FROM TECHNOLOGICAL HUMANITY
TO BIO-TECHNICAL EXISTENCE

SUNY series, Intersections: Philosophy and Critical Theory

Rodolphe Gasché, editor

FROM TECHNOLOGICAL HUMANITY TO BIO-TECHNICAL EXISTENCE

Susanna Lindberg

SUNY
PRESS

Published by State University of New York Press, Albany

© 2023 State University of New York

All rights reserved

Printed in the United States of America

No part of this book may be used or reproduced in any manner whatsoever without written permission. No part of this book may be stored in a retrieval system or transmitted in any form or by any means including electronic, electrostatic, magnetic tape, mechanical, photocopying, recording, or otherwise without the prior permission in writing of the publisher.

For information, contact State University of New York Press, Albany, NY
www.sunypress.edu

Library of Congress Cataloging-in-Publication Data

Name: Lindberg, Susanna, author.
Title: From technological humanity to bio-technical existence / Susanna Lindberg.
Description: Albany : State University of New York Press, [2023] | Series: SUNY series, intersections: philosophy and critical theory | Includes bibliographical references and index.
Identifiers: LCCN 2022035144 | ISBN 9781438492582 (hardcover : alk. paper) | ISBN 9781438492599 (ebook) | ISBN 9781438492575 (pbk. : alk. paper)
Subjects: LCSH: Technology—Philosophy. | Human-machine systems—Philosophy. | Biotechnology—Philosophy. | Transhumanism. | Philosophy, Modern.
Classification: LCC T14 .L53 2023 | DDC 620.8/201—dc23/eng/20221026
LC record available at https://lccn.loc.gov/2022035144

10 9 8 7 6 5 4 3 2 1

Contents

Acknowledgments	vii
Introduction	1
Chapter 1. What Is the Human Being?	15
Chapter 2. What Is Called Technics?	39
Chapter 3. The Originary Technicity of the Human Being	79
Chapter 4. De/constructing Humanity	109
Chapter 5. Humanity and Inhumanity of Technical Communities	197
Chapter 6. From Technological Humanity to Bio-technics	259
Notes	279
Bibliography	331
Index	345

Acknowledgments

I thank very warmly the Helsinki Collegium for Advanced Studies where this book was written. The inspiring company of its fellows certainly made the research work much lighter.

I also thank the unknown referees of this manuscript for their encouraging words and perspicacious comments, which improved the work as a whole and, most importantly, opened interesting avenues for future work.

Finally, I thank Aengus Daly for his invaluable help with the English language. Whatever faults remain are due to my own choices.

A passage of the introduction on post- and transhumanism summarizes a more detailed explication published as "On Prosthetic Existence: What Differentiates Deconstruction from Transhumanism and Posthumanism," in *Humanism and Its Discontents: The Rise of Transhumanism and Posthumanism*, ed. Paul Jorion (New York: Palgrave Macmillan, 2022).

Introduction

What does contemporary technology make of the human being?

This pressing question may sound like a lament over the defiguring of the beautiful, classical image of humanity. It can also sound like a jubilant cry of delight at the prospect of finally overcoming the obsolete figure of humanity and moving toward a better trans- or posthumanity. The question of the fate of technological humanity has not disappeared over the course of the last century, and this seems to reaffirm the need to ask ever anew what the marvelous new technologies make of the humanity that made them in the first place.

The starting point of the following work, however, is the claim that such considerations tend to move in circles in which it is impossible to see how technology, interpreted as a neutral instrument of human intentions, could actually amount to anything other than a reaffirmation of the very humanity that employs it. As long as technology is considered in an instrumental framework, the most marvelous and the most monstrous technologies that could be imagined still bear the (wise or unwise) intentions of the human being. They thus retain the human even when they promise or threaten to exceed its image.

The very question of technological humanity is marked by a fundamental inadequacy when it is based on the circle of intent and instrument. The instrumental conception of technology downplays the ambiguity that results from the fact that technology tends to escape its user's intentions by following its own internal logic (its *pharmakon* effect discovered by Jacques Derrida and further developed by Bernard Stiegler in particular). It overlooks the consequences of how technology innervates the human world (its *environmentality*, whose precise character was already a matter of dispute between Martin Heidegger and the German Philosophical anthropologists in 1920s and remains so today between thinkers like Peter Sloterdijk and Erich Hörl). It ignores the intertwinement of technology with the very fabric of the nonhuman earthly world in general (as seen in the phenomenon of the anthropocene first designated by Paul Cruzen). This blindness arises from a failure to ask what technology is in itself. In chapter 2 of the present work, I will attempt to shed light on this obscurity not by asking what technology is in the narrow sense (an instrument coupled to human science) but what *technics* is in the most general possible sense, a sense that also includes human (and perhaps other living beings') skills and techniques, singular artworks and industrial production systems, and seemingly immaterial codes and programs. I will show how this general sense of technics has evolved from initially being seen as an instrument of human skill, then being interpreted as autonomous mechanism, before finally being seen as lifelikeness in contemporary bio- and information technologies. The last stage makes apparent what was to some extent already there at the outset: technics has a quasi-autonomous, quasi-living, quasi-intelligent way of being that I designate using the term *bio-technics*, a notion illustrated by contemporary bio- and information technologies but that I will use in a much more general sense. Bio-technics is not a simple projection of human intentions. It has a quasi-life of its own insofar as it has its own nonconscious directedness, which can be called "intentionality" only if the term *intention* is severed from its original meaning such that it can also be attributed to the programs that animate the simplest living entities, such as cells, as well as to technical beings.

Technics was always constructed in order to prolongate human (or other living beings') capacities, and in this respect it has been constructed in the *image* of human beings (and other living beings). Technics also has always had a speculative role, acting as a mirror of their human constructor, hence the fears of human degradation and the dreams of human enhancement that emerge from this technical mirror image. Whatever appears alien (monstrous/marvelous) in a technical context seems to reflect some alienness (monstrosity/marvel) in the human being itself so that the inevitable distortions in the technical image of the human become the origin of distortions in the human. I will present the complexities technics introduces to the study of the human in the first two chapters of this book, giving an introductory overview of the problem field. In chapter 1, I will show how a utilitarian conception of technics projects a utilitarian vision of the human being who restricts itself to being a subject of technics and an object of anthropotechnics. I will also show how the limits of this perspective become visible in the view of the human as an educable, free, and creative being that prevails in the human sciences. The idea of anthropotechnics presupposes this idea of the plasticity and transformability of the human, which ultimately undermines the very possibility of showing where the image of the human has been destroyed or overcome. There is no pre-given figure of the human because its very nature lies in its capacity to reinvent its figure. Finally, I will point to the discrepancy between technological humanity and the subject of philosophy, a discrepancy that makes the philosophical questions surrounding technics and humanity so difficult. Classical philosophy thinks that insofar as technological humanity is a historical figure of the human, and even one possible figure of the historicity of the human, it is distinct from the subject of philosophy, who is ultimately only the self-consciousness that belongs to the act of pure thinking. But at least since Heidegger contemporary philosophy has emphasized that philosophy always takes place in a given historical situation that it needs to account for, which means that today the subject of philosophy must also account for itself in its technical situation. The subject of philosophy cannot be limited to a pure reason that thinks itself untouched by its technical supports, but on the contrary, it must question technicity as the fundamental relation that ties humans and nature together. The core of the book is an inquiry into this question. In further chapters I show how Martin Heidegger and Philosophical anthropologists such as Helmuth Plessner, Michel Foucault, Jacques Derrida, Bernard Stiegler, and Giorgio Agamben discover and deconstruct the question of technological humanity. These chapters are particularly concerned with indicating how the figure of humanity has tended to fade away as technics became a philosophical question and how the idea of an originary technicity that coincides with life itself gradually emerged in its place. In the end, the bio-technicity that characterizes contemporary technics is also an image of the bio-technicity that turns out to be the condition for any figure of humanity (anti-, trans-, post-, non-in-human humanity). This bio-technicity is ultimately shared by humans, other living beings, and technical beings: bio-technics is how existence is given and gives itself under technological condition.

At first glance, the deconstruction of technological humanity and the emergence of a general bio-technics seems to follow from a very particular diagnosis of the contemporary world, from a metaphysics of the present era, as Michel Foucault would put it. We do find ourselves in an unprecedented technological situation, its ground lies in modern thermodynamic and nuclear technologies and its present state lies in information technologies and bio-technologies. It is necessary to ask in all manners how this technological situation has affected human existence and human self-understanding. To ask what contemporary technologies make of the human being is not just to echo yet again the age-old lament of departing generations who are puzzled by the latest inventions nor the jubilation of new generations spurred on by new hopes, for there is something specific to contemporary technologies that puts the human being at stake in a new way. What is novel in contemporary technologies is their reflexive character: they are directed back onto the human beings who have built them, instead of primarily cultivating, modifying, and exploiting external nature, as earlier technologies did. The most characteristic contemporary technologies are bio- and information technologies, where the *bio* is also the human body and the *info* is also human thought. Progress in these domains seems to have an unprecedented impact on the human being's own mind and body. This is not to say that older technologies did not affect the human being. All technologies mark their users. For example, medicine and the printing press have long contributed to the development of the human mind and body. But while older technologies have generally been seen as enhancing or contaminating the human situation, there is today a widespread foreboding that contemporary technologies aim to change what the human being itself means, such that with the progress of technology, the human species feels that it runs the risk of divinizing or defiguring itself.

Are those forebodings justified? What do they really mean? These questions are much more complicated than they appear at first sight, for not only does the nature of contemporary technologies change extremely rapidly, thus obscuring the meaning technologies may have, but also and in particular the nature of the "human being" that they are suspected of transforming is itself obscure. Moreover, the question of the human being is reflected back into the philosophy that seeks not only to inquire into the anthropos but also to investigate human existence as the site of the subject of philosophy itself.

Before going any further, I need to digress for a moment to remark on my linguistic choices, and these can actually be read as another metaphor for the contemporary difficulty of knowing what is happening to the human being. The reader may have noticed that I have taken the liberty of referring to the human being with the pronoun *it* and with its derivatives (e.g., *itself*). I have accustomed myself to this use of *it* and I have also extended it to all instances where one would normally expect to find a third-person singular pronoun, with the exception of references to specific persons whose gender has hitherto been designated unambiguously (Plato is *he*). This choice, as well as this digression justifying it, are of course due to the crisis surrounding the third-person singular pronoun that has gripped the English language for decades now and that is by no means unrelated to social and to anthropo- and bio-technical progress. I explain my choice in a footnote,[1] and I simply hope that the reader will be comfortable with this choice, which is not meant to convey any value judgments but rather to avoid common pitfalls surrounding gender and grammar. I suggest that you read this book as a test of what happens if *he/she/they* singular is replaced by *it*. After reading these pages you will know whether the experiment is successful.

Now, the primary aim of this book is not to make a cultural diagnosis or to critique the contemporary world that is shaped by the aforementioned technological and even linguistic factors. I am neither a technician capable of evaluating concrete technologies nor an anthropological, social, or political scientist trained in the evaluation of their psychological, sociological, or political consequences. Instead, this book focuses on the philosophical concepts of technics and humanity; it attempts to trace the unraveling of an idea of "technological humanity" in contemporary philosophy and the emergence of a bio-technical conception of existence beneath it.

I will start this philosophical investigation by discussing in this introduction the contemporary intellectual currents of transhumanism and posthumanism that focus on the idea of technological humanity.[2] They keep open the question of whether intense technological transformations can ultimately extend to fundamental anthropological transformations and furthermore of whether technological changes can become so significant that they actually put an end to humanity and change it into something else, into a condition sometimes described as "transhuman" or "posthuman." I outline these positions because they provide the most striking illustration of different possible perspectives on the idea of technological humanity today. However, like most "isms," these intellectual currents express general world views rather than precise philosophical positions. This book does *not* aim at a continuation of post- or transhumanist projects but instead at uncovering a field of philosophical problems underlying and conditioning such projects, albeit often unthought by them and really much older and more general than these discussions. The philosophical core of this book lies in this general question: *How does technics configure human existence?*

We shall see that this question belongs on a very different level to post- and transhumanist reflections, but the radicality of the latter makes manifest the necessity of the former.

Transhumanism and posthumanism are not fixed notions but historically variable positions. They share a positive attitude toward technology, which is why they are sometimes taken to be synonymous. Stefan Lorenz Sorgner even suggests that trans- and posthumanisms could be united into a "metahumanism."[3] Some commentators claim that both trans- and posthumanism are continuations of postmodernist and poststructuralist ideas.[4] Yet when we look more closely, these two currents turn out to be contrary to each other in many ways. Although the transhumanist Nick Bostrom has identified himself with posthumanism,[5] transhumanism generally tends to use the tools of analytical philosophy and is opposed to poststructuralism,[6] while many versions of posthumanism see poststructuralism and postmodernism as their forebears. These posthumanists often refer to Michel Foucault, Jean-François Lyotard, Gilles Deleuze, Jacques Derrida, Jean-Luc Nancy, and Bernard Stiegler. The approaches of these thinkers provide useful conceptual tools insofar as they all think that technics is a distinctive feature of humanity (*"propre de l'homme"*). But they also emphasize that this very technicity undermines the possibility of any proper characteristics (*"propre"*) as well as of the very notion of humanity (*"l'homme, l'humain, l'humanité"*). They think technicity as an originary mode of existence but do not proclaim a new humanism, transhumanism, or posthumanism.[7] However, the terms *poststructuralism* and *postmodernism* are so vague that the thesis of a rapprochement between posthumanism and poststructuralism holds of some authors, for example Rosi Braidotti and Stefan Herbrechter, but not of others, as Cary Wolfe and Frédéric Neyrat have shown.[8] Once we admit the impossibility staking out definitive positions, we can give a schematic characterization of transhumanism and posthumanism.

Transhumanism is an intellectual position that advocates the radical transformation of the human being's biological, mental, and social conditions by means of technology. The first article of the World Transhumanist Association's *Transhumanist Declaration* states, "Humanity stands to be profoundly affected by science and technology in the future. We envision the possibility of broadening human potential by overcoming ageing, cognitive shortcomings, involuntary suffering, and our confinement to planet Earth." An earlier version of the *Transhumanist Declaration* declared: "We support the development of access to new technologies that enable everyone to enjoy better minds, better bodies and better lives. In other words, we want people to be better than well."[9] The transhuman philosopher Max More specifies that these "radical alterations in the nature and possibilities of our lives [result] from various sciences and technologies such as neuroscience and neuropharmacology, life extension, nanotechnology, artificial ultraintelligence, and space habitation."[10]

Transhumanists perceive such transformations as human *enhancement*, desirable as such, even if it results in a form of life that is no longer human. Robert Ranisch and Stefan Lorenz Sorgner explain that "the result of such a technologically induced version of evolution is referred to as posthuman. However, there is no commonly shared conception of what posthumans are, and visions range from the posthuman as a new biological species, a cybernetic organism, or even a digital disembodied entity. The link between the human and the posthuman is the transhuman, an abbreviation for a transitional human, to which transhumanism owes its name."[11]

Although the "posthuman" may refer to the product of such technological enhancement, the term also has another use, which allows Cary Wolfe to claim that it is the opposite of transhumanism.[12] Posthumanism denotes a break with the traditional humanism perceived as an ideological construct that paternalistically imposes Western "phallo-logo-centrism" on the entire world (and indeed, the transhumanist conception of "better minds, better bodies and better lives" reproduces the most classical Western ideas of human excellence). Many feminist, postcolonial, and ecologist authors call themselves posthumanists in order to demarcate themselves from and oppose humanism in this sense (and its ideals of mind, body, and life). This is why the term has also been used to characterize authors such as Donna Haraway who see the other of "man" in the feminist figure of a technologically mediated cyborg, authors such as Cary Wolfe who associate posthumanism with ecology rather than with technology, or authors such as Jane Bennett who are interested in the materiality of human and nonhuman life.[13] To sum up a complex situation in a slogan, we can say transhumanism is situated in Silicon Valley and posthumanism is in departments of art and literature, or we can think of transhumanism as a liberal-capitalist view of human becoming and posthumanism as a leftist one.

The origin of the term transhumanism is generally located in Julian Huxley and Pierre Teilhard de Chardin's writings.[14] Today's transhumanists are engineers, scientists, futurist artists, and utilitarian philosophers. Their ideas are defended by the World Transhumanist Association (WTA), which was founded by the philosophers Nick Bostrom and David Pearce in 1998, whose declarations can be found on the Humanity+ website. Transhumanists think that biomedical technologies should be used nontherapeutically, for example to enhance human health and longevity, emotional and cognitive capacities, physical traits, and behavior. Some transhumanists dream of mind uploading and expect the advent of Singularity, or the becoming self-conscious of the infosphere.[15]

The term *posthumanism* also has several origins. Its first occurrence has been traced back to the Macy conferences on cybernetics (1946–1953), but it was first used in the contemporary sense by the postmodern philosopher Ihab Hassan in *Prometheus as Performer* (1977).[16] Hassan thought that the change of the human form has now become so radical that "humanism may be coming to an end, as humanism transforms itself into something that we must helplessly call posthumanism." While this idea seems similar to the visions of the transhumanists, Hassan has a different reference point, namely the idea of the "end of man" formulated by Michel Foucault in 1966.[17] Foucault does not mean that the biological species *Homo sapiens* would change in any way but rather that the idea of humanity is a historical construct that is gradually becoming obsolete. Foucault's idea of the end of man is not the end of the human animal but of a particular self-image we have of ourselves. The authors who first studied technological posthumanity, especially Donna Haraway and N. Katherine Hayles, agree with Foucault rather than with those transhumanists who take the notion of the transformation of the human species literally.

What is the role of technics in the self-constitution of the human being according to these intellectual currents? The transhumanist position is quite clear. In most cases, transhumanist authors see technics as a means for human self-enhancement and enhancement as desirable as such. For instance, Max More affirms that "transhumanists regard human nature not as an end in itself . . . we can learn to reshape our nature in ways we deem desirable and valuable . . . we can become posthuman. . . . Transhumanists refer to 'technology' as the primary means of affecting changes to human condition."[18] The aim of technics is thus the production of the human being by itself. In familiar Aristotelean terms, in transhumanism, the human being sets the final cause (itself) as well as the formal cause (also itself) and it is the agent of this production and the matter formed. Technics is the instrument of this production, which is valuable if it effectively realizes human intentions but is not problematized in itself.

Correlatively, the critical discussions of transhumanism tend above all else to evaluate its aims. For example, commentators and science fiction authors have shown why the *final causes* of transhumanist production, such as immortality, absence of pain, and an ultrarapid and tireless brain could even be undesirable. When we look more closely at the transhumanist ideal of "better minds and better bodies," we realize that instead of challenging traditional enlightenment humanism, transhumanism actually adopts and reinforces its most productivist and dryly intellectualist elements.[19] Together with the enlightenment ideal, it thus tends to reproduce the flip side of this humanism: a harsh productivism and a hostility to incarnated existence. We can further question whether it is really so obvious that the human being is the agent of transhumanist self-production. For who is that agent? An individual could choose to improve itself in a certain way, but in so doing it could only choose among the technical *dispositifs* that are already available in society, which the latter it cannot choose. And finally, we can wonder whether human beings can ever be seen as simply the material (*"hyle"*) for transhumanist production. Harnessing the body and the mind for preset aims might well destroy or detrimentally impact important aspects of their own development.

Such questions and misgivings mainly arise in bioethical approaches to transhumanism. A very different, and in my opinion more fundamental, question emerges when we interrogate technics itself instead of just focusing on how humans use it. When technics is understood simply as means, it suffices to evaluate the ends human beings make it serve. But this perspective remains blind to the effects of technics, which cannot be simply reduced to human intentions (pollution) and it does not heed how technics formats human intentions (the creation of new needs). Martin Heidegger famously calls the fundamental structure of the modern technical world the *Ge-stell*: it disposes of the human being and of nature before it comes to be at the human being's disposal.[20] Michel Foucault's term *dispositif* reinterprets *Ge-stell* with regard to social techniques in particular. Both show how the general structure of the contemporary world is a technical framework that predetermines what we can be and do and what we deem desirable. If technics is such a general framework, its effects can never be delimited as the effects of simple means: the very existence of means is already a consequence of the framework that conditions and surpasses their use. The posthumanists who distinguish themselves from transhumanists by developing Foucault's (and Derrida's) idea of "the end of man" accordingly see technics as a dispositif of a finite historical situation that conditions the human (and nonhuman) beings who live in it. But at the same time, technics is also something that the human (and nonhuman) beings can invent and use in different technics of self and of the world. Technics is really the endless mediation between existents and their situations.

One could perhaps say that most versions of posthumanism are existential and political interpretations of this situation. Who are we if we are inextricably bound to technics, if we are formatted and formatting beings? What do we make of ourselves? For instance, Haraway's "cyborg" is a possible name for the existential situation of a being that through endless technical mediations is profounded intertwined with nonhuman life. In a parallel fashion, Hayles locates posthumanity in the situation where human beings have given over one part of their thinking to technical systems and then identified themselves with the resulting hybrids. Does such posthumanity amount to an overcoming of humanity? It at least heeds the extension of humanity into domains ignored by classical humanism. But then again, what does "humanity" mean? In *A Thousand Plateaus*, Deleuze and Guattari note: "Why are there so many becomings of man, but no becoming-man? First because man is major-itarian par excellence, whereas becomings are minoritarian; all becoming is a becoming-minoritarian."[21] This is why there is no becoming-man but only becoming-woman, becoming-child, and becoming-animal. If humanism just means becoming-man in the majoritarian sense, then these posthumanist intertwinements of human with nonhuman technicity represent the overcoming of humanity. But if the idea of the human being can accommodate its minoritarian becomings as well, then the hybrids of human/woman/man and machine reconfigure "humanity" instead of abandoning it.

Trans- and posthumanism fundamentally ask whether and how technological progress changes humanity into something that comes after "humanity." The motivation of this book is different and (it seems to me) more fundamental because it aims to show that before evaluating such a change we should first decide what we mean by "humanity" and "technics." The figure of humanity can be changed only if it is a *figure* in the first place, but if it is, as I believe, a plastic capacity for transformation, its only possibility of being fundamentally changed would be the cessation of this capacity and its fixation into an unchanging figure. On the other hand, if technics is only the instrument of predetermined change, it does not really induce change but only realizes preprogrammed possibilities and thereby reproduces the extant figure of the human. But if it is the very situation in which change takes place—the bios used by technics and the new form of life created by technics—then it is difficult to know when it is the effect and when it is the agent of a change whose origin lies in the situation itself.

Instead of defining the notions of humanity and technics, I am interested in their relation, in the *originary technicity* whose effects are both humanity and technics and which calls for a constant adjustment of these terms. In Agamben's terms, this relation happens as the reciprocal *use* of humans and technics; in deconstructive terms, this *use* becomes thinkable as the reciprocal *mimetic* relation between life and technics; and in Heideggerian terms this reflects the *in* of the existential situation of being-in-the-technical-world. What this means should become clear in the following chapters.

In chapters 1 and 2 of this book I will present the notions of humanity and technics. In chapters 3, 4, and 5 I will show how the complex question of technological humanity has gradually been deconstructed, first in German philosophy from the 1920s (Martin Heidegger and the Philosophical anthropologists Max Scheler, Helmuth Plessner, and Arnold Gehlen), then in French "poststructuralist" philosophy that began in the 1960s (Michel Foucault and Jacques Derrida), and finally in continental philosophy of the 2000s (Bernard Stiegler and Giorgio Agamben). This is, admittedly, a selective history and many other important authors could also be included. However, it seems to me that these authors present a sufficient variety of perspectives on the question of the human/subject. Through reading these authors, I will show how an idea of bio-technics gradually emerged in the place of technological humanity. I do not take the terms *technological humanity* and *bio-technics* from the authors treated in this study, but my discussion of them takes place in the context of an investigation of a field of problems that these authors had in part discovered but did not treat in detail because they had other objectives in view.

In the chapter 3 I will show how technics (instead of reason) appeared as a determining feature of human existence for the first time in Heidegger's existential analytic and in the simultaneous development of so-called Philosophical anthropology by Scheler, Plessner, and Gehlen. By paying attention to the technological imprint on the human existence, these authors thought of the core of human existence in terms of a negativity. Negativity and nothingness were not only a fundamental metaphysical force but also a sign of the hollowing out of traditional humanism that called for different ways of thinking of the human, such as in terms of life or *Dasein*. Sometimes this negativity was seen as reflecting the nihilism of the modern age, for example in Heidegger's texts on the technological era, and sometimes as acting as the source of a creative liberty that enabled the creation of a technological world, for instance in the works of Philosophical anthropologists. In all these cases, technics was much more than an instrument: it was the form of a world in which humanism has lost its substance.

In chapter 4, I will show how so-called poststructuralist thinking developed an *antihumanism*, which thought of the human being not as the origin but as a simple effect of prevailing systems of significations. Foucault shows how the human effect is produced by different social power technologies to which technologies of the self can respond. Derrida studies signification itself as a technics and not as the expression of a logos or of a meaning intention. He also studies the consequences of the parallel between technics and life, both of which are thought of in terms of codes and programs, and thus preparing the possibility of thinking of "bio-technical existence." However, Derrida himself does not affirm such a theory. He instead develops a kind of a phenomenological ontology of spectrality and of the *khora* that frames the tele-technological existence that comes forth today as a fundamental condition of human existence.

In chapter 5, I will show how, instead of asking what anti/humanism consists of, thinkers as different as Bernard Stiegler and Giorgio Agamben focus on the markedly ethical and political investigations of the conditions of resistance to *inhumanity*. Interpreting human existence in terms of concrete technics, these thinkers draw attention to the way in which technics belongs to communities and attaches individuals to collectives. This is why technics is not only shared by but also accompanied by the possibility of alienation. In the concluding chapter, I make a brief inquiry into the possibilities of resisting such alienation through a free use of technics that allows ways out of its toxicity.

I also draw together the idea of bio-technical existence implicit in the readings presented in this book. Examining humanity in the mirror of technics thus shows how technological humanity is really a deconstructed humanity: a life. Examining the effects of technics on existence attracts attention, not to consciousness that defines classical humanity but to nonconscious aspects of existence that coincide with life. At the same time we see that this "life" is by no means a natural given. It is thoroughly marked by technics and actually *is* the technical agentivity of a skilled body, of a mechanically supplemented mind, of calculus that supports thinking.

Existence is technicity—not a product that a technical program could predetermine but a life that forever escapes technical formatting. Life and technics do not coincide into one general thing; *bios* and *techne* remain separated by the very hyphen that connects them. One cannot display the reason of the other. They mime one another, invent one another, use one another, and together display the fundamental structure of our world.

Chapter 1

What Is the Human Being?

How does the present "technological condition," as Erich Hörl puts it,[1] modify the human condition? Does it really modify the human condition? The human being is an extremely complex research object because it implies diverse and even incompatible perspectives, and these themselves often awaken strong passions, for whatever is said about humanity reflects back upon the speakers and the listeners themselves. In order to avoid running together incompatible levels of inquiry, we can divide the problem area into three different, albeit not distinct, groups. The first studies the human being from a natural scientific viewpoint, the second, from an anthropological perspective, and the third examines the human subject philosophically. This book is focused on the third level of inquiry, but we first need to delimit this from from the other two.

Anthropotechnics

Anthropotechnics means the transformation of human individuals and collectives by technical means.[2] Such techniques can be social, such as in the case of disciplinary techniques used in schools and prisons that were famously described by Michel Foucault; cognitive, as in the case of digital technics studied by N. Katherine Hayles, Bernard Stiegler, and Giorgio Agamben in particular; or medical, in the general sense of direct interventions into the human body, including the brain. Let us start by casting a glance at the various questions that arise in the last of these cases. The human being is there investigated after the fashion of the natural sciences and is manipulated by scientifically developed technology. This natural scientific and engineering perspective on the human serves as the main inspiration for transhumanism, but it has also been investigated by bioethicists who do not share and often oppose transhumanist ideology.

The natural scientific and medical perspective evaluates technologies with a view to their capacity to affect the biological processes of human life and human thought as neurological operations. What characterizes many contemporary technologies is their increasing intimacy and their ubiquity, which gives the false impression of their immateriality. They become ever more intimate with our bodies, which are not only equipped with new smart devices but almost inhabited by ever more discreet medical aides that function as internal prostheses (e.g., insulin pumps, organ transplants, electronic stimulators, and pharmacological regulators located deep within the organism). In principle, with the gene scissors CRISPR-Cat9, some elements of human bodies might soon be produced technologically even before the persons to whom they "belong" are born. On the other hand, the disembodied (not to say spiritual) dimension of intellectual and social life is increasingly affected by the impersonal ubiquitousness of digitalization that mimes and is parasitic upon thought and communication. While yesterday's science fiction thus becomes reality, today's science fiction imagines technological futures in which today's humanity itself gradually becomes obsolete. Philosophers have responded to these developments in many ways: while transhumanists have asked when the transformations of human mind and body will reach a tipping point beyond which the human is no longer human, bioethicists stick closer to existing technologies in order to explore the frontier between enhancing people and creating monsters.

By displacing the sense of normality and even of humanity itself, medical technologies have contributed to human self-understanding and stimulated human self-reflection. Medical technologies become philosophically relevant when they affect the material conditions of existence such that they have existential consequences. I want to briefly outline the main types of existential questions raised by recent technological developments as they have been treated in bioethics in particular. Bioethics will not be the topic of the later chapters of this book, but it will be useful to see how existing technologies have been discussed in this field to provide some background to the later discussions.

1. One of the most existentially important bodily events is death. The development of medicine has increased human longevity significantly

(when other factors such as poverty do not cancel out this progress). As Martin Heidegger says, "Being-certain with regard to death . . . will in the end present us with a distinctive certainty of Dasein [*ausgezeichnete Daseinsgewißheit*],"³ but whether death comes somewhat sooner or a little later does not change the fundamental role of death as the limit condition of human existence. For some people, however, human mortality seems so scandalous that they imagine various ways of circumventing death by hiding it either behind the promise of an eternal afterlife or rebirth (many religions) or behind a promise of a technological prevention of death (yesterday's alchemists and today's transhumanists). In the light of existing medicine, the idea of canceling death technologically is merely speculative. However, the attraction of this idea is surprising, not so much because an individual's fear of death is incomprehensible but because when we envisage the social consequences of the disappearance of death we can see it leads to a dead end. A deathless society would have to be a childless society, unless humanity is ready to suffocate from sheer overpopulation or unless deathlessness is reserved for just a tiny wealthy elite that takes its pleasures regardless of the cost to the rest of humanity. In either case, a deathless society would no doubt soon become a gerontocracy where, in the absence of the novelty introduced by children, the same people—you know who—eternally retain power. Speculative fiction has showed time and again why there is nothing utopian about these visions.

In reality, more important than such speculations on the disappearence of death are changes of the significance of death brought about by contemporary medical technologies.⁴ Among the strangest consequences of the present-day medicalization of death are the uncertainties surrounding the exact criteria and therefore the exact moment of death as well as the change in the status of the cadaver. From an eerie non-thing with a lingering proximity to a beloved person's specter that therefore needs to be treated with special respect and piety, the cadaver today becomes a simple material object, a resource, with some zombie-like features.⁵ This transmutation follows from certain technical-legal changes. While the cessation of the functioning of the heart used to be the criterion of death, today's medical institutions prefer to define death as brain death, which itself has many different definitions, the main one being that legal death can now be said to occur while the heart is still beating. This change of definition was necessitated due to developments in organ transplantation technologies. When the heart stops, the body is so to speak too dead with respect to the possibility of organ harvesting. But if death is defined as cessation of brain activity, the organs can be cut away while the body is still somewhat alive, irrigated by the heart and other bodily functions. At that moment the body is still warm, its breast still expands thanks to artificial ventilation, and it can still react to the operation. These technological possibilities have changed the significance of death in a disturbing way. Death is no longer a natural and social event. It is a medical and juridical event in which a human body, which is strongly protected by law as a person's most inalienable property while that person is alive, instantaneously becomes a medical resource that belongs unconditionally to the medical institution that then uses it to treat other patients. In more and more countries this right is now legally enforced and strengthened to the degree that it overrides considerations of piety toward the deceased and the mourners. Influenced by campaigns depicting organ donation as a "gift of life," the general public is, by and large, overwhelmingly in favor of these legislative changes. However, it is disturbing to think that because the brain-dead person is juridically dead, its reactions during the operation are not interpreted as pain and protestation (and yet, to lessen the unease of the operating personnel, the dead bodies are actually anesthetized before the operation to remove the organs despite the paradoxes involved in such a procedure). It is also disturbing to think that the medical institution's right to a dying person's body is deemed superior to the right of family and friends to spend the last moments with the passing person undisturbed and of taking the time needed to mourn with the body. The moment of death and the dead body itself were once considered sacred and entitled to a special respect, but this is no longer the case when bodies are taken for organ harvesting.

    2. Another existentially central bodily event is birth. Birth control and abortion are very ancient techniques that merely choose whether or not a child is born, but if the child is born, it is welcomed as the stranger that it is. Today's reproductive medicine opens very different questions since it has developed the means to control what kind of a child is born, little by little transforming the actual birth from chance to choice.[6] In industrial countries, the development of human fetuses is already subject to monitoring and various diagnostics in the uterus. In artificial insemination procedures, it is technically possible to select and even to enhance a human fetus before the placement of the embryo in the uterus. Today embryo diagnostics and selection are carried out very rarely and only in order to avoid serious pathologies. Tomorrow, the development of reproduction technologies could open up the question of whether not

only monitoring but also selecting and even manipulating embryos ought to become common practice. It is not technically possible to determine a child's future life (an embryo cannot be genetically programmed to grow into a talented mathematician),[7] but it is possible to eliminate targeted genetic flaws and even add new genetic features (e.g., the babies born in 2018 whose embryos were genetically manipulated by Dr. He Jiankui in order to give them protection against HIV—without noticing that the manipulation is likely to increase the probability of dying younger).[8] Certain analytical bioethicists, such as Allen Buchanan, Dan W. Brock, Norman Daniels, and Daniel Wickler, insist that as the state should not forbid individuals' projects of self-transformation (plastic surgery, gender reassignment, learning drugs, brain stimulators, etc.), and it should not meddle with their reproductive dreams either. They go as far as maintaining that parents should not lessen their progeniture's chances in life and thus that embryo selection should be an obligation (this idea is, of course, anathema to most defenders of the rights of people with disabilities). Embryo selection is an eugenistic practice, but Buchanan, Brock, Daniels, and Wickler claim that it differs greatly from the totalitarian state eugenics of the twentieth century[9] because in contemporary consumer eugenics the state does not control natality and education but rather individuals make the choices. The question is much more complicated than this, however, because the freedom of human enhancement is not being claimed here for oneself but for one's children. An individual can never choose to be born, but birth is an event of a community (of the baby, the parents, the eventual donors, the medical staff, and the entire community that will care for, educate, and in manifold ways welcome the newcomer). Ordinary birth control and even embryo selection do not produce "designer babies," but the genetic manipulation of the embryos does, especially since the invention of the genetic scissors CRISPR-Cas9, which make the fabrication of chimaeras a less chimeric idea. If the artist Eduardo Kac could place an order to INRA laboratories to insert medusa DNA into a rabbit embryo in order to produce a rabbit that glows green in the dark,[10] it is theoretically possible for avant-garde parents to place an order for a green glowing baby or another kind of a human chimaira.[11] Genetic manipulation does not follow from the proper and autonomous choice and consent of the future child but from adults who impose their will on the child's body, such that it becomes a little less a stranger and a little more a product. When a future child is the result of other people's projects, does it really make a difference if it is part of the state's project, say, of having better workers or of the parent's project, say, of having the smartest children in the neighborhood? I admit that I quite cynically doubt the intelligence of average parents as much as I do the goodwill of the average states.

The most well-known critics of eugenics have paid particular attention to the ethical and political problems connected to this problem. Andrew Pilsch has labeled critics who have warned against the consequences of a new eugenics as "bioconservatives."[12] He quotes Francis Fukuyama, but one should instead probably mention Hans Jonas,[13] Jürgen Habermas, or Bernhard Waldenfels. The term *conservative* is misleading because it suggests that the critics of modern consumer eugenics are critical on religious grounds, but it is generally the other way around. Jonas and Habermas are following Enlightenment principles when they warn against both consumer and state eugenics: their aim is not to sanctify the present-day human genome but to protect the freedom and the autonomy of the child from eugenist utilitarian reification.

3. There is an entire field of philosophical discussion concerning the effects of new information technologies on personal identity. The philosophers who ask if I would still be myself if my mind were "uploaded" to a computer formulate the question very simply.[14] Apart from the purely speculative character of the question, this way of studying human identity is naive: it assumes that mind and body are two separable entities instead of aspects of one living being, it attributes personal identity to only one of these sides (namely to the disincarnate mind), and then it equates the "life of the mind" with computable operations. Instead of supposing that "mind" is something independently of its "supports," it would be more interesting to see how psychophysical identity also gradually evolves with its different technical prostheses. Is my being me altered if I am enhanced by some new cyborg organs, like Stelarc's robot arm or Neil Harbison's third eye, or by advanced medical interventions, like sex reassignment surgery or transplantation surgery, the latter of which has been described very well by Jean-Luc Nancy and Francisco Varela?[15] Is my being me altered if I entrust my memories and my social life to social networks operated by means of digital devices? Ultimately, these cases may not be so different from technically simpler cases. For instance, is my being me affected if I get new eyeglasses, strong medicine, or fancy drugs—or when my bodily state is otherwise altered, for example by getting pregnant, falling ill, working out, or getting old? I believe most of us would say that in all of these cases I remain me, but who I am is enriched by

new experiences—and eventually deprived by some others. As a whole, personal identity is so much more than a mind-body with more or less sophisticated extensions, above all because it is a capacity (of feeling, thinking, acting) instead of a stabile form.

A richer way of studying the effects of digital technologies on personal identity focuses not directly on the individual but on the world that digital technologies set up and to which individuals must adapt. Important philosophers of technology such as Heidegger, Jacques Ellul, Max Horkheimer, Theodor Adorno, and Félix Guattari[16] have already shown how profoundly, since the mid-twentieth century, industrialization has changed the Western world in particular and, little by little, almost the entire global lifeworld. Industrialization has severed life from its natural cycles and submitted it to technological rythms: it has provided a plentiful though standardized material culture but also created the monumental problems of pollution and waste that exacerbate the development known as anthropocene. Today technological progress is generally thought to have reached a new phase with digitalization that provides social products and pollution. Especially in industrial countries today, ordinary life is increasingly mediated by multiple algorithms such that it has become an "algorithmic life" (as Éric Sadin puts it) formatted by "algorithmic governmentality" (as Antoinette Rouvroy, Guido Berns, and Bernard Stiegler call it). Whether material or social-psychological, industrial technology functions by arranging an environment so that today "the mode of participation of technology is fundamentally environmental while at the same time transforming the environment."[17] Our personal identity reflects our environment, and as this is increasingly digitalized, so too is our identity. The recent experiences of confinement in a highly digitalized environment during the COVID-19 pandemic have shown very concretely how digitalization affects personal and collective identity by increasing both serenity and stress.

4. By providing the environment in which many—most—people live, digital technologies have a strong existential impact because they contribute so significantly to what is called thinking. Since their invention, computers have regularly been presented not only as handy tools but as existentially important machines insofar as they extend and mimic the human brain or thought. When Alan Turing wrote his seminal—and speculative—text "Computing Machinery and Intelligence,"[18] he saw the computer as a double of the human mind: if the imitation game between mind and machine shows that the machine yields similar results to the mind, machines can be considered as similarly intelligent. The imitation

game is still used today to test so-called artificial intelligences. The underlying cognitivist hypothesis that claims thinking "can be accounted for in terms of manipulation of symbols according to explicit rules"[19] more or less defines the research program nicknamed GOFAI (Good Old-Fashioned Artificial Intelligence).

Today this research program has been challenged both by analytical philosophers, such as John Searle, and by researchers inspired by Heidegger and Merleau-Ponty, such as Hubert Dreyfus,[20] but also by those analytical philosophers interested in *embodied cognition* and *situated cognition* who argue that, prior to abstract symbolic processes, cognition is an embodied agent's purposeful encounter with the world and furthermore an interaction with the objects, other subjects, and symbolic structures of an entire situation.[21] A still more radical position is Andy Clark and David Chalmers's idea of an *extended mind*: they claim that cognitive processes are not all "in the head" but that human reasoning leans heavily on external supports such as instruments, language, pen and paper, books, computers, and finally the entire culture so that "the individual brain performs some operations while others are delegated to manipulations of external media."[22] It is not just that not all cognitive processes are conscious, but they do not all happen just in the human "mind" (or brain) but also in objects (in paper on which we write, in books that we read, in computers, and more generally in all kinds of technical and cultural instruments). Just as there is no private language, there are no private technical and cultural objects either. This is why cognition is extended into cultural objects and into the human community whose objects they are.

In continental philosophy, the idea that thinking happens in the field consisting of both people's brains and their technical supports was formulated by the paleoanthropologist André Leroi-Gourhan in works published back in 1943.[23] Leroi-Gourhan shows how as soon as hominids became "humans," part of the evolution of their brain happened "outside of the head," in the tools that retain knowledge of how external nature functions (a silex axe "knows" how wood breaks when you hit it) such that the individual who wields the tool does not need to know this explicitly (it's enough to know how to use the tool) and also such that the knowledge can be shared by many individuals (when I give my axe to you, you can cut wood as easily as I can). If tools retain knowledge of natural processes, writing is another technique that preserves knowledge in a more subtle form, for it preserves linguistic messages. As Plato has shown in the *Phaedrus*, whose importance Derrida reveals in his now classic text "Plato's Pharmacy" (1972),[24] writing also conserves memory outside of the human mind, such that the individual can forget something but still possess it in writing and also such that it can be shared by many human beings (anybody who reads it, whether the original writer is present or not). Derrida shows that writing (any writing from handwritten notes to huge digital libraries) is a prosthesis of thinking that supports thought by conserving memories and by allowing their distribution far from their original inscription. But writing also handicaps thought because a reliance on written notes actually encourages forgetfulness and thoughtlessness, and writing detached from its author is defenseless against erroneous, sloppy, and malevolent readings. The philosophical consequences of Derrida's observations will be explained in chapter 4.

Today's digital supports of thought combine all these features and intensify them exponentially because digital and algorithmic life determines the entire lifeworld to such a large degree. Enormous digital archives not only conserve intentionally produced messages, preventing them from being forgotten or ignored, but they also conserve traces of all kinds of unthinking activity on the Net. Thanks to sophisticated programs, digital processes can function on their own without human intervention, and thanks to machine-learning technologies, they are to some extent capable of evolving without human intervention. This does not mean that they have become conscious. As N. Katherine Hayles has shown, computing machines belong to an *unthought* element of human thought/society that functions differently to the human mind but that nonetheless are an indispensable complement and supplement of human thought.[25] As Bernard Stiegler has shown in and after his seminal *Technics and Time* trilogy, technical supplements replace neither the morphogenetic memory of the body nor the epigenetic memory of the individual's experiences. This is why they are not necessarily conscious. Instead, they are an epiphylogenetic memory situated in tools, that is, a memory external to individual memory and precisely for this reason shared by entire communities.[26] This is why, as the historian Yuval Noah Harari notes in his bestseller *Homo Deus*, the digital supports of thought make us stronger as a species but also weaker as individuals.[27]

5. New technologies also modify our *affective* states in multiple and diffuse ways. Affectivity is a peculiar area of existence that cannot be attributed to either mind or body because it affects them both and actually shows their originary unity. It also cannot be attributed to either the individual or the collective because it often takes place on an impersonal

level where they cannot be easily distinguished (think of sporting, artistic, or political events). The technics that contributes to affectivity is also very diverse and reveals the originary unity of technologies, technics, and techniques. For example, living in a given technological context affects our feelings and rhythms (in a modern industrial city most bodies are well fed but chemically intoxicated; in a digital environment most minds fall prey to its enticements). There are also technically produced substances that are meant to affect our moods and affective states (pharmaceuticals, especially psychotropic drugs, recreational drugs, and alcohol). On the other hand, media, entertainment, and art also contribute to our affective states. On a mechanical level, this is because digital technologies (including machine-generated surveys of the suspectable public) make them increasingly omnipresent, but on a psychological level their ever finer techniques of expression can impose messages ever more efficiently.

These five types of reflections—death, birth, identity, thought, affect—on the effect of technological progress on human existence are the subject of intense and widespread discussion. Philosophy confronts them in deliberations on bioethics or transhumanism. However, neither bioethics nor transhumanism ask what the humanity is in its own right that the different technological supplements modify or overcome but take the meaning of this word for granted. On the contrary, a second—anthropological or more generally humanities—perspective on the human being thematizes this question.

Overcoming Humanity

The second perspective on the human being can be called anthropological in the very general sense of the science that asks what the human being is. This perspective sometimes has the same motivation as the posthumanist inquiries into the limits of humanity. It is informed by different human, cultural, and social sciences whose research object is the human being and which are reflected in philosophical anthropology. From a very general point of view, anthropological inquiries take the human being as their research *object* whose nature or essence they strive to define. Heidegger rejected anthropology in "The Age of the World Picture" because he considered that a science asking *what* the human being is cannot understand what he took to be the fundamental ontological question, namely *who* Dasein is: "Anthropology is that interpretation of man that

What Is the Human Being?   25

already knows fundamentally what man is and hence can never ask who he may be. For with this question it would have to confess itself shaken and overcome."[28] But Heidegger's rejection is undoubtedly too hasty. For if anthropology is indeed fundamentally an inquiry into human nature, at least modern anthropology shatters claims to a definite human essence rather than making and fixating them, and it studies humanity in its variations. This is why already in Heidegger's time, the so-called Philosophical anthropologists—especially Max Scheler, Helmuth Plessner, and Arnold Gehlen—defined the human being as an openness to and a capacity for change. In the middle of the twentieth century, the so-called poststructuralists (Foucault, Derrida, Deleuze, etc.), and later certain gender theorists (Butler, Haraway), interpreted the human being in an analogous fashion as a capacity for plasticity and figuration, as we shall see more precisely in chapter 4.

Historically, the simplest approaches to the "anthropos" have been quasi zoological, examining the human species with regard to its biological features (phenotypical typologies based on features such as skull shape or genetic definitions of the human species in terms of human descent or in terms of the human's demarcation from other species). Hegel had already rejected such biological reductivism (e.g., manifest in his time in phrenology) by remarking ironically that in such approaches "the being of spirit is a bone," which is for him the ultimate misinterpretation.[29] Modern anthropologies, on the contrary, are cultural and social sciences that do not build on human biology but are interested in the cultural and societal variations of humanity. Should we say, following Heidegger, that such anthropological variants nonetheless presuppose an unquestioned human essence? Or that the study of these variants aims to discover universal structures defining humanity? Examples include Sigmund Freud's theory of incest prohibition as developed in *Totem and Taboo* (1913),[30] which supposedly thematizes the fundamental matrix of human psyche, or Claude Lévi-Strauss's work, which illustrates the universal phenomenon that marks the transition between nature and culture in *The Elementary Structures of Kinship* (1949),[31] thus allowing the explanation of these elementary structures of kinship. Or should we affirm that the study of anthropological variants aims at understanding other ways of being human and even forms of humanity that include nonhuman beings like animals and spirits? The anthropologist Philippe Descola shows in *Beyond Nature and Culture* (*Par-delà nature et culture*, 2005) how the identities and differences between human beings and other entities have been explained (and are still explained) using four different ontological regimes (animism, totemism, naturalism, and analogism), of which only naturalism, the historical foundation for the Western scientific world vision, establishes a strict distinction between humanity and the rest of beings and defines humanity in terms of a distinctive subjectivity that accompanies language and culture.[32] Contrary to this, in animism, as in the Amerindian perspectivism studied by Eduardo Viveiros de Castro, "the common condition of humans and animals is not animality but humanity."[33] As against naturalism, which assumes all beings share a common nature but that different species are separated by different "cultures," animism proposes that all beings share a common humanity but they are separated by their different bodily natures. Furthermore, in totemism, there is a deep physical and psychical affinity between humans and their totems[34] and in analogism numerous general aspects are combined in a singular manner in each individual, whether human or nonhuman, such that all individuals are distinct but allow the discovery of different analogies between individuals of different kinds.[35]

Understanding what "humanity" might be does not mean neglecting animism, totemism, and analogism and regarding them as primitive world views that were happily overcome by naturalism, but rather it means understanding why all of these ontologies are not only still of contemporary relevance but also comprehensible by and even acting within all of us. When we take these perspectives into account, we see that the distinction between humanity and other beings is not as self-evident as the naturalist tradition would have it and that understanding the human condition might actually require understanding other beings' conditions as well. Or to put it in the words of the French philosophers Sophie Gosselin and David Gé Bartoli: "Contrary to the universality postulated from the unifying metaphysics of the European city, the experience of ontological duplicity opens human existence to *sylversatility* that it has been unceasingly suppressed. Contraction of 'sylvestre' (savage, forest) and of 'versatile' (capable of reversal, ontological duplicity or ambivalence), we call sylversatility the element of an experience of the world that exposes us to non-conventional, non-social, non-human, marginal or liminal entities, that is, to diverse manifestations of metamorphic powers of 'naturing' from whence a world can occur."[36]

Beyond the limits of the science of anthropology itself there are a number of human, cultural, social, and political sciences and special studies that struggle with the question of the idea of humanity and its limits.

The question is a sensitive one because it has political consequences: any idea of the human can become a normative paradigm and any norm can turn out to be biased and to generate exclusionary practices. The classic example of such an implicit bias occurs when the ideal type of human is tacitly or unconsciously thought to be male, such that women appear to be atypical, deviant, and in the worst case inferior variants of this type (and people who do not fit into either of the two main sexual categories are judged to be even more deviant). Thanks to the political and theoretical work of several generations of activists and researchers, not only in gender studies but of a variety of sciences that have learned to take into account the question of gender, it should by now be evident that these typologies are scientifically unfounded and that the exclusionary practices founded on them are unjust—although they still exist and are even given pseudo-justifications in pseudo-scientific discourses. Another classic example of the political consequences of human typification is the racism that defines the ideal type of humanity as white (and sometimes assumes that whiteness characterizes Europeanity, Americanity, and Christianity, which is of course false). It is obviously absurd to use skin color, facial features, or the like to justify political, cultural, and economical inequalities, but unfortunately this still happens. Today the principle of evaluating prevailing ideas of humanity in terms of discrimination that they may cause has become almost a method in certain branches of social studies that then use this principle as a starting point in their defense of the rights of different underprivileged groups (the disabled, the fat, etc.). For me, the most interesting of these discussions concerns the question of infancy because on the one hand it touches on a condition that all people have shared in their lives, and on the other hand it impacts the most longstanding traditional definition of the human being as an animal endowed with logos. Supposing that infants are as yet deprived of logos or that children have a lesser access to logos, is their humanity diminished by their linguistic incompetence?[37]

If one way of understanding humanity is to study its variants, another is to study its opposition to what is not human, to what can initally be called *animality*. Descola showed that the strict opposition between the human and animal is only a fundamental ontological characteristic of naturalism—but this is of a paramount importance as naturalism is the most influential of all contemporary ontologies because it is the ontology of the scientific worldview. Derrida and Agamben in particular have studied the division between human and animal not as a scientific belief but as the underlying metaphysical division that innervates the entire history of philosophy: "It is as if determining the border between human and animal were not just one question among many discussed by philosophers and theologians, scientists and politicians, but rather a fundamental metaphysico-political operation in which alone something like 'man' can be decided upon and produced."[38] "Animal" is by no means a biological entity, but the word designates a vast quantity of different forms of existence that only share the condition of not being human. Derrida inscribes this duplicity in the quasi-plural word *animot*.[39] As a classical philosophical concept, animality designates a condition that is common to humans and certain other living beings but from which humanity distinguishes itself through some specific capacity that has generally to do with logos. Man would thus be *zoon logistikon*, where *zoon* refers to animality defined as a capacity for autonomous movement and sensibility, and *logistikon*, to whatever differentiates humanity from the rest of animality, such as reason, speech, politics, and so forth. This is how humanity defines animals as its other with a view of knowing who itself is. Asking what animality is is an essential part of asking what humanity is, such that I ask who I am by asking who is the animal that I follow (mimicking Derrida's book title *L'Animal que donc je suis*).

However, the question of the relation between humans and other animals is not merely a metaphysical one: it is also an existential and a political one. Existentially, humans have almost always lived together with animals, and companionship with animals is an essential part of our self-understanding.[40] Other species contribute to who we are. Actually, even on a genetic level, the human genome is not hermetically sealed off from other species, our DNA contains elements at least of our Neanderthal and Denisovan cousins and our organisms contain nonhuman microbiotic organisms without which we would not survive.[41] However, when it comes to what are evidently other animal species, the modes of coexistence with wild and domestic animals have changed enormously since the advent of industrial agriculture. Especially in rich countries, the only domestic animals left will soon be companion animals like cats and dogs, while animals kept for food or clothing are hardly companions anymore but resources exploited in industrial settings. At the same time, this brings about the "sixth extinction," caused by humanity as it wipes out numerous wild species and rustic variants. As what used to be a coexistence thus becomes pure and often cruel exploitation and even extermination; the place of the animal in human society and the political community changes and it becomes a marker of human cruelty and culpability.[42]

An analogical but opposed "other" to the human is the machine. The question of the proximity of the human and the machine has an equal speculative weight to the question of the relation between the human and the animal.[43] Today the speculative machines that the human evaluates its own humanity against are generally either robots or artificial intelligences. If animals were supposed to share the condition of life with us but not the intelligence, robots are on the contrary supposed to share the condition of intelligence with us but not life. Contrary to intelligent machines, humans can investigate the nature of their own intelligence and more generally of logos. Like animals, intelligent machines too pose existential and political questions regarding the tasks that can be entrusted to machines or the rights and responsibilities of intelligent machines.[44]

All of these different lines of inquiry that investigate the sense and the limits of the human—different ways of being human, the human's animal, and machinic others—seem to make manifest a general suspicion concerning the pretensions of "humanity": isn't the idea of humanity ultimately just a huge prejudice? Shouldn't it be surmounted, isn't it a "phallo-logo-centric" ideology translated into an anthropocentric ideology? Shouldn't anthropocentric humanism be overcome in the name of a larger posthumanist vision of life, such as the ones defended by Braidotti, Wolfe, Sorgner, and many others? Or should we instead avoid defining humanity by means of any definite "others" of any putative "normality" and think of it as a plastic and versatile category that is forced to reflect upon itself when it encounters *inhumanity*, either in the world or in itself? As Jean-François Lyotard says in the introduction to *The Inhuman*:

> "What if human beings, in the humanism's sense, were in the process of, constrained into, becoming inhuman (that's the first part)? And (the second part), what if what is 'proper' to humankind were to be inhabited by the inhuman? Which would make two sorts of inhuman. It is indispensable to keep them dissociated. The inhumanity of the system which is currently being consolidated under the name of development (among others) must not be confused with the infinitely secret one of which the soul is hostage."[45]

We shall return to the question of the inhuman in the chapter 5.

Humanity and the Subject of Philosophy

The third perspective on humanity that will be treated in this introductory chapter is the philosophical question of the human being. It is by far the most complicated of the perspectives presented so far, not only because the general term *human* remains ambiguous and the perspectives on it are so varying, but as we shall soon see, because in philosophy the human being is not an unequivocal research object or a clearly determined concept but a fundamental problem that splits and divides as soon as philosophy looks at it. Perhaps it is nothing but the difference between two (or more) poles.

Terminological difficulties presage the philosophical problems attached to the term *human being*. Kant deems the human the center of philosophy when he lists the three defining questions of philosophy—What can I know? What should I do? What may I hope for?[46]—with a fourth one that grounds them all and that defines anthropology: What is a human being? (*Was ist der Mensch?*)[47] Translating *Mensch* as "human being" is somewhat misleading, insofar as Kant's *Mensch* denotes the human animal in all its primitive crookedness or warpedness,[48] whereas the "human being" evokes the humanist ideal of humanity. In French, *Was ist der Mensch?* was translated as *Qu'est-ce que l'homme?* and in English it was until recently translated as "What is man?" As the terms *Mensch*, *homme*, and *man* are becoming obsolete because they equate humanity with masculinity, they are now generally replaced by the term *human being*, which is also the term that I use. However, it is useful to note that this change to some extent dilutes the difference between the Greek *anthropos* and the Latin *humanitas*, which underlies the difference between theoretical and practical questions.

The complexity of the problem of the human being can be seen in this division between the scientific *anthropos* and the moral *humanitas*, but it becomes much more intricate when the notion of the human being is associated with the philosophical notion of the *subject*. The juxtaposition of the anthropological being and the philosophizing instance begins with Descartes, who, deciding to doubt all knowledge imposed by authority, founded philosophy anew starting from himself such as he was, a man like any other seeking calm and clarity in a warm room, as he recounts in the *Discours de la méthode*. But although Descartes re-inaugurates philosophy starting from himself (and claims that everybody else can do this too), he also immediately abstracts from his empirical self, preformatted as it is by education, beliefs, sensations, and passions, and goes on abstracting until he finds a truth so certain—I think, therefore I am—that it needs no empirical confirmation but is universally true. The evaporation of the empirical subject actually guarantees the truth of the cogito. "Who am I? The thinking thing!" is the unique, universal subject that everybody shares with Descartes at all times and places. It is this empty self-relation that assures the certainty of the self. Still this could never have been ascertained without the anthropological moment that Descartes discovered and discarded in the same gesture (for how could human animals, perhaps mere "hats and coats which may cover automatic machines,"[49] produce the sublime thought of cogito unless by suppressing themselves? Today we tend to ask, however, whether any being capable of the cogito, and not just the human being, can be the subject of philosophy).

The split between the empirical and the transcendental subject opens up almost unnoticed beneath Descartes's feet and spreads secret fissures throughout modern philosophy. It provokes a mighty earthquake in Kant's philosophy, which seems to speak of nothing else but the subject—although its declared theme is the human faculties rather than the subject and although what it affirms of human subjectivity is far from a simple thesis.

The central question of the *Critique of Pure Reason* is that of human knowledge, *menschliche Erkenntnis*.[50] Kant shakes up philosophy so intensely because he does not study objects of knowledge but subjectivity as an objective condition of objectivity. He calls such philosophy transcendental: "Ich nenne alle Erkenntnis transzendental, die sich nicht so wohl mit Gegenständen, sondern mit *unserer Erkenntnisart* von Gegenständen, *so fern diese* a priori *möglich sein soll*, überhaupt beschäftigt."[51] The true subject of critical philosophy is this transcendental subjectivity. The empirical consciousness only encounters a scattered multiplicity, transcendental subjectivity brings the unity of experience to this. But the transcendental subjectivity also brings unity to its own internal division between the incommensurable faculties of sensibility and understanding. Transcendental subjectivity is the capacity for creating unity by synthetizing the diverse and by schematizing the divided faculties. The subject is not a substance (like Descartes's "thinking thing") and the unity is not given, but the production of unity is a *pure activity*. Furthermore, the subject is not only divided into sensibility and understanding. There is also a more complicated division within the subject itself as shown especially in section sixteen of the transcendental analytic of the second edition. When speaking of the "I think" that must accompany all my representations, Kant distinguishes between the "I think" that accompanies the multiplicity of the intuition (pure or originary aperception) and my knowledge of this "I think." Pure aperception itself cannot be intuited, it can only be thought by means of transcendental philosophy, which shows that it is the transcendental unity of self-consciousness of which nothing further can be said.[52] This is why the transcendental dialectic criticizes the Cartesian idea that "I think" could be an object of experience, an idea that underlies the endeavors of rational psychology to define "I think" as a subject or as a substance.[53] Following the fine analysis of Jacob Rogozinski, the "I think" of *Critique of Pure Reason* is so pure that it has no face and no name, and "even though in my pure consciousness of myself, 'I am being itself,' in this opaque night 'nothing is given for me to think,' and the being that I am is nothing."[54] But on the other hand, in the *Critique of Practical Reason* the ethical law makes us capable of determining our existence, "as if the subject's non-permanence and non-identity did not forbid it to pose itself, to determine itself, to give itself a Law in an autonomous manner."[55] Although Hegel, in *Phenomenology of Spirit*, denies that Kant's categorical imperative ever has the capacity to command even the least concrete historical action[56] and therefore never allows the acquisition of a determined existence, many later commentators have tried to discover the possibility of historically determined existence in Kant's philosophy. While Rogozinski attributes this capacity to pure practical reason, others have related it to Kant's philosophy of history, which is based on his philosophical anthropology,[57] or directly to his anthropology.[58] We need not solve this problem that belongs, properly speaking, to Kant studies, but we do need to pay attention to the multiplicity and the split nature of the Kantian subject that is also revealed in this discussion. As we saw, the Kantian subject was divided first into sensibility and understanding, second into the pure synthesizing activity and the subject's knowledge of this very activity, and third into theoretical and practical activity (which can be synthetized in the *Critique of the Power of Judgment*, as Deleuze suggests). Each time the subject splits into incommensurable powers whose common ground, if there can be said to be one, is invisible to any kind of intuition and even understanding—but we can also see that the subject is the secret power that creates the unity of what is so profoundly divided.

Finally, as we already saw, the multilayered subject revealed by the critical project is, and is not, the same as the anthropological subject discussed in the *Anthropology* and also in Kant's various writings on philosophy of history. As Foucault observes in his introduction to his translation, the lecture course published as *Anthropologie in pragmatischer Hinsicht* (1798) appeared after the three Critiques but it was written for teaching purposes during the writing of these major works. As such, *Anthropology* is only a collection of empirical and necessarily incomplete series of facts about the human soul (neither *Seele* nor *Geist* but the untranslatable *Gemüt*), but it is nonetheless a philosophical anthropology. Christian Krijnen says that anthropology is for Kant a philosophical discipline, albeit not a pure critical one but only an applied, pragmatic philosophy.[59] Foucault, who in his introduction to his translation of the *Anthropology* looks laboriously for parallels between the critical project and the anthropology, later presents Kant as the thinker who sketches the possibility of a philosophical analysis of the present time, for example in *Was heißt Aufklärung?*,[60] which develops an anthropological view of man. Hence we can see that the question "What is the human being?" can refer to either anthropology as concrete empirical-historical knowledge of the human being or the critical project and the secret unity of the subject split into those three functions that provide the defining questions of the three critiques.

The aim of this short Kantian digression was not to present all facets of Kant's concept of the subject but to show why it is both so difficult and so fundamental. Like the subject of philosophy, which is split into its anthropological and philosophical components that are both incommensurable and must go together, the subject of philosophy is in itself split into sensibility and understanding, into theoretical and practical reason, into reflective and determining judgment, and into consciousness and self-consciousness, and it is split still further into other even finer distinctions. None of these distinctions can be explained by resorting to a stable common ground or result (as in scholastic dialectics where the opposition of thesis and antithesis is overcome in the synthesis). Instead, they are grounded in the activity of synthetizing that cannot itself be brought to a halt and rendered visible because it is the condition of all visibility.

Fichte's *Tathandlung* sums up and completes Kant's idea of the subject as pure activity, and afterward all German idealists strove to reconcile the purity of the philosophical subject with the reality of life. Hegel thematizes the philosophical concept of the subject much more explicitly than Kant. But Hegel's subject, brought to completion in the *Science of Logic*, is the Absolute and not an anthropological being. In *Phenomenology of Spirit*, where he shows the birth of the Absolute from concrete historical human experience, he certainly relates the finite and the infinite consciousness most vividly, but finitude there is not that of the anthropological being as such but that of different forms of historical consciousness that are not the singular *I* but are always already the *we* and already marked by the cunning of the Absolute. With respect to the present study, the divisions indicated by Kant are more interesting than the reconciliations effected by Hegel, for they show where the deepest problems lie.

The problem of the relation between the subject of philosophy and the anthropological subject remained subterranean in Descartes but became implicitly present in Kant. It moves to the philosophical foreground in phenomenology of the twentieth century. Husserl argued that the possibility of the new philosophical science of phenomenology depends on the overcoming of the natural attitude in epoche. As detailed in *Ideas*,[61] the epoche reveals the pure transcendental consciousness that consists primarily in the pure act of the cogito understood as directedness (toward phenomena and not just toward the thinking subject itself as in Descartes's cogito) and secondarily in the study of the structures of the intentional consciousness using the method of eidetic variation in particular. Husserl does not distinguish the transcendental subject from the empirical subject as confidently as Kant. He rather returns to the act of epoche ever anew, forever trying to defend transcendental subjectivity from the possibility of fading into mere empirical experience. Many later phenomenologists sought other ways of dealing with the split between empirical and transcendental consciousness: instead of protecting the purity of the theoretical consciousness they sought to conceptualize the impure structures of concrete practical existence. This characterized Heidegger's approach in particular. In his reading of Kant, he stressed the role of the fourth question, "What is a human being?" and interpreted Kant's transcendental subject as the movement of temporalizing, thus as pointing toward his own thinking of Dasein. Even so, Dasein's temporality differs from the Kantian subject because it is not the temporalization of experience but the temporalization of the existent Dasein itself. In a parallel fashion, Heidegger's interpretation of Dasein's intentionality as care differs from the Husserlian subject because intentionality is not that of a theoretical consciousness but that of (mostly unconscious) practical activity. This attention to concrete existence attenuates certain divisions of the subject (particularly between theoretical and practical) but it brings forth other divisions (particularly between authentic and inauthentic modes of existence). However, what is methodologically interesting is the emergence of a new relation to divisions and obscurity in philosophy. Instead of hiding them or trying to overcome them, they are now regarded as essential parts of subjectivity that demand specific approaches, for example, through receptiveness or indirectly.

Heidegger is important in the present context for another reason, however. He is the first to interpret human existence explicitly in terms of technics, which cannot be reduced to either theoretical or practical stances. (After Heidegger, it is possible to notice in German idealism and even in Kant certain indications of the possibility of thinking life in terms of technics. These signs are very interesting but they did not constitute the center of the thinking of these philosophers in the way that they constitute a very essential thread of Heidegger's philosophy.) In *Being and Time*, care is mediated by the tools used in the world, which are themselves interpreted from the practical totality in which they make sense. In Heidegger's later work, the world Dasein exists in is studied by drawing on two distinct interpretations of the ancient *techne*, namely art and especially poetry (*The Origin of the Work of Arts, On the Way to Language, Elucidations of Hölderlin's Poetry*) and technology (interpretations of the contemporary industrial technologies as the *Ge-stell* of the modern epoch).[62] Because he sketches out a then-unprecedented description of being-in-the-world as a technical activity, Heidegger is a necessary starting point for the problematic of this book as well. I will show how impulses from Heidegger's work were further developed by later philosophers (such as Foucault and Stiegler) into a thinking of the human being, not as self-consciousness but as self-technique in which technics constitutes the past in which the human finds itself thrown, the present as the non/technique of pure reflection, and the future as the art of inventing the future of existence. The present needs to be called a non/technique, for although reflection is also a technique of self-knowledge and self-constitution, reflection cannot be reduced to a technique only because that would amount to thinking the human being solely as its own construction. But reflection also includes an encounter with what escapes construction—a genuine *surprise of self*. Reflection also needs to encounter total otherness—exteriority, intimacy—that eternally flees technics' power, whether this exteriority is that of the world or that of the self itself. In this way, the study of the human technicity seeks to do justice to constructivist conceptions of human existence while at the same time showing why they undo and deconstruct themselves.

Technical Humanity?

In the introduction, the intellectual currents of trans- and posthumanism were used to thematize the contemporary urge to think humanity in terms of technics. Now we have seen how the naturalist conception of humanity, which transhumanism also presupposes, and the anthropological conception of humanity, which posthumanism also mobilizes, fall short of explaining technical humanity. They presuppose it but do not really say what humanity is if it can be so deeply affected by technics. Such a question needs philosophy, but the philosophical question of humanity is not untouched by the question of technics either.

How does technics affect the empirical aspect of humanity that was at the same time presupposed and yet transcended in the classical philosophers discussed above? If technics can go as far as modifying "humanity" (as body, mind, or existence), then the "humanity" that is defined by these elements must somehow be capable of transformation and is not an unchanging essence. It cannot be just transformable—a receptive *hyle*, a substance waiting to be formed—because it is also a transforming activity—a *dynamis*, a capacity for technical activity. What needs to be emphasized here is that technical humanity is not a form-giving mind acting on a form-receiving body, it is the entire mind-body, the comprehensive capacity that is capable of giving form because it was already capable of being formed. It is formatted and form-giving, bound and free, determined and inventive. Regarding a philosophical grounding of such a view of human existence, we can take recourse to its interpretation as plasticity by Catherine Malabou, as *formation des formes* by Juan Manuel Garrido Wainer, and as metamorphosis by Boyan Manchev[63] and to its interpretation as a capacity for transformation that emerges where forms are not fixed.[64]

Philosophical reflection on technological humanity is therefore not enacted by a mind examining its bodily support. It is another way of looking at technical humanity as a whole, at the entire comprehensive capacity for transforming and being transformed. It is a reflection that becomes possible when this capacity is somehow interrupted and transcended, such as happens in another epoche that turns around to look back at the technical human being in order to ask after the condition of its possibility. Before seeing how such philosophical reflection has been carried out and how it has been further developed in recent philosophy, we need to pay attention to the fact that what technics means is by no means evident. Among the different points of view described above, all of which are contemporary, the anthropotechnical viewpoint tacitly assumes technics to be mainly a mechanical procedure, while the anthropological viewpoint takes it as a skill and even a mode of creativity. This yields two contradictory ideas of what technics is. For philosophy, technics has

What Is the Human Being? 37

long since been an almost invisible, unthought domain. Undoubtedly some philosophers of technology have taken technics as their object, but philosophy has difficulties in thinking the role of technicity in philosophical subjectivity itself. This is why this book aims to study technics as a constitutent of the philosophical subject itself. But in order to do this, it first needs to untangle the multiple meanings of technics. This term can be used as a heuristic means of shedding light on the obscure regions of the subject by articulating and operating some of its constitutive splits differently. So while this chapter has examined the concept of the human being, the next chapter will examine the concept of technics. I will not seek to contribute to the philosophy of technology (e.g., the one developed for example by Don Ihde and Peter-Paul Verbeek) nor to study techniques and methods of thinking. I will instead show how technics contributes to human self-reflection. In so doing, I want to show how technics is not a simple object present to human beings and how the human is not the simple product of technics but rather to see how they each produce one another in a singular movement of originary technicity.

Chapter 2

What Is Called Technics?

In order to see how technics contributes to the self-understanding of humanity, we need to determine what is meant by technics.

On the simplest level, this is a question of terminology. The term *technics* is itself characterized by rupture and multiplicity.[1] With *technics* I refer to the phenomenon that ancient Greeks designate with the singular word *techne*, that Germans call *Technik*, and that the French call *technique*. This book is written in English, which has as many as three different terms for techne: *technology, technique,* and *technics*. Perhaps because English is not my mother tongue or perhaps because this English terminology is tailored to technological rather than philosophical needs, I find this multiplication of terms to be a cumbersome complication. In what follows, I wish to speak about the unitary phenomenon of techne and I will usually refer to it with the most abstract of the available English words, *technics*. This choice also indicates that I do not aim to contribute to what is generally called the *philosophy of technology* but to investigate the way in which the concept of technics affects and even constitutes the concept of humanity.

The concept of technics has a history and a geography and these naturally reflect technological evolution. This book is no more a history of technology than it is an anthropology, and I will accordingly only refer to this history when it bears upon the argument. The history of technics is not restricted to the history of the occurrences of this word and its derivatives, but it is also associated with the things referred to under the names of tools, instruments, organs and machines, arts and artifices, systems and mechanics, and so on, most of which cannot be referred back to specific historical situations but keep cropping up in different contexts.

In what follows, I will not investigate this multifaceted historico-semantic field for its own sake but I will consider the phenomenon of technics in terms of three distinct layers that contribute particularly to the concept of humanity: *techne, machine,* and *bio-technics.* Each has been the core phenomenon of technics at a given moment in history that, in its turn, has left its mark on the history of philosophy. Over the course of history, however, these different interpretations of technics did not occur in succession but coexisted, although their relative importance has changed over time.

These three perspectives also do not correspond to the three senses of humanity discussed previously, but they continue the investigation of the last, philosophical issue of how technics contributes to humanity. These considerations treat the question of technical humanity in terms of three layers. Firstly, technical humanity is form-giving because it is humanity endowed with techne (i.e., it has handy, crafty, skillful, inventive, and creative ways of knowing and forming the world and the human being itself). Secondly, technical humanity is formatted because humankind is adapted to a technical situation. When technical equipment acquires functional autonomy, it is called a machine. The fully developed industrial machine creates a situation in which the worker is submitted to mindless work not chosen by the worker. This situation creates the alienated humanity analyzed particularly in Marxism. Thirdly, before any definite type of technology and before any determinate figure of humanity or even of life, life and technics share a common material condition to which they can come together and fall apart in the first place. This materiality is nothing substantial: it is a dynamis, a power, or a capacity that can be partly formalized in codes and programs but that is not reducible to them. This is what I call general bio-technics. These three perspectives demarcate an epochal history. The first of these moments marks the conception of technics in antiquity; the second, the modern; and the third, the contemporary.

Heidegger famously interpreted technics as a mode of knowing. Here we shall also see that all versions of technical humanity have their rationality. Its bio-technical and especially cybernetical versions even create the illusion that they consist of pure rationality. But the rationality that belongs to technics is never the same as philosophical reason: it is at most the law governing bio-technical matter. In order to reflect upon technical humanity philosophically, its bio-technical rationality must be interrupted and transcended. Only in this way can we confront the multiple senses of technical humanity.

Antique *Techne tou biou*

In contemporary philosophy of technics, the most frequent understanding of *technics* is in the sense of technical objects and systems. However, originally the Greek word *techne* did not refer to things but to the subject who uses them and also to the subject independently of any tools and instruments. In ancient Greece, techne thus opens up the question of technics as a dimension of human existence. Techne is the form-giving know-how of the subject who gives form to its world and ultimately to itself. As Heidegger says in *The Question Concerning Technology*, techne was first and foremost a form of knowledge, which for him means a way of discovering the world. But as Foucault points out in his later lecture courses, the aim of this knowledge is to live a good life in the polis. Let us now sum up the fundamental sense of techne in antiquity.

According to the standard reading, the ancient Greek *techne* is knowledge as know-how, which implies knowledge of the world but focuses on the human activity of producing. It is generally translated into English as craft, skill, art, or a form of expertise, such as carpentry, medicine, music, or dance. Although Plato and Aristotle do not dedicate entire treatises to techne, both use the term frequently in comparison with the higher forms of knowledge such as *episteme* and *sophia* on the one hand and *phronesis* on the other.[2] Their question does not really concern techne in and of itself. Instead, they ask if there can be a philosophical techne or if dialectics surpasses the given rules and if there can be an *ethike techne* or if *phronesis* is higher than any techne. At stake is the possibility of education: if wisdom and especially virtue follow a rule, they are techniques that can be taught, and if not, they cannot (and the ill-behaving sons of Pericles[3] are incorrigible). While Plato, in the *Protagoras* and the *Republic*, was at least intrigued by the possibility of an *ethike techne*, Aristotle rejected it clearly in the *Nicomachean Ethics*[4] and keeps the *technai*, skills in determinate areas, apart from higher forms of knowledge that are not limited to particular areas but encompass thought and good life in general. After all, techne is the knowledge simple craftspeople have and not that of the free citizens who take charge of the polis.

As Bernard Stiegler points out in the introduction to his *Technics and Time 1*, since the classical texts only consider techne in its relation to other forms of knowledge, their considerations did not develop into a theory of technics as it is in itself. Nonetheless, the characteristics of techne as defined in Greece remain determinative for all subsequent theories of technics. In everything that follows, it will be useful to bear in mind the fundamental distinctions Aristotle makes in *Nicomachean Ethics*. For Aristotle, techne is the virtue of productive intelligence (*poiesis*), which applies intelligence to particular cases. Both *poiesis* and praxis are productive sciences, but poiesis produces a work that is exterior to the agent whereas praxis aims only good action itself, *eupraxia*, and thereby produces the agent itself (Aristotle, *Eth. Nic.* IV, 5, 1140b). Aristotle understands techne as a form of knowledge that differs from science. The latter's objects are necessary and eternal (*Eth. Nic.* 6.3, 1139b); however the objects of production (*poiesis*) and action (*praxis*) are variable. Every art is concerned with becoming (generation and corruption), that is, with "bringing something into being, i.e. with contriving or calculating how to bring into being some of those beings that can either be or not be, and the cause of whose production lies in the producer, not in the thing itself which is produced" (*Eth. Nic.* 6.4, 1140a). As a form of knowledge, techne has an intermediary status. On the one hand, it is more than mere experience (*empeiria*) because it is a form of intelligence based on reflection and especially on calculation. This is why it can also be taught. But on the other hand, it is knowledge concerning changing natural things instead of eternal and necessary principles, hence the art of techne is not infallible and its workings can be undone by contingent occurrences. "This is why its domain is the same as chance or fortune, as Agathon says—art waits on fortune (*tyche*), fortune waits on art" (*Eth. Nic.* 6.4, 1140a). In the *Metaphysics*, Aristotle further examines techne as a dynamis. "All arts, i.e. all productive forms of knowledge, are potencies; they are originative sources of chance in another thing or in the artist himself considered as other. And each of those which are accompanied by a rational formula is alike capable of contrary effects [; this is how] the medical art can produce both disease and health" (Aristotle, *Met.* IX, 2, 1046b). In the whole history that we will follow in subsequent chapters, we will constantly see technics associated to dynamis (capacity or power), which is neither universal necessity nor random occurrence but relative chance that can be calculated. This is why it belongs to the specific, historically changing domain of finite functionality between truth and truthfulness.

I would like to point out in passing that techne's kind of knowledge, which is neither knowledge of universal forms (like mathematics) nor knowledge of of simple passing impressions (*empeiria*) but means making calculations about changing, contingent realities, was rediscovered by some mid-twentieth-century continental philosophers who were interested in minor forms of knowledge that were neglected and even repressed by mainstream epistemology. Jacques Derrida in particular examined this in his studies of writing, the exemplary technical means of language, and thereby also of thinking. I will treat Derrida's thinking about writing in more detail in chapter 4, but I want to mention here that it is indebted to the ancient idea of techne. Like techne, writing does not inscribe the logos itself but only particular cases (linguistic expressions). This means it does not deal with eternal ideas but with changing contingent content. It is a rule that cannot be deduced but only repeated. It thus does not preserve a strict identity but merely iterates what is similar. As similarity implies difference, writing also allows both conservation and loss of signification, both truth and falsity. For a long time, Derrida's reflections on writing provoked indignation on the part of epistemologists who equated knowledge with the science of eternal necessary truths and saw Derrida's claims as a Trojan horse for relativism. Yet not all knowledge is of universals and especially not technical knowledge, which has now become an omnipresent feature of human life and science. Derrida's studies of writing are therefore important sources of inspiration to contemporary theoreticians of technics.

For the ancients, techne was also a secondary form of knowledge because its value lies outside of itself, in its product or work, *ergon* (sometimes translated into English as "function"). The value of the carpenter's craft lies in the house, the value of the doctor's skill lies in the health of the patient. In principle this distinguishes techne from nobler forms of knowledge such as prudence, which has its aim in itself, in the good life it produces. However, this distinction does not apply to all technai; for example, the product of music is the music itself, the product of dance is the dance, and so on.

Techne is motivated by its *ergon*. The ancients did not regard concrete things as being as valuable as the pure universal principles found in mathematics; to the moderns, techne did not appear as worthy as the laws and phenomena of nature such as gravitation or the movement of the planets. The *ergon* of techne did not appear as an esteemed object of consideration prior to being interpreted as a work of art. It was first in the Renaissance art and then in philosophy, especially in Kantian and post-Kantian philosophy of art, that art was seen as the reflection of nature's own creativity (Kant), the sensible manifestation of the idea (Hegel), or even the highest expression of philosophy itself (Schelling). In the eyes of an ancient Greek philosopher, the elevation of the work of techne into an expression of the Idea would not necessarily have been a good thing, since it implies a loss of control on the part of the technites over what is being produced. Thus, according to Plato, the power of the rhapsode Ion is not a real techne—a knowledge—because it is dependent on a kind of loss of knowledge characteristic of the rapture brought about by the divine gift of inspiration (Plato, *Ion* 533d).[5] But on the other hand, as Tom Angier notes, in *Phaedrus* Plato regards divination, *mantike*, as possibly being a techne even though it does not master its object but involves a certain loss of control, a kind of folly, namely the visions sent by God (Plato, *Phaedrus* 244b–45a).[6] Thus some exceptional technai, especially divination and poetry, are not simple techniques but depend on inspiration (enthusiasm)—or maybe even consist in a paradoxical art of enthusiasm. "If anyone comes to the gates of poetry and expects to become an adequate poet by acquiring expert knowledge of the subject without the Muses' madness, he will fail, and his self-controlled verses will be eclipsed by the poetry of men who have been driven out of their minds" (Plato, *Phaedrus* 245a).

The manner in which the work of art escapes from the strict rules of techne illustrates the character of all techne insofar as techne does not have total control of its *ergon*. In *Nicomachean Ethics* Aristotle said that the domain of techne is the same as that of *tyche*: techne does not control its object with the unyielding necessity of scientific logos but only with a relative rigor that can be affected by *tyche*, good or bad chance. What this means is clarified in Book II of *Physics*. Aristotle there defines natural beings (*physei onta*) by comparing them with technical products (*technai onta*), such that we can infer the character of the latter from the former.[7] Natural beings have the principle of their movement and their coming to be in themselves, whereas technical beings have it in another being, namely in the human being that produces them (Aristotle, *Phys.* II 192b). Now, first, as "art imitates nature" (*Phys.* II 194a), both natural and technical beings can be explained in terms of the four causes (material, formal, efficient, and final). However, through a curious inversion, Aristotle's explication of nature often imitates the explication of technics, which the latter is easier to grasp; for example, wood is the material cause of a house, the architect's plan is its formal cause, the activity of construction is its efficient cause, and sheltering humans is its final cause. Aristotle also explains on a more general level how *technai onta* are produced in aiming toward an end or goal, meaning that *physei onta* must be explained in terms of finality (*Phys.* 199a). Second, Aristotle notes that *tyche* (luck) and *automaton* (chance) are also called causes (*Phys.* II 195b) of physical and technical beings. Tyche is luck or fortune dependent on human deliberation, including its faults and mistakes. *Automaton* is chance or hazard dependent on unreasoning agents, such as monstruous births (*Phys.* II 199b). Although physical and technical objects are subject to necessity to a great extent, especially the necessity residing in the material cause, chance, especially the chance affecting the final cause, also contributes to their becoming what they are. This also explains why technical objects are never simple reflections of the technician's intentions. They must realize nature's own dynamics and develop possibilities first given by nature itself: "Art either executes what nature is incapable of doing or imitates nature" (*Phys.* 199a). Art seizes nature as its hyle, material cause, and it can bend hyle to forms that nature cannot produce alone, like transforming wood into a bed. But hyle itself is part of the physis, which can no doubt be used as material but which also has its own dynamis, its own productivity that enables many different developments. For example, we can think of Odysseus and Penelope's bed, carved by Odysseus from a unique living olive tree that still had its roots in the ground and around which their palace was built (Homer, *Odyssey* book 23). One could imagine that this tree invited Odysseus to change his ideas of what a bed can be.

We saw how the divine inspiration that alters the artist's aims is one opening that *tyche* can take to interfere with the work. There is another opening to be found here, albeit one Aristotle does not really elaborate on, namely, that the potentiality for physis in hyle can allow spontaneous strokes of luck (*automaton*) to sneak into the work, for example the accidental flaws, faults, exceptions, and surprises that arise from the material and undo the craftsman's original plans, sometimes even contributing to the discovery of new ways of doing things. The technites can produce his work because he knows how to use the element of the physis as the work's hyle, but he cannot master the entire physis. This is why his control of the work is only ever partial. Aristotle's notions of luck and chance explain, from the restricted point of view of individual beings, the phenomenon that pre-Socratics explained in cosmological terms: human craft and cunning are bound to fail before the disquieting forces of physis and moira.[8] Or, as Heidegger puts it in his famous interpretation of a chorus of Sophocles's *Antigone*, human techne (which Heidegger insists in translating only as knowing and not as craft or skill) is in an infinite conflict with *dike* (the law of the totality of being) such that *dike*, as "the most violent," always limits and overwhelms the violence of human techne.[9]

In antiquity, techne was a craft whose value could lie in its result but whose reality nonetheless resides in the person who exercises the skill. This is why the Greeks saw the question of the techne as essentially a question about the technites, that is, the craftsperson. Once again, the craftsman is not really made into a question in and for itself: different texts refer to the practitioners of different skills—carpenter, navigator, doctor, musician, dancer, midwife[10]—but not with a view to knowing what their skill consists in but in order to provide metaphors that help to understand the higher skills of philosophy and ethical life. We know that midwives assist parturients, so we understand what Socrates means when he says that he wants to help people to deliver their thoughts; we know that doctors heal bodies, so we understand that philosophers ought to heal souls. What really interests the Greeks are these general skills of life that go beyond all particular crafts: Can living in a truthful and virtuous manner be taught, is there a *techne tou biou*, an art or a skill of life itself?

As we saw, Plato and Aristotle were intrigued—and hesitant—about the possibility of the human being's being its own work. On the one hand, the *ergon* of praxis is praxis itself, and of course, whatever a man does makes him into what he is. On the other hand, a virtuous man shows excellency in his praxis, and he would probably be even more excellent if he was educated to virtue already in his youth. Education would be easier if there was an *ethike techne*, an ethical technique that could be taught and that could be applied in real-life situations. But is there a *techne tou biou* that produces a virtuous person?

This question, for a long time marginal among historians of ancient philosophy, was brought to the fore by Michel Foucault,[11] especially in his last works where he examined techniques of the self (*techniques de soi*),

> which is to say, the procedures, which no doubt exist in every civilization, suggested or prescribed to individuals in order to determine their identity, maintain it, or transform it in terms of a certain number of ends, through relations of self-mastery or self-knowledge. In short, it is a matter of placing the imperative to "know oneself"—which to us appears so characteristic of our civilization—back in the much broader interrogation that serves as its explicit or implicit context: What should one do with oneself? What work should be carried out on the self? How should one "govern oneself" by performing actions in which one is oneself the objective of those actions, the domain

in which they are brought to bear, the instrument they employ, and the subject that acts?[12]

In setting out to write a history of the techniques of the self, Foucault found that this idea was richly developed in antiquity but then marginalized and even suppressed after what he calls the Cartesian moment in philosophy that postulates that knowledge is the only subjective condition of truth.[13] For Foucault, the matricial presentation of self-technics can be found in Plato's first *Alcibiades*, which opens up the problematic, unfolds it in its complexity without yet closing it off by providing a rigid solution. In the *Alcibiades*, Socrates engages a dialogue with the young, beautiful, noble, and wealthy Alcibiades who has mighty ambitions for governing his city and perhaps other cities and countries as well. Socrates shows the young man that his ambition of giving counsel to the city of Athens has a weak basis because he does not know what justice and injustice are as he was never educated to virtue (the youth's foster father, the wise and virtuous Pericles, was decidedly incapable of giving a proper education to his sons and wards). Socrates then proposes something that Alcibiades, too, becomes aware that he desires, namely an education in virtue and justice. But unlike the sophists, Socrates does not sell to the young man any *ethike techne*, as if there were a rule, a method, or a neatly formulated guidebook that could be passed on. An ethical agent is not a substantial *ergon* of a techne, he is a subject. A man skilled in the good life has a similar relation to himself as the technites has to his work (the dialogue is full of craftsman metaphors), but there is also something more to the former. As Foucault says, there is a difference of end, object, and nature between the techne of the doctor healing himself and the techne which allows a person to take care of himself.[14]

Socrates tells Alcibiades that he should learn to take care of himself, *epimeleia heautou*. He should thus follow the Delphic advice "know thyself," *gnothi seauton* (Plato, *Alcib.* 119, 124, 127, 128c–29a). The proper relation to truth and justice presupposes a proper relation to oneself. Foucault draws attention to the fact that the knowledge of oneself Socrates recommends to Alcibiades in this dialogue is not a theoretical knowledge of ideas (e.g., knowledge of geometry and of other theoretical sciences required of the future philosopher-king in Book VII of the *Republic*). It is rather a practical knowledge of justice needed to govern the city but especially to govern oneself. And how—through what techne—does one learn justice and virtue? There are no formal rules of the art that are to be observed, as in music and gymnastics, which Alcibiades has already mastered. The only way get to know oneself such that one can take care of oneself is dialogue—the very dialogue with the lover-mentor to which Socrates has invited Alcibiades. Dialogue is a technique without technique, a constant self-discovery in interrogative but kindly discussion with a friend-master. The dialogue also has an instrument that is nothing like an ordinary technical tool: it is the eye of the master, the very pupil of the eye, which functions as a mirror in which the pupil can see itself (*Alcib.* 132c–33c). This takes place in a wonderful movement: no eye can ever see itself seeing, but each eye can see itself seen by another eye and this is how the look of the other helps each to see the self. The dialogue's conclusion is an *ergon*, which is nothing like a thing. It is the good life in the political community enabled by the knowledge of what is just and what is not, or better, by the capacity for looking for it in each situation.

And how do the mentor's kindly looks and questions form the pupil to virtue and justice? Unlike doctors and teachers of gymnastics, the philosopher-mentor does not aim at an ideal form of a substance, such a healthy, skillful body. The philosopher-mentor aims to form a person to act well, that is, to foster in the student a dynamis, a capacity for acting virtuously. Good action means that the soul governs the body and furthermore that it uses (*khresthai*) the body in the best way (*Alcib.* 129d–30b). Like the carpenter who knows how to use tools but in a superior sense, the soul must learn to use the body and more generally all possibilities of life. The body is the universal instrument, the instrument that uses instruments, and more importantly, it is the instrument of life. It is not a substance but a dynamis, a capacity to produce and to act, and the soul must learn to govern it to act virtuously. We will see later why *khresis* became important to Foucault[15] as well as to Giorgio Agamben, who developed the concept of use in *The Use of Bodies*.[16]

In his lectures, Foucault shows how this idea developed in the Hellenistic and Roman periods, especially in Stoicism. Rather than make the considerable detour of going through those texts, I will simply enumerate Foucault's main observations concerning them.[17] Firstly, like in Plato, the aim of the Stoic *techne tou biou* is man himself, whereas in later monastic Christianity the aim of the exercises of the soul is to learn to renounce oneself. However, contrary to Plato, the Stoics do not think that the main motivation for the techniques of the self is the governing of the city but rather the individual, eventually living with friends but in seclusion from political life. Contrary to Plato, the Stoics also think that the *techne tou biou* is the task of a whole life and does not just pertain to education in youth.

The object of the good life still has a connection to the polis, but Stoic cosmopolitanism equates the polis with the cosmos, meaning that life according to nature should also be characteristic of the rule of political life as well (although often it is not). More important, however, is the formation of oneself by breaking bad habits, by hardening oneself for combat, and by purging oneself of bad passions and especially of the fear of death. The means used in this care of the self is still other people, but instead of living dialogue with a master this is now frequently mediated by writing. For instance one writes personal notes that help to reflect upon oneself, such as Marcus Aurelius's *Meditations to One's Self* (*Ta eis heuton*), or one writes letters to friends, such as Seneca's correspondence with Lucilius.[18] The care of the self becomes an exercise with more formal instruments than in Plato, involving for example the reading of texts of important writers, memorizing and repeating their sentences, writing carefully about what one has done and thought, and rereading one's texts and meditating upon them. Maybe one could say that these more formal instruments of self-technique produce a more disciplined, less creative self. This does not mean that the self could ever be just an instrument with a stable form, a definitive product of self-producing technologies, as "constructivist" interpretations of Foucault claim he aims at. The true potential of self-techniques is rather revealed by the *Alcibiades*, which shows how the self is gradually formed in dialogue and in community, formed by dialogue and dialectics that are never fixed doctrines but always ways of opening up new questions. In this sense, the *Alcibiades* already foreshadows what Foucault takes to be the self-techniques of our time: invention, creation, experimentation, and production of ways of life.[19]

This is how the first sense of technics culminates in the technics of the technician itself, as concretized first in the art of education and ultimately in the paradoxical skills of life and of philosophy. As we have seen, in the antique context technological humanity is hardly anything other than humanity itself: the ideal of a virtuous and excellent life. Antiquity is conscious of the technical means of ethical life, but it regards them as simple means that disappear before the aim and ignores its eventual negative effects that come to the fore in modern philosophy.

Modern Machines and Instruments

But what about the technical object itself? Antiquity did not pay much attention to it, but it sowed nevertheless the idea of a lucky chance provided by nature, *automaton*, that the artisan might be lucky enough to seize (*tyche*) in a clever manner (*mehane*).[20] The dawn of modernity is marked by the desire to control that chance better thanks to science. The technical object had always been thought as an instrument, *organon*, of the human body (rather than its mind). Better instruments could be built by looking at the organism of which the instrument was the extension (as Ernst Kapp puts it) and by imitating its structure. The instrument develops in imitation of the organ (like the telescope imitates the eye). The *nec plus ultra* of the instrument is the automat, which does not evoke the stroke of good luck anymore but the self-moving mechanism built in imitation of an entire organism. Mechanism is no more a simple means-of (a telos external to it); it becomes the arrangement of moving parts that appears to have a telos of its own in the minimal sense that it has a movement of its own. Machine is built in imitation of the human body or of life more generally. At the same time it becomes the model against which modern medicine tries to understand the workings of the human body and of living organisms in general. Twofold relation of imitation thus connects the organism and the machine, each one imitating the other, and each one also turning out to be different from the other. That is the law of imitation: it connects by similarity where identity cannot be reached.

The properly modern interpretation of technics culminates in the idea of the machine. The era of the machine is not a precise historical period. Machines have existed well before philosophical antiquity. But it is only in the modern era that machines seem to organize and orient an entire civilization. In *Technics and Civilization* (1934), Lewis Mumford dates the beginning of the Western machine era to the thirteenth century, when the mechanical clock was introduced into the monastic life, not in order to know time (clock-time is not a natural time) but in order to organize time into regular units that allow the organization of an entire life into regular actions.[21] These are the first premises of a modern time-reckoning civilization that now grids the entire world. The logic of the machine pervades the entire intellectual life of the seventeenth century. The invention of the experimental method in science is based on the same principles as any clockwork, namely abstraction from reality, neutralization of the observer, and repetition of the same operation.[22] Experimental science also relies on machines used as scientific instruments. As Koyré says, the *perspicillum*—the telescope Galileo describes in *Sidereus Nuncius* (1610)— can be considered as the first truly scientific instrument, and it actually inaugurated a new phase in science, the instrumental.[23] The development of the printing press was also central to the development of sciences. But over and beyond any singular technical invention, Mumford shows that "the machine" is a general term that can be used as shorthand for an entire technological complex that was born at the beginning of the modern era[24] as well as for the mindset that underlies it. The foundations of this mindset are laid out in the modern science (Galileo, Newton, Bacon), in modern philosophy (Descartes, Hobbes), and in new art (Leonardo da Vinci), whose works launched the mechanist worldview that has transformed the face of the earth and society.

Although recurrent in all these works, the idea of the machine does not have the clear and distinct contours of mathematical truths from which it is supposed to stem. It seems to me that it is above all the imitation of a living body with its imperfections and irregularities rather than a perfect incarnation of pure mathematics. Mathematics is one tool for the better design of machines, but in addition to this, the material element always remains submitted to chance. The science of the modern era takes technics to be primarily a means and not an end, and this is why it tends to leave the essence of technics unquestioned. When it pays attention to the technical object itself, it generally runs together instrumental and mechanist interpretations and thus creates results such as the following:

(1) The machine is frequently initially interpreted in an instrumental setting. From this point of view, the machine appears as an extension of the instrument such that all machines are instruments although not all instruments are machines. The instrument is thereby seen as a means by which human intentions can be realized. These intentions can be subject to moral evaluation but the means can only be evaluated as to whether they serve these intentions well or not. From another perspective, however, the instrument is also a thing in its own right: it is a tool that extends and enforces the activity of the working body (and mind).[25] The Greek word *organon* captures the connection between the body and the instrument because the organon means the instrument and the tool but it also came to signify the organs of the body. This is how the organon brings in the mimetic relation between the organic and the artificial: the instrument imitates (prolongates) the organ, but on the other hand the organ is understood by way of comparing it with a mechanism. This

comparison becomes particularly important in early modern mechanistic science, which explained the organs of the body as instruments and the entire organism as a mechanism.

(2) When considered in more detail, the machine differs from the instrument because it can function without the organism that wields it. Mumford defines the machine in contrast to the instrument by saying that unlike the tool, the machine is not only manipulated but it tends toward autonomy, such that it does not merely mime an organ but an entire organism.[26] Like the organism, the machine works on its own and thus can serve as a metaphor for the living organism.

(3) Of course, in the end the machine is not a real organic whole but an assemblage of moving interconnected parts reliant on an external source of energy that repeats the same operations indefinitely and that can only tend toward autonomy. A well-made machine fits tightly together and repeats exactly the same movement tirelessly (like the clock Mumford describes). It works so effectively because it abstracts from contingencies irrelevant to its function (like the darkness of the night). But it alienates the beings that are submitted to it from their natural course precisely because of this abstraction. Thus the introduction of the clock alienates people from the natural rhythms of their minds and their bodies as well as from environing nature. Criticism of the machine is based on the fact that the it creates an artificial totality that is at odds with its organic counterparts, as already described in Hobbes's *Leviathan* and in Descartes's reflections on man-like automations: the machine is destructive (of the natural state of things) precisely because of the effectiveness of its capacity for abstraction.

The machine metaphor contributes to our question concerning technological humanity both on the level of thinking (i.e., the *differentia specifica* of the human being) and on the level of the understanding of the human body. The ambiguity of the metaphor suggested above is manifest on both levels.

For the ancients, thinking was a pure intellectual activity, but at the outset of the modern era, thinking discovered the utility of instruments and machines, which two terms it at first uses interchangeably. The paradigmatic early modern study of instrumentality is Francis Bacon's *Novum Organum* (1620), which means precisely *new instrument*. As is well known, the new instrument proposed in this treatise is the new experimental science—the *Novum Organum* counters the old *Organon*, Aristotle's logical treatises. Bacon criticizes the reduction of science to logical deductions and sophistry, which he sees as the Platonic and Aristotelian heritage, and he calls for a philosophy of nature that only the pre-Socratics had a notion of but that has since then largely gone to waste. Against the "errors of false philosophy," based on pure reasoning, Bacon's new instrument for knowledge should take the reader "close to things themselves."[27] The instrument of the new natural science is first of all the experimental method itself: "The true order of experience . . . first lights the lamp, then shows the way by its light, beginning with experience digested and ordered, not backwards or random, and from that it infers axioms, and then new experiments on the basis of axioms so formed."[28] Natural philosophy progresses with order and method, it moves back and forth from experience to axioms that might explain them and from the new axioms to new experiences that might validate or invalidate the axioms. But in order to do this, one needs to pay heed to the mechanical arts—symbolized by the "lamp" that must be lit and directed. Mechanical arts intervene here in several ways.

On the one hand, mechanical arts provide instruments that assist the senses, for instance the telescope. Bacon emphasizes that the simple use of senses does not suffice to make a scientific experience, but experiences must be worked over and evaluated methodically: "The subtlety of experiments is far greater than that of the senses themselves even when assisted with carefully designed instruments; we speak of experiments which have been devised and applied specifically for the question under investigation with skill and good technique."[29] Bacon is adamant that empirical findings must then be arranged, coordinated, and put into tables and other written documents.[30] This is another way in which the methodical study of the results derived from experiences requires the use of technical aids, namely documents written in order to preserve, share, and allow consideration of experiments, and ultimately also the use of the printing press, which enables the findings to be widely disseminated. Finally, science needs abstract mathematical instruments, as shown particularly by the success of the differential calculus invented by Newton and Leibniz.[31] In sum, while the instruments of ancient science were only the operations of Aristotelian logic, the instruments of the new science include instruments used in experiences, written documents, and mathematical operations.

On the other hand, nature must be examined in its relation to mechanical arts in general:

> And as for its composition, we are making a history not only of nature free and unconstrained (when nature goes its own

way and does its own work) such as a history of the bodies of heaven and the sky, of land and sea, of minerals, plants and animals; but much more of nature confined and harassed, when it is forced from its own condition by art and human agency, and pressured and moulded. And therefore we will give a full description of all the experiments of the mechanical arts, all the experiments of the applied part of the liberal arts, and all the experiments of several practical arts which have not yet formed a specific art of their own (so far as we have had an opportunity to investigate and they are relevant to our purpose). Moreover (to be plain) we put much more effort and many more resources into this part than into the other, and pay no attraction to men's disgust or what they find attractive, since nature reveals herself more through the harrassment of art than in her own proper freedom."[32]

This does not necessarily mean that man's sole task is the domination and harrassment of nature, as Bacon's twentieth-century critics have often maintained. It means that nature only reveals itself through the mediation of art (technics) and art functions only if it conforms itself to nature's own processes. "What is absolutely needed, is to do a thorough survey and examination of all the mechanical arts, and of the liberal arts too."[33] Given that men had hitherto invented many useful things in the mechanical arts by chance, Bacon hopes that "when they do so with method and order, not impulsively and desultorily, many more things are bound to be uncovered."[34] Mechanical arts are thus an essential part of Bacon's program. However, as Sophie Weeks has shown, mechanics is a complex notion for Bacon. It means (1) the empirical work of artisans, unreflective and based on chance; (2) *Experientia literata*, which "proceeds by putting together former inventions"; and (3) philosophical mechanics, which is "connected with physical causes."[35] Only the last really transforms mechanical arts into tools of inquiry that are not simple *experimenta fructifera*, fruit-bearing experiments, but *experimenta lucifera*, light-bearing experiments.[36]

In Bacon's philosophy, the mechanical arts contribute to natural philosophy. But at the same time, although arguably not in his own work,[37] nature itself comes to be seen as a mechanism. The great early modern thinkers—Newton, Hobbes, and Descartes—depict both the system of the world and the system of human life as mechanisms. The machine is not considered as a concrete instrument of experimentation but used as a representation that helps to clarify natural processes. "Machine" is a structuring metaphor that helps to conceptualize the workings of the world, of the animal body, and even of certain features of the human mind. Moreover, the machine's logic of cause and effect became an analogon of scientific and philosophical activity.[38] Of course, at that time when religion controlled what science could say, it was safer to present natural mechanisms as built by God, represented as a celestial artisan. However, in the last instance the machine metaphor was meant to represent nature as an autononmous system that does not need the hypothesis of the maker. It therefore led naturally to the eighteenth-century materialist and atheist science and to the twentieth-century idea that God is not a scientifically or philosophically valid hypothesis. This is how, on the basis of materialism and mechanism, the scientific world view ended by standing on its own feet, which is itself a remarkable achievement. Today, thinking—and especially scientific thinking—relies heavily on all kinds of technical aids, as mentioned below: scientific instruments, abstract techniques of guiding thought, and machine-based metaphysical models. The question does not concern whether or not to use these aids but to what extent their contribution to thinking can be limited to an obedient instrumentality and to what extent the instruments of thinking have grown into automatisms that escape human control.

As we noted, the machine metaphor also contributed to the question of technological humanity by providing a mechanical model of the human body (and mind).

Descartes was the first to introduce the machine metaphor into the biological sciences. He uses it most extensively in his *Treatise of Man* in which he gives a totally mechanical explication of the human body and of mental functions such as the senses, imagination, and memory. He invites the reader to compare man with the machine in order to understand the functions of man's body more easily as follows:

> *Introduction:* These men will be composed, just as we are, of a soul and a body. . . . I suppose the body to be just a statue made of earth, which God forms with the explicit intention of making it as much as possible like us. . . . We see clocks, artificial fountains, mills, and other similar machines which, even though they are only made by men, have the power to move of their own accord in various ways. And, as I am supposing that this machine is made by God, I think you will agree that

it is capable of a greater variety of movements than I could possibly imagine in it, and that it exhibits a greater ingenuity than I could possibly ascribe to it.

Conclusion: I desire that you consider that all the functions that I have attributed to this machine, such as the digestion of food, the beating of the heart and the arteries, the nourishment and growth of the bodily parts, respiration, waking and sleeping; the reception of light, sounds, odours, smells, heat, and other such qualities by the external sense organs; the impression of the ideas of them in the organ of common sense and the imagination, the retention or imprint of these ideas in the memory; the internal movements of the appetites and the passions; and finally the external movements of all the bodily parts that so aptly follow both the actions of objects presented to the senses, and the passions and impressions that are encountered in memory: and in this they imitate as perfectly as is possible the movements of real men. I desire, I say, that you should consider that these functions follow in this machine simply from the disposition of the organs as wholly naturally as the movements of a clock or other automaton follow from the disposition of its counter-weights and wheels. To explain these functions, then, it is not necessary to conceive of any vegetative or sensitive soul, or any other principle of movement or life, other than its blood and its spirits which are agitated by the heat of the fire that burns continuously in its heart, and which is of the same nature as those fires that occur in inanimate bodies.[39]

Descartes thus explains the functioning of the body by way of comparison with a machine. Georges Canguilhem says that the machine can explain everything but "it cannot account for the construction of machines." This is why the machine metaphor leads Descartes to infer the creator God from the existence of the body just as we infer from existence of the machine that of the artisan who built it.[40] However, if the metaphor is understood as a metaphor and not as a model, the hypothesis of the creator God is not inevitable. Such a reading was developed somewhat crudely by La Mettrie in *Machine Man* (1748)[41] and in a more sophisticated manner by D'Holbach and Diderot, whose *D'Alembert's Dream* (1769) is an elaborate textual construction representing the entirety of human sensibility through the charming metaphor of an affective harpsichord.[42] Today's medicine is possible only because body is regarded simply as a mechanism.

Between the general machine metaphors that open and close his treatise, Descartes gives detailed mechanical explications of the different functions of the body and of the mind, adding many drawings and diagrams to illustrate his points. Although the mental faculties are explained mechanically, Descartes maintains the idea of a soul that is not reducible to mechanics. He explains this more clearly in the fifth part of the *Discourse on the Method*, where he likens humans and other animals to automatons, which are made by the hands of God and therefore are far more sophisticated than the automatons made by man. However, he also mentions two things that even the finest automatons cannot do, namely speak (instead of just imitating sounds like parrots do) and reason:

> And the second difference is, that although machines can perform certain things as well as or perhaps better than any of us can do, they infallibly fall short in others, by which means we may discover that they did not act from knowledge, but only from the disposition of their organs. For while reason is a universal instrument which can serve for all contingencies, these organs have need of some special adaptation for every particular action. From this it follows that it is morally impossible that there should be sufficient diversity in any machine to allow it to act in all the events of life in the same way as our reason causes us to act.[43]

Descartes does not ask what a machine is. He takes readers' familiarity with machines for granted: their world was one that was increasingly populated by different mechanical masterpieces whose ingeniosity stimulated the imagination. The automatons in theaters and in public spectacles were not really meant to be taken as real humans and animals, except momentarily and by surprise. The marvelous machines whose lifelikeness proved their ingeniosity were finally, obviously, only just machines to be admired and toyed with. In the same way, the machine metaphors in Descartes's texts are not meant as statements of how the world and the human body really are but instead as providing an image that makes their functioning more intelligible, that makes their invisible springs visible—and the scientific text more agreeable to read. Like the numerous drawings and figures in the *Treatise of Man*, the machine metaphors are an essential part of the explanations. But at the same time, Descartes emphasizes that the mechanical images are just images, that the automaton is not a real human being (because it is morally unthinkable that it were so). Always accompanying the idea of the machine is this ultimate difference between the machine and the living being.

In his rich article "Legitimating the Machine: The Epistemological Foundation of Technological Metaphor in the Natural Philosophy of René Descartes," Andrés Vaccari provides a detailed examination of Descartes's use of the machine metaphor. He stresses two important things. Firstly, the machine metaphor is, as we already saw, an epistemic tool. Although Bacon had banned analogies and other rhetorical extravagances in his description of the "idols of the market" in *The Great Instauration*, Descartes trusts that his readers know how to read metaphors and uses them freely in order to make visible elements and mechanisms that would otherwise be invisible. In *Principia*, he says: "I do not recognize any difference between artefacts and natural bodies except that the operations of the artefacts are for the most part performed by mechanisms that are large enough to be perceived by senses."[44] The use of mechanical metaphors is all the more consequential since the aim of knowledge of nature is manipulation and intervention—the betterment of the conditions of human life. Secondly, Vaccari shows that "what matters to Descartes is not this or that machine, but the laws of all machines, the very ontology of machines."[45] The machine is not just a metaphor among others but is the "meta-analogy," an overarching image that sustains Cartesian metaphysics. Etymologically, the very notion of machine supports this understanding: *machina* also means "framework, scaffolding."[46] Contrary to the antique techne that dealt partly with potentialities that it could not entirely rationalize, modern mechanics makes the world rational without remainer by "breaking things down into elements and mechanical action."[47] It provides a clue, a model for the functioning of science. It is a metaphoricity that is "constitutive of thought itself, essential to the very possibility of science. . . . Yet metaphor can be conceived as the very condition of the possibility of thought—suggesting the prospect that all logos, in the ends, amounts to nothing but techne."[48]

The Cartesian machine epistemology and ontology helped in understanding the human body and mind by likening their obscure functioning to the more abstract functioning of machines. While Descartes, La Mettrie, or Diderot's machine metaphors were only philosophical fictions, the same deepfelt analogy between living beings and machines drove the most ingenious artisans of the seventeenth and eighteenth centuries to build

What Is Called Technics? 59

automatons that were as lifelike as possible. This dream of constructing a machine that imitates life lives on right up to today's efforts to build anthropomorphic robots (e.g., Hiroshi Ishiguro's *Geminoid HI2*). This is how the theoretical fiction of the analogy between life and machine, and even between mind and machine, remains an important tool of the human being's speculative autoreflection.

The machine is not only a theoretical tool and a speculative construction, but concrete machines have very concrete effects on human beings, their societies, and their environments. The idea of the machine has had epoch-making practical, moral, and political consequences, for technics emerged not simply as the instrument of human intentions, but the technical system emerged as an autonomous machinery in which human beings are but powerless cogs. Since industrialization became firmly established in the nineteenth century, many authors have focused on the dehumanizing effects of the mechanical civilization. They are so well known that a brief reminder suffices for our purposes.

1. When a human body is seen as a machine, it can be treated as one. This has undoubtedly enabled spectacular progress in medicine, but it has also enabled the emergence of disciplined and ultimately alienating industrial work in which the human being is robbed of freedom of movement, imagination, and skill.

2. Like the thirteenth-century monks who became adapted to the rhythms of the clock and thereby to monastic discipline, the nineteenth-century factory worker became adapted to the machine and thereby to the entire socioeconomic system that organized production mechanically and ultimately adapted to the political systems reliant on industrialization—which came to include both the capitalist West and the socialist East. Both systems saw human beings as a workforce to be disciplined but also reproduced and hence, at least to some extent, even to be cared for.

3. Today, as the theoreticians of the anthropocene point out the most forcefully, it is obvious that not only human society but also its natural environment, indeed the entire planetary nature (animal and plant species, ecological systems, the world climate, water, and even geological systems), carries

traces of the machine civilization, the worst of which are climate change, the sixth extinction, and the different forms of pollution (nuclear, plastic, chemical, etc.).

While these givens are obvious (although some have heeded them more than others), the diagnosis of the underlying causes have varied. The main divide is between those who attribute responsibility to machine logic itself[49] and those, mainly following Marx, who attribute the chief responsibility to capitalism and call for a different use of those same machines.[50] It is Heidegger who best articulates the philosophical stakes of the epoch of technics. He affirms that the rule of modern technics coincides with the accomplishment of the epoch of metaphysics, which changes the human being into subject, nature into object, and thinking into calculative rationality.[51] The transcendental horizon of the epoch of technics is the *Ge-stell*, which changes humans, nature, and thinking alike into resources. Alain Badiou is a good example of a philosopher who contests Heidegger's diagnosis and who instead understands the modern devastation of humanity—atomization of society, rupture of bonds between people, equalization of everything by money—to capitalism.[52] Both, however, refer to the logic that we have pointed out, namely the machine's capacity for abstraction, atomization, calculation, and the use/exploitation of nature.

Be that as it may, most regard technics as an ambivalent and essentially amoral reality that can potentially be better applied or transformed to be capable of a better application. Badiou ridicules the Heideggerian rejection of machines and calls instead for better machines that serve people, not capital. Mumford gives voice to a dream formulated in the era of romanticism when he rejects the cold mechanical use of machines and calls for a different use, one in the service of life and of an organic society. But the solution is undoubtedly not so easy. An organic society can be as toxic as a machinic one, for this idea underlies all twentieth-century ethnic nationalisms, most especially in their totalitarian forms. More fundamentally, how can we recognize and maintain the difference between the mechanical and the organic given that since the very beginning of modernity the one has always been the image of the other? As in Aristotle, *technai onta* are different from *physei onta* because the latter are autonomous while the former realize externally imposed ends and are reliant on external raw material and produce external products—except that the machine tends toward automatism and autonomy and its aim is to rival life. And inversely, the *physei onta* might appear to be self-moving and autonomous, but this is so only within the limits of their finitude, which makes them dependent on their milieu, including their technical surroundings. The opposition between living and artificial beings is much less firm than we might wish. This has lead to a contemporary inversion of the question: instead of thinking life as the image of a machine, technics is now thought of in the image of life.

Contemporary Information Technologies and Biotechnologies

Doesn't criticism of a machine society ring true but also seem somewhat out-of-date? Hasn't the technical paradigm shifted with the invention and spread of so-called new technologies? Surely the machines and industrial complexes characteristic of high industrialization are still omnipresent such that, as Bernard Stiegler says, our society is hyperindustrial rather than postindustrial.[53] But hyperindustrial machines are increasingly automatized and operated by computational technologies whose fundamental logic cannot be adequately grasped by the modernist idea of the machine. This is why Bernhard Waldenfels can claim that the modernist paradigm of technology has been overcome by a hypermodern paradigm that, as we shall later see, aims instead at operating like a biological organism.[54] One could say that hypermodern technologies show an inversion of modern technologies: while modernism constructs nature in the image of a machine, hypermodernism constructs technics in the image of life, thus realizing Canguilhem's suggestion of thinking technics as a "universal biological phenomenon and no longer only an intellectual operation of man."[55]

In contemporary life, so-called computational or information technologies are everywhere and they are by no means limited to industrial contexts. Far more omnipresent than the machines, they have infiltrated all areas of the society from commerce to education, from warfare to administration, from social relations to logistics, from research to entertainment and art. The expression "new technologies" refers not only to information technologies but also to less visible but equally important biotechnologies, as well as to certain emergent forms of technologies, for example nanotechnologies. Sometimes the newest technologies are referred to by the acronym NBIC, nano-bio-info-cogno-technologies.[56] Have these technological evolutions brought about a paradigm change, or have they realized a paradigm change whose principle was already sketched out in the late nineteenth century?[57] In history, one can rarely see an epochal change as it is taking place: history is made afterward, and so too is philosophy. It would be presumptuous to declare the arrival of a new era just because of some new inventions, however impressive (the Internet), when in other respects civilization functions in very much the same way as it did in the nineteenth century (the use of fossil fuels continues to grow even though their dangers should be well known by now). However, whether a historical tipping point has been reached or not, it is possible to distinguish a new paradigm unlike the modern mechanist one.

Waldenfels clarifies this paradigm change by distinguishing between classical, modern, and hypermodern paradigms of technics. According to him, while classical technics aims to bind nature's force and modern technics aims to master it, hypermodern technics aims to liberate technics itself as a quasi-nature. While the classical tool was used by the artisan, the modern machine used external forces. Hypermodern technics, on the contrary, consists in automatons capable of autoregulation and systems capable of autoorganization. Hypermodern technics does more than imitate living beings: it "lives" and constitutes a new kind of a *techno-nature* or *nature-technics*, or *bio-technics* and *techno-bios*.[58] This means, firstly, that nature is today deeply marked and pervaded by technics (as can be seen in phenomena like climate change and the sixth extinction that would not have occurred in the absence of technological industrialization)[59] and secondly that our knowledge of nature is profoundly mediated by technics.[60] But on the other hand, this does not mean that nature and technics are simply identical, as if nature were nothing but a technological projection or as if technics were nothing but a realization of nature's potentiality. Nature and technics mime one another, but they also threaten and surprise one another in ways that are often impenetrable and obscure.[61]

Both modern and hypermodern technics refer to life as the model for technics. However, what has changed is not only the sense of this comparison but also the understanding of biological life, which is no longer thought of as just an organism but essentially in terms of processes. In modern era the determinate machine imitated a definite organism. In the hypermodern era it is the principle animating machines that imitates the principle of life *et vice versa*: imitation is no more between definitive structures but between different forms of bio-technical logics. While the mechanist machine was a relatively closed structure, hypermodern bio-technical processes are dynamical processes that do not simply keep structures in movement but traverse, transform, and finally constitute their very matter. Bio-technical processes are lifelike insofar as they are autopoietic and technical because they are cybernetical. Instead of working only in terms of externally set aims, they realize an internal finality whose aim is themselves. The internal finality of bio-technical beings is not rational in the sense of self-consciousness, but it follows a more elementary reflexive logic.

This kind of logic was actually described in terms of the "machine" by Gilles Deleuze and Félix Guattari in their *Anti-Œdipus* (1972). Their use of the term *machine* is different from the previously discussed modern one because it does not denote a given structure but is on the contrary a counter-term to the structuralist idea of structure and to the metaphysical idea of subject.[62] Deleuze and Guattari renounce the opposition between vitalism and mechanism[63] and claim instead that both organisms and machines are "machines" with two different states: on the one hand, the machine as structural unity and the living being as an individual are "molar" phenomena; on the other hand, both are constellated by "molecular" phenomena. "But in the other more profound or intrinsic direction of multiplicities there is interpenetration, direct communication between the molecular phenomena and the singularities of the living, that is to say, between the small machines scattered in every machine, and the small formations dispersed in every organism: a domain of nondifference between the microphysical and the biological, there being as many living beings in the machine as there are machines in the living."[64] The "molecular machines" are "desiring machines" that should not be taken as organized structures but only as a functioning constituted by "flows and cuts" that ends up being productive, formative.[65]

Deleuze and Guattari do not introduce their idea of machines in order to develop a contemporary theory of technics but in order to provide an alternative theory of the human unconsciousness and also of political drives. These do not represent, they produce, as the famous opening of *Anti-Œdipus* puts it:

> It is at work everywhere, functioning smoothly at times, at other times in fits and starts. It breathes, it heats, it eats. It shits and fucks. What a mistake to have ever said the id. Everywhere it is machines—real ones, not figurative ones: machines driving other machines, machines being driven by other machines, with all the necessary couplings and connections. An organ-machine is plugged into an energy-source-machine: the one produces

a flow that the other interrupts. The breast is a machine that produces milk, and the mouth a machine coupled to it. The mouth of the anorexic wavers between several functions: its possessor is uncertain as to whether it is an eating-machine, an anal machine, a talking-machine, or a breathing machine (asthma attacks). Hence we are all handymen: each with his little machines. For every organ-machine, an energy-machine: all the time, flows and interruptions. . . . Something is produced: the effects of a machine, not mere metaphors.[66]

In Deleuze and Guattari's text, the machine is nonetheless mainly an image illustrating a certain logic. The image is so telling because it echoes the general intellectual tendency to explain phenomena through an underlying *immaterial technicity* that was characteristic of many twentieth-century sciences, from psychoanalysis and psychology to anthropology and social sciences, but also extending from mathematics and logics to cybernetics and engineering, from linguistics to biology. In all these very different areas of research, one can distinguish an interest in subrational processes, an interest that—to first state it schematically—focuses, in the animal rationale, on the animality, whose materiality is reinterpreted as a biological operativity, and in rationality itself, on the elements that are below consciousness, figured as unconscious or formal operations that constitute the matter of thought. What interests me here is that both animality and nonconscious rationality were then modeled in terms of a certain abstract bio-technicity.

As so many sciences were at some point in their development marked by this general intellectual tendency, retracing their history in any detail lies beyond the scope of this study. Both N. Katherine Hayles's *How We Became Posthuman*[67] (1999) and Erich Hörl's *Sacred Channels*[68] (2005, translated in 2018 as *Sacred Channels*) make insightful historical inquiries into this epistemic revolution that occurred in many areas, which were often unaware of or could not recognize each other's merits. In what follows, I refer to these works for the necessary intellectual history but concentrate, for my part, on the general movement of thought that emerges from these different inquiries into immaterial technicity. Although for good reasons this immaterial technicity or technical materiality is not presented in these terms by scholars themselves, the following theories tend toward a theorization that is more general than any individual science and even tend toward an ontology modeled on technics. This is not an ontology of code, like the one that is postulated more or less implicitly by those who refer both the brain and the computer to a common computational principle. Codes are certainly used to describe all kinds of processes from biology and brain research to computer science, but they are not the ultimate structure of reality, the ontological ground on which everything rests. They are only a way of modeling phenomena. Codes—not some specific programs but the general codeability—may be the a priori structures of our knowledge so that they constitute the transcendence of our time, but this is only an artificial and historically changing quasi-transcendental horizon. In chapter 4 we will see with Derrida what this means for philosophy. To end the present chapter, we will limit ourselves to considerations of modern bio-technics on a more phenomenological level.

First, the focus is shifted from positive phenomena and from signifying thinking, not toward anything like a substantial ground but nonetheless toward an underlying logics that provides an explicative background or horizon of the phenomena. Second, this background is described in terms of quasi-technical processes that appear sometimes as mechanical and repetitive, sometimes as autopoietic and recursive. Third, the ground cannot be thought in terms of ideas but only in terms of contextuality or environmentality.

First, then, the focus changed from positive phenomena and signifying thinking toward their inapparent, non-signifying background. The novelty of the approach lies in the realization that this background, even though non-signifying, does not need to be irrational, for it can have its own formally valid operativity that research can strive to uncover. Or like Aristotle said, techne is not an episteme but it can still be taught and is therefore a form of knowledge. This realization takes place in many domains. One of the best known is psychoanalysis, which is practically born as an inquiry into the mechanisms of the unconscious that do not appear rational to consciousness but that are not random either for they have their own objectives. And while Freud studies unconscious desires as desires repressed by consciousness, Deleuze and Guattari detach them more radically from consciousness and describe unconsciousness as an autonomous productivity of "desiring machines." Curiously, during the same period, mathematicians and logicians who often loathed psychoanalysis were preoccupied with a program that to some degree paralleled that of psychoanalysis. Erich Hörl reveals how around about 1900, the French philosopher and mathematician Louis Couturat dealt with the question of whether thinking was grounded intuitively or symbolically by showing how symbolic thinking could be described in purely formal terms such that "what had previously been unthought and unthinkable about thinking itself, indeed about the rationality of mind—pushed the autonomization and ultimately the machinization of the symbolic."[69] For Couturat, consciousness is but a screen behind which and unbeknownst by which is the intellect, a purely formal operativity that makes thought possible. The operations of the intellect cannot be reached by psychological study but only by logical analysis that discovers symbolic calculi.[70] Drawing from Leibniz, Couturat showed how these fundamental operations of the intellect are, on the one hand, beyond the reach of intuition, such that consciousness is blind to them, and yet, on the other hand, it is possible to discover their logic and establish a general science of calculus, the Leibnizian *characteristica universalis*, in which reason is guided by signs alone and not by evidence.[71] Hörl contrasts the mathematician's reflections to those in linguistics, namely Saussure's reflections on language at the turn of the twentieth century. Saussurean linguistics is reminiscent of Couturat's "logistics," which discards intuition in favor of pure calculation, insofar as in Saussurean linguistics sense is not based on real-world references either but on differentiation on a pre-signifying level. However, for the logician, the symbolic can be reduced to pure calculation, whereas for the linguist, language must use signs that are necessarily marked by the contingency of their birth and thus fundamentally by time (which calculus cannot take into account).[72] Hörl also shows how the ground prepared by logics, mathematics, and physics was later reoccupied not only by linguistics but also by ethnology and especially by the structural anthropology formulated by Claude Lévi-Strauss.[73] He shows that Lévi-Strauss's re-elaboration of the theoretical grounds of anthropology was actually informed by the birth of cybernetics, which was then understood as a "logic of machines,"[74] and it thus continued work that began in pure logic. Like symbolic logic and cybernetics, structuralist anthropology turns its attention to phenomena below the level of intuition, to the elementary structures and operations that regulate human societies. The anthropologist then articulates these in differential systems that are neither visible nor even comprehensible as such but which nonetheless account for the structure of the society.[75]

To sum up, in many sciences we find a parallel movement of looking for a non-signifying but formally valid operativity beneath the manifest or signifying reality. My aim, of course, is not to claim that these all amount to the same thing but to draw attention to this parallel. What interests me here is the fact that, by and by, the model that helps in grasping the fundamental operativity in the "unthought"[76] background of phenomena turned out to be technical rather than purely logical. Two technical models soon became particularly important, namely, the calculator and writing, the computer combining both (it operates through calculus but one of its most important functions is conserving and distributing documents written in different signs such as words, images, or codes). The calculator is generally thought to instantiate calculus: it realizes it flawlessly, rapidly, and with no loss of information. Writing, on the contrary, is more ambivalent. As we shall see in chapter 4, in his move beyond structuralism Derrida has shown that writing is a supplement that is both a remedy and a poison to the memory it is supposed to assist and it is also a disseminative source of both understanding and misunderstanding.

What is the essence of technicity manifest in these examples? It is not the calculator, the writing, or the computer as such. These blocks of matter are just supports that could very well be replaced by other supports. Technics itself appears to be somehow immaterial: it is the information, the code, the program, the pattern, or the message in which the procedure has been encoded. It is also the procedure that unfolds the pattern by repeating it, but further by introducing differentiation and even modifications to it. Because technics appears immaterial, it can be thought of as instantiated indifferently in different supports. This is why, for instance, Alan Turing and Norbert Wiener thought that men and machines could be seen as fundamentally the same and that the computer could be used as a model for human cognition.[77] But as we shall see in a moment, there are others who have always claimed that actually all machines, and not just writing, are material and contextual, and that this contributes to their functioning and sense.

Second, what is the exact nature of the (quasi-)technical processes that constitute the non-signifying and nonetheless formalizable ground of manifest, signifying phenomena? This is the question that most interests researchers and this is what engineers need to know in order to reproduce these phenomena artificially.

Examined through the lens of the older machine paradigm, a technical process would appear to consist in repetition. A technical object is a structure that instantiates a program or a code that it repeats identically, for example Leibniz's, Pascal's, or Babbage's calculating machines. The idea of technicity as repetition reveals its insufficiency when it is applied to living or signifying structures. One cannot explain an individual living being solely in terms of the replication of its form (the improbable idea of a dog) or, in the modern context, only as the deterministic unfolding of its DNA for each living organism realizes its form/code differently in function of many contextual factors. The same goes for many acts of reading the same book, acts that see the same signs but experience the same stories differently.

In the mid-twentieth century, the idea of creating artificial life or thinking culminated in cybernetics. This is a vast multidisciplinary research program whose theoretical core is generally referred back to the Macy conferences (1943–1954) but which developed just as essentially from out of concrete laboratory and engineering work. In *How We Became Posthuman*, Hayles gives a clear account of the development of the central ideas of cybernetics. Cybernetics is a general theory of communication and control that was meant to apply to animals, humans, and machines alike. As Rosenblueth, Wiener, and Bigelow state in their groundlaying article "Behavior, Purpose and Teleology" (1943), these all can be described in terms of causal determinism, but this does not capture their specificity, which lies in their being purposeful beings. This does not of course mean that they all have some exterior aims, but rather that they act following their own intentional aims and behave in a voluntary manner and not randomly. Furthermore, all purposeful action requires some feedback, which makes it teleological as well. The question of the unthought ground of living, mechanical, or intellectual entities amounts to asking after the exact logic of this purposeful action.

In the first phase of cybernetics, the organism's action was conceived as aiming at homeostasis. Hayles explains this concisely by saying that while a living organism's homeostasis consists in seeking to maintain a steady state when it is buffeted by its environment, in cybernetics this idea was extended to machines that were expected to maintain homeostasis by using feedback loops (which gradually came to be thought of in terms of information).[78] Cybernetics could equate living beings and machines because it blackboxes their internal structure and concentrates only on communicative input and output, on control, and on information systems, as Rosenblueth, Wiener, and Bigelow emphasize in their article. According to Hayles, second-order cybernetics emerges when reflexivity is taken into account such that the cyberneticians themselves are included in the feedback loop. The apex of second-order cybernetics is Humberto Maturana and Francisco Varela's *Autopoiesis and Cognition* (1972), where the starting point lies in biology and not in mathematics, the starting point for Wiener's and many other earlier cyberneticians' work. Maturana and Varela thought it was impossible to explain living organisms solely in terms of purpose or function. For them, life is cognizing, however elementary, and through cognizing the living being makes itself into what it is. It is autopoietic because what it is reflects what it learns of the exterior world. Moreover, its autopoiesis is self-reflexive because it acquires knowledge not only of its environment but also of its own experience of the environment: autopoiesis is self-reflectivity. This is why, instead of asking how the organism obtains information about its environment, one should instead ask how it attained a structure allowing it to operate in the medium in the first place.[79] Finally, as Hayles puts it, the third wave of cybernetics emerged when "self-organization began to be understood not merely as the (re)production of internal organization but as a springboard to emergence."[80] Hayles attributes this change to Varela's own critique of autopoiesis in his work on embodiment and enaction in particular. "Enaction sees the active engagement of an organism with the environment as the cornerstone of the organism's development. The difference in emphasis between enaction and autopoiesis can be seen in how the two theories understand perception. Autopoietic theory sees perception as a system's response to a triggering event in the surrounding medium. Enaction, by contrast, emphasizes that perception is constituted through perceptually guided actions, so that movement within an environment is crucial to an organism's development."[81] In other words, the living organism is not only a circular attempt to maintain itself in function of the milieu, but it needs the milieu's intervention in order to evolve. It can be modified in function of its "program" but it cannot modify its own program, and this is why such modifications—real novelty—can only come to the organism from the outside.[82]

The theories of autopoiesis and enaction were developed especially in the context of biology. But within the general context of cybernetics they also reflected on other domains and especially information technology, where they helped generate hypotheses on artificial life. Could life be constructed using sufficiently clever programs that not only reproduce themselves but also evolve in function of their situation? Today these questions are studied in particular in the context of research in so-called artificial intelligence, which consists in reality in machine learning programs that not only reproduce their program but also use it to examine available data, "learn" from it, and modify their own functioning, even programming, in function of this data. As Yuk Hui shows, machine learning is thus characterized by a recursivity that must be distinguished from simple repetition: recursivity is "characterised by the looping movement of returning to itself in order to determine itself, while every movement is open to contingency, which in turn determines its singularity."[83] The stakes in a third-order cybernetics view of living organisms are thus the same as the stakes in artificial intelligence engineering: the technical-biological entity is conceived of as an autopoietic whole that aims not only at maintaining itself against the environment but also at opening itself to novelty that enables evolution.

Some theoreticians (like Norbert Wiener) saw the possibility of a general theory of all kinds of animal, human, and informational behaviors in cybernetics. But they did not go as far as interpreting this general theory as the ontological basis of reality. Gilbert Simondon's theory of individuation is a critical response to cybernetics that can be understood as a full-fledged ontology of the contemporary scientific world. Simondon was inspired by the program of cybernetics, but he also criticizes several of its fundamental ideas, most notably the idea of the homeostatic/autopoietic/embodied organism's autonomy and the analogy between machine and organism. This is why he sketched his own theory of "allagmatics" (theory of changes) against cybernetics (theory of control).[84]

The roots of Simondon's objection to cybernetics lies deep in his ontology of individuation that aims to pose the problem of the individual anew. In the autopoietic and cybernetic theories that we examined previously, the individual was given in reality but not interrogated. The question was how the individual maintains and produces itself, not how it came to be in the first place. For Simondon this is lazy thinking, for in philosophy the individual should not be a given starting point for inquiry but its very question. What is the individual and how did it become what it is? For Simondon, there is nothing like an individual that could be understood as a substance or as a subject. Or to put it another way: we should ask after the genealogy of the apparent individuals. The real question is not the individual but individuation. Everything that is, for Simondon, is in the process of individuation, and the real question is, How does individuation occur? This question also leads him to reinterpret cybernetics.

According to Simondon, one cannot explain individuation in isolation from the milieu from which and *in which* and *as which* individuation actually happens. Individuation is really a process in the preindividual milieu. The preindividual milieu is never stable, but it is an undifferentiated, unstructured field. This field is nothing like a homogenous primary matter, it is full of tensions, heterogeneities, and incompatibilities. In

Simondon's terms it is a *metastable* situation, not a stable one in which everything is fixed and permanent and change is impossible. Neither is it a total dispersion in which individuation does not happen, but rather it is a metastable situation in which many possibilities for a provisional stability—metastability—can be found but have not yet been found, such that the system remains in a mobile state, full of potentialities.

Simondon's main example of individuation in a metastable preindividual field is the process of *crystallization*.[85] Crystallization happens when a saline solution can function as a supersaturated mother liquid. When a germ is dropped in this liquid, a crystallization process begins from the germ and gradually spreads out into the solution, which thereby acquires a structure that it did not have previously. The formation occurs in accordance with a logic that Simondon calls "transductive"[86]: neither induction nor deduction, a transduction passes from next to next, it communicates information (here, the crystalline structure) to the next, from there to the next, and so on. Unlike deduction and induction, transduction is not a process of knowledge but a process in reality itself. This is why it necessarily runs up against contingencies (here, the impurities of the liquid) and is modified in function of them. Transduction describes a formation process that does not occur as the hylomorphic imposition of a form on to passive matter but rather as the communication of information from one point to another in a metastable situation in which the information is constantly reinterpreted in each new situation.

The physical process of crystallization is a very simple case of individuation. But individuation can happen in all levels, for example on the physical, vital, social, transindividual, or technical level. Each type of individuation happens in a different type of preindividual field; for example, for a physical individual such as a crystal the preindividual field can be saline solution, for a living individual such as an animal it can be the ecosystem in which it lives, for a technical object it can be the totality of technical elements and industrial dispositifs within which the object can be built. Individuation is not separable from the preindividual field: it is not a process in the preindividual field, it is the process of the individual field itself. In all these cases—but differently in each case—individuation takes place as a mediation between incompatible elements. It happens as the development of possible solutions to the problem field that the preindividual field is. Such a solution is never definitive or completely stable; it is always a provisional, metastable solution that brings the present force to one possible state of balance.

Individuation is not a principle (*principium individuationis*) but a singular operation each time that structures a metastable preindividual field where the problems that it contains find a provisional solution that itself leads to new problems, and so on. Individuation is not a realization of preexisting virtualities. It happens because new solutions are invented; individuation is driven by "resolving invention,"[87] and it is always inventive. Individuation does not lead to a new fixed identity either. It is in a state of constant mobility and transformation where the individual can only be characterized by an internal resonance. It is not a "subject" that could be separated from the milieu, such as an interiority from an exteriority, it is nothing other than the relation between interiority and exteriority or the constant passage from the exterior to the interior and the other way around.

Life is one form of individuation. It is a vital individuation that emerges as a solution to a problem in purely physical field.[88] A living organism is not just a metastable solution to a physical problem: it is the constant invention of solutions to the problems of its milieu and this could even be described as a constant birth. According to Simondon, a machine can ever only function and reproduce itself, it cannot invent itself. This is why a living organism and a machine are different. The analogy postulated between them by cybernetics is false. Simondon sees cybernetics as the science of direction whose more or less phantasmatic aim is to build the perfect automaton, which Simondon calls the "robot." For him, the robot in the sense of a totally autonomous machine is a purely mythical and imaginary being that "does not exist."[89] The robot, in this sense, is the real focus of cybernetics and the objective of artificial intelligence.

However, Simondon also points to another sense of cybernetics. He actually rethinks it in the sense of allagmatics, by which he means a "science of change."[90] This allagmatics does not belong to one regional science at a given time, but it is the art of passing from one region to another—from physics to chemistry, from chemistry to biology, from biology to human culture, and so on. Allagmatics shows how the metastabilities of one milieu can come forth as an individual on another level. For example, life is a result of insoluble tensions in a physico-chemical situation.

The allagmatic cybernetics in Simondon's sense differs completely from Hayles's cybernetics. Traditional cybernetics explains how individuals constitute themselves, whereas Simondon's allagmatic cybernetics is a technics of provoking individuations, which can only ever happen in metastable milieux that do not consist of individuals but of an entire field of elements in which different individuations can take place. In Simondon's sense, individuation never happens just to the individual, it happens in a multiplicity of elements, as multiple, mobile relations between elements. What the previous three theories of cybernetics saw as the point of encounter between the individual and a milieu—an encounter that immediately triggered the individual's return to itself—is in reality an encounter between several elements of an open milieu. The individual is not a closed system that touches the exterior world only to withdraw into itself. It is the trace of encounters between many tensions and elements. This is why, instead of individuals, there is really the milieu.

Moreover, why should the ground of bio-technical entities be thought of in terms of *environmentality*?

Let us recapitulate. We have seen how several generations of thinkers tried to articulate the hidden logics of the inapparent, non-signifying background of manifest, signifying beings. By and by, the abstract formalist and structuralist logics gave way to a more refined bio-technical logics capable of explaining the behavior and evolution of living, technical, and signifying entities. These bio-technical logics aim to describe the behavior of real living and technical beings, instead of pure ideal entities. They describe reality not as an impenetrable substance but as the abstract logics of its constant production. These logics are real insofar as they say how entities come to being and become what they are—or, as Simondon puts it, how they are individuated—and in this sense they have ontological force. The abstract term *bio-technics* designates the immaterial materiality of the biological, technical, and signifying beings that we are and with which we exist.

These bio-technical logics do not describe universal truths (like mathematics) but the behavior and becoming of real singular beings (like a rat in a maze or a cruise missile seeking its aim). One cannot explain the life of such singular beings simply by recourse to a code that the entity would replicate and instantiate. As the theories of enaction in biology and recursivity in artificial intelligence research made clear, the explication of the becoming of such singular entities requires the explication of their coming to be (through external agents) and of their transformations (triggered by external hazards). As Simondon puts it, their individuation always happens in a concrete preindividual milieu. Singular bio-technical beings are always to some extent miraculous: their existence is not necessary and their becoming is surprising, and yet they are there. In order to account for the novelty that their existence, transformation, evolution, and destruction represents, we need to refer to the external factors that could only emerge from their environment. Environmental factors provide the contingent chances that give the impetus to bio-technical entities and supply some of the matter to evolution.

Biotechnological codes are abstract, but they also exist only if they are embodied and if they evolve in factual contexts. They do not need this double factuality after the fashion of a form that needs matter (this is the classical hylomorphism Simondon strongly criticizes) but as a code needs contingent obstacles, ruptures, exceptions, and problems in order to evolve. These moments of decoding and re-encoding constitute the bio-technical entity's environment. This means that environment is not a homogenous dimension in which entities move like animals across the savanna. It is a preindividual milieu traversed by tensions and incompatibilities that individuation processes try to solve. Bio-technical entities are each others' environment, and this environment constantly changes as the entities shift in relation to one another. They are not related as substantial billiard balls on a table nor even like the interference of one code with another but like elements of each others' preindividual milieu.

In the end, this amounts to a general metaphysical position. I propose to call the abstract dimension in which instantiated bio-technical logics evolve the dimension of techno-nature, quite simply because it includes both intermingling dimensions without melting them down into a third, common ground. It is the metaphysical dimension or the transcendental horizon for conceiving bio-technical entities. It is not unitary, coherent, and true enough to constitute a general ontotheological substance or idea: it is the impossibility of such an overarching position. It is the general contextuality, the necessity of being many, the necessity of the contingency of the many being there. Its mode of being was beautifully described by David Gé Bartoli and Sophie Gosslin in their important book *Le toucher du monde*, which describes nature not as an observable totality but as the coemergence of corpora through their technicity.[91]

The condition for bio-technical entities is environmentality—or ecology, as Hörl says in his "Introduction to General Ecology: The Ecologisation of Thinking,"[92] which also stresses that contemporary ecological thinking is increasingly denaturated as ecology also appears as a technological condition or as a techno-natural situation. On this level of reflection, environmentality does not mean environmentalism, which is a political position—and surely the most decisive one today—although environmentalism could of course relate to philosophical environmentality. The philosophical environmentality is the metaphysical—that is, transcendental—condition of bio-technical reality. This means that all entities need to be thought of in context and that contexts are not fixed places but ever-changing constellations of other bio-technical beings.

And as Heidegger in particular has shown, contemporary transcendence is articulated in technical terms, but it is not the singular *Ge-stell* that still bears the shadow of the earlier mechanical universe. It is a techno-natural environmentality in which entities endlessly produce and transform other entities.

Many Layers of Technics

This chapter presented the phenomenon of technics in terms of three core terms: techne, the machine, and bio-technics. Although we can find the technical phenomena referred to by these terms in any epoch, each is distinctive for a specific epoch: technics is thought of in terms of techne in ancient Greece, as the machine in the modern era, and as bio-technics in the contemporary world. Each of them gives a distinctive picture of the human being: techne is the human being's capacity for giving form and this can also lead to its capacity for being formed to good life; the machine is the rationalized image of the human body and mind that is thus also capable of subjecting the human being to an alien rationality; and bio-technics describes the immaterial materiality of nature, body, and thought.

Today's idea of technical humanity includes all the senses of humanity and technicity presented in the previous two chapters. The humanist perspective considers the human being as a subject of techne, as a technician with knowledge and skills that can also be used on the human being itself, as in the examples of education to the good life and philosophy. The anthropotechnical view of humanity sees the *anthropos* mainly as a machine or a mechanism that can be manipulated at will. If the ideas of techne and machine reflect aspects of the classical conception of humanity, the idea of bio-technics tends to blur and overcome the limits of this conception. This is why it provokes further "metahumanist" speculations (Stefan Lorenz Sorgner combines trans- and posthumanism under this term).[93] Bio-technics does not conform to the rationality that defines the classical subject of philosophy, although it calls to a reflection upon the technical conditions of rationality. In this era, philosophy needs to reflect upon the concrete anthropological situation opened by new bio-technics, but over and beyond this, the subject of philosophy needs to rethink itself in a reflective movement in which it asks: What are the ontological conditions of bio-technicity? How is philosophy itself affected by its being reflected in the mirror of biotechnological existence?

What are the philosophical conditions for bio-technical humanity? And what is philosophy in the era of bio-technics? It is easy to see that in order to think of bio-technical humanity one has to stop modeling human nature on the philosophical subject of the cogito type or the *I* that accompanies pure reason, insofar as these share reason's eternal and unchanging character. It makes no sense to speak about transformations of human nature if this nature is not transformable to start with: the form of human is transformation (capacity to change and to be changed). Reason is an essential part of the human but it does not determine the aspect of the anthropos. It only says that the human is something that reflects upon itself in the light of reason. The idea that the nature of the human is technics and artifice was already outlined by Rousseau and Nietzsche and it has become particularly manifest in the era of bio-technical humanity. The idea that human nature is in technicity is related to the Hegelian idea that not only is the essence of the human being freedom, but also, precisely because it is freedom, there cannot be any essence of the human being. The essence of humanity lies, on the one hand, in the free action by which it makes itself into what it is not, and on the other hand, in the historicity that bears the traces of this action. Similarly, if human nature lies in its technicity, "humanity" cannot be a figure that fixes, say, the morphology or the physical capacities required of beings who claim to be human. The nonhuman is not a disfigured human being but just a thing incapable of freedom, which can lead to self-reflection, to changes of self. Technical transformations do not alter humanity until it is no longer human because the essence of humanity is precisely its capacity for technical alterations.[94] The criterium of humanity is not an anthropological figure, it is the moral sense of humanity that enables the choice for and against inhumanity, as Jean-François Lyotard might put it. Thus the properly existentialist sense of humanity in the era of bio-technics would be that there is no definable aspect or figure of humanity, but there is humanity as long as rational beings reflect upon the sense of their humanity and decide about inhumanity. Because of its transformative character, human nature is constitutively open—open to betterment and disfiguration, to excellence and inhumanity, to all kinds of transformations that are human as long as they called upon to judge about their humanity. But if the human being is open to transformations, it is because its nature is not fixed but remains a question to itself. As Arnold Gehlen explains Nietzsche's famous expression *Mensch ist das nicht festgestellte Tier*: "This expression is true and and it has exactly two meanings. It means firstly: there is no determination of what the human being is, and secondly: the essence of the human being is somehow 'incomplete,' not 'fixed,' both expressions are exact and can be adopted."[95]

Let us now see how the idea of the originary technicity of the human took form in early twentieth-century German thought.

Chapter 3

The Originary Technicity
of the Human Being

The nickname *bio-technical humanity* invites us to think of human *bios* (life, essence, being) as essentially determined by *technics* (as product or activity). We saw that thinking of the human bios as the trace of technical activity requires thinking of it as a plastic capacity for change and transformation instead of as a fixed form. But doesn't such a versatile and multiform figure of humanity finally amount to the fundamentally nihilist idea of a formless humanity? Or does formlessness on the contrary presuppose a metaphysical negativity indicative of a capacity rather than an incapacity, a power rather than a powerlessness?

Such questions require an investigation into the metaphysical grounds of bio-technical humanity. The philosophical investigation of philosophical bio-technics is much older than this word because it draws from a long line of studies of human nature in terms of originary *artifice* or *technicity*. The genealogy of the originary technicity of humanity bypasses the idea of the artificial nature of the human being that takes form well before Nietzsche, especially in the works of Diderot, Rousseau, Herder, and Hegel. But as Bernard Stiegler has shown, the idea of technics as the originary supplement to the lack of a human nature can be traced back as far as the myth of Prometheus and Epimetheus recounted by Plato in the *Protagoras*.[1] However, it seems to me that the first systematic philosophical inquiries into technics as the metaphysical question of the essence of the human being, and not simply as an accessory remark, takes form in an interesting debate in German philosophy at the end of the 1920s between

Heidegger's phenomenology, especially *Being and Time* (1927), *The Fundamental Concepts of Metaphysics* (course taught in 1929–1930), and "The Question Concerning Technology" (1953) on the one hand and on the other hand the research program of so-called Philosophical anthropology, whose most important works were Max Scheler's *Die Stellung des Menschen im Kosmos* (1928), Helmuth Plessner's *Die Stufen des Organischen und der Mensch* (1928), and Arnold Gehlen's *Der Mensch: Seine Natur und seine Stellung in der Welt* (1940).[2] It was in 1928 that Scheler wrote that "in no other historical era has the human being (*Mensch*) become so much of a problem to himself as in ours."[3] At that time, the loss of evidence concerning the human being was already brought about by scientific and technological progress in industrial modernity, which avant-garde art and science fiction (Huxley, Čapek) also investigated. Philosophers faced the mystery of the human being as a philosophical problem that required a fundamental rethinking, not just of one philosophical object among others (*Mensch*) but of philosophy itself as one of the ways of being of this problematic entity. (It's likely that the uncertainty surrounding the human is one reason for the unparalleled philosophical inventiveness of this period.) Although Philosophical anthropology and Heidegger's existential phenomenology were not the only approaches in which interesting reformulations of the question of the human being took place at that time, they were the ones that thought of the human being not in function of rationality (that reinvented itself as logical calculus or as phenomenology) but in function of technics. Furthermore, they think of technics itself not as applied science but as an outgrowth of "life," of prescientific biological or everyday existence. The human is for them a living being endowed with technics, more precisely, a being whose life is technicity. Although many other species use technics as well, the human use of technics appears different from that of any other species because the human being's entire world, and therefore its own self, is a technical construction.

In this chapter, I will compare Philosophical anthropology and Heidegger's fundamental ontology because they present similar interpretations of the place of the human in the world but explain it using two different philosophical frameworks. Philosophical anthropology, more precisely Plessner and Gehlen, develop a theory of the *artificial nature* of the human being. They think the human being in terms of a philosophical biology, and furthermore, as life that continues and overcomes itself in technics. The human is a being without a proper nature because its nature is precisely its capacity for technics and artifice. Like Husserl,

Heidegger is opposed to the reduction of the philosophical question of the human to biological explanations of it as a living being. Contrary to this, he formulates his thinking of existence against both scientific and philosophical biology and anthropology.[4] But at the same time, Heidegger too thinks of human existence essentially in function of technics. In *Being and Time*, Dasein's everydayness is interpreted through the equipment. This takes on a wider meaning in its activity of "dwelling and building," and finally human existence in the contemporary world is thought of in function of the *Ge-stell* of the epoch of technics. Technicity, not reason, is the key to understanding Dasein's world—and reason itself is thought of as logos interpreted as technical calculus or as poietic thinking (*Denken und Dichten*). Technics is not the "human nature," but it determines the human way of being by providing its privileged relation to the world: technics forms its world and thereby itself.

I shall now present these two descriptions of the originary technicity of the human being. I shall show that Philosophical anthropology and Heidegger share an analogous idea of this being's fundamental constitution. Both think of human existence as a homelessness or unhomeliness, the core of which is a form of negativity (lack, nothingness) that exposes the human being to alienation, falseness, and hollowness. But in both works, negativity turns out to also be the space of freedom and potentiality, without which technical invention would not be possible. Both think of technics as a supplement or a complement to this originary negativity. The human being *is* nothing but it builds itself a world that it calls home. The finite, historically changing world built by humans in turn forms human beings, not directly as individuals but as communities, cultures, peoples, and any kind of group that then form individuals. The human being is a nothingness that supplements its intimate negativity by technics. It also sees the negativity of technics, which is due to the fact that technics, thought of as a continuation of nature's own poiesis and not as a materialization of natural laws, cannot be definitive (like a truth) but it is always a finite, provisional construction. Because of this finitude, technics is necessarily infested with obsolescence, is blessed by invention, and always undergoes change.

Both parties to this debate think of the human being as a nothingness. In order to supplement its lack of essence, the human being makes the world of technics that makes the human being in turn. Both present technics as a point of reflection by which the human being finds itself in a world. Yet while Heidegger's Dasein finds itself thrown into a technical world, the Philosophical anthropologists' man builds itself a world, such that Heidegger emphasizes human situatedness and the Philosophical anthropologists human agency. In the end, Heidegger and Philosophical anthropologists do not so much think different things but they think of them *differently*, and it is these alterations in ways of thinking that will be important for us.

Philosophical Anthropology: On Natural Artifice

While philosophical anthropology is a general philosophical discipline, the name Philosophical anthropology with a capital *P* is used to group together an approach developed from the late 1920s onward by Max Scheler, Helmuth Plessner, and Arnold Gehlen. All Philosophical anthropologists aim to rethink the philosophical constitution of the human being in the modern situation in which "humanity" can no longer be explained by theology or philosophical idealism nor by a simple mechanistic materialism. They underscore the fact that the contemporary worldview must be based on science and that philosophy needs to take this into account. Philosophical anthropologists must especially heed the discoveries of biology, psychology, and sociology. They particularly insist on thinking of the human being as a biological being. It is not that they take biological findings for philosophical truths, but they nonetheless develop a bio-philosophy that is compatible with biology and that can also provide a philosophical grounding for the science of biology.

Following Darwin, the Philosophical anthropologists take the descent of man from animals and simpler living beings for granted—which also means they describe the gradual emergence of the human from lower levels of life. However, these German thinkers were also critical of Darwin's explanation of life as a struggle for survival, instead of also including sympathy and cooperation, and of simple adaptation to the general environment instead of a more open and even creative relation to the world. Unwitting inheritors of a romantic philosophy of nature, they saw the theory of natural selection as a purely utilitarian and politically harmful idea. They were more influenced by the biological theories of Jacob von Uexküll, who focuses on the behavior of the individual organism in its milieu. Such an approach also better answers the needs of the philosopher, who works on the scale of the individual subject, rather than Darwin's studies that focus on entire species or molecular biology's interest in an infra-individual living substance. Uexküll studies the organism, and more precisely the animal, by asking in a Kantian manner what are its conditions of experience.[5] He invites to us ask after not how we see the animal but how the animal sees its own world. "To the physiologist, every living creature is an object that exists in the human world. He investigates the organs of living things and the way they work together, as a technician would examine a strange machine. The biologist, on the other hand, takes into account each individual as a subject, living in a world of its own, of which it is the center. It cannot, therefore, be compared to a machine, but only to the engineer that operates the machine."[6] One can only understand an organism in relation to its *Umwelt*, which is not simply its immediate physical environment but its subjective world: it is not the organism's objective space-time but the particular space-time synthetized by this animal.[7] The *Umwelt* appears to the animal as elementary sensations (*Merkzeichen*) that can become objects for it as meaningful perceptions (*Merkmal*) and finally as impulses to action (*Wirkzeichen*). An *Umwelt* is thus the totality of markers that are significant to an animal; the animal not only perceives these markers but can also change its behavior in function of them (in this regard Uexküll's bio-semiotics prefigures later biological theories of autopoiesis and cybernetic ideas of feedback[8]). Uexküll studies the animal's behavior and explains it in a phenomenological and semiological manner. He thinks that the animal's interaction with its world can result in melodic and rhythmic patterns that individualize it,[9] such that it is, so to speak, its own work of art. As the romantic philosophy of nature had noted and later ethological research confirmed, not only human beings but also many other animals relate to their environment in a technical manner. It is hardly possible to distinguish the animal's life from the activity of building a nest, using tools (many simians), "dressing up" (elephants cover themselves with dirt in order to protect themselves from the sun), and of course communicating. While Darwin describes the animal's relation to its environment in terms of adaptation, Uexküll thus develops the idea of a constructive world-relation.

The idea of the artificial nature of the human being is properly developed by Helmuth Plessner, but as his thoughts reflect the general constitution of Philosophical anthropology first delineated by Max Scheler in *Die Stellung des Menschen im Kosmos* (*The Human Place in the Cosmos*), I will start by outlining the latter's main ideas.

Scheler is arguably the first to grasp the philosophical implications of Uexküll's theory.[10] Although Uexküll is not quoted in *The Human Place in the Cosmos*, Scheler's definition of the living being clearly echoes his work: "A living being is always an *ontic* center that forms 'its own' spatiotemporal unity and individuality. . . . A living being is that X which limits *itself*, and which has individuality."¹¹ Examining, like Uexküll, the living being as the subject of desire and perception, Scheler further explains that the living being goes toward its environment because of an "ecstatic impulsion" (*ekstatische Gefühlsdrang*)¹² through which the organism is in contact with something other than itself. This contact is not a simple causal relation but is really an "openness" to something exterior to which the organism reacts (like a plant reacts to light or an animal to the smell of edible things). Biopsychical organisms (especially animals) also have an instinct (*Instinkt*), associative memory (*assoziative Gedächtnis*), and practical intelligence (*praktische Intelligenz*). Animals do not act following mechanical impulses but they behave (*Verhalten*). They are not bound to a singular mechanical model of action but they can actually learn, get used to, and invent new ways of behavior. Things are significant for them, and they live in specific ways with particular rythms, behaving in singular manners. Intelligence is associated with dynamic life process belonging to both humans and animals.

While the plant's drive-feeling simply goes outward, the animal's drive begins outward but when it meets obstacles, it turns back inward and sometimes makes the animal change its behavior (this reflection is the animal's "consciousness"). The animal's interaction with its environment is a functional circle consisting of both instinctual activity directed at the environment and reactions to resistances in the environment. Scheler traces the human being's metaphysical origin to the moment when the circle of life is broken in an act of negation. Thanks to negation the human being can objectify both its environment and also itself. It can rise above the opposition between organism and environment and assume an eccentric position from which it can contemplate the entire relation (this is its self-consciousness).¹³ What sets the human being apart from other living beings is this eccentric position from which it can say "no" to the environment and even to life itself. All life can to some extent turn against inorganic nature, but only the human being can say "no" to the environing world, turn against its instincts in a way that makes it "world-open" such that it can "have a world."¹⁴ Because of negation and openness, the human being is capable of ideation, that is, it can ask what the world is, what it is itself, why, and whether it could be otherwise. These *differentia specifica* of the human are not ideal a prioris but the a posteriori features of life's own development.

These differences between the plant, the animal, and the human being are not facts of evolutionary biology. They are metaphysical differences that reflect Aristotle's *De anima*, where "soul" is defined as the *entelechy* of living beings.[15] *Entelecheia* as psyche is not *entelechy* as life force postulated by nineteenth-century vitalists (whom the Philosophical anthropologists opposed). Aristotle's psyche is nothing other than the very movement of life that comes forth in the three specific forms of nutrition, sensation, and thought, such that plants have only a nutritive soul, animals also have a sensitive one, and humans are distinguished by their noetic soul. The *differentiae specificae* fixed by Aristotle are reflected in most bio-philosophies of the early twentieth century, not only in Philosophical anthropology but also for instance in Heidegger's phenomenology and Hans Jonas's metaphysical biology.[16] Today we have good reasons to disagree with the strict distinctions drawn between plant, animal, and human.[17] But at the beginning of the twentieth century, these distinctions allowed thinking of the continuity of the phenomenon of life. These distinctions also add little to the study of plants and animals, their main objective being the clarification of the "special position of man."[18]

Unlike Scheler, Helmuth Plessner was a trained biologist, and the major part of his magnum opus *Die Stufen des Organischen und der Mensch* (1928) develops a biological philosophy of plant and animal life. What does he mean by life? Not the lifeless matter of physics, nor any of the principles postulated in earlier scientific biology, such as preformation, teleology, series, or creative evolution. When Plessner defines life as *positionality*,[19] the term is neither biological nor historical but philosophical. Positionality is a "mode," a modality of being of organic life, a kind of an ultimate a priori form of life.[20] Plessner is much closer to Hegel than to Kant, for he does not study our experience of the living being but the mode of being of the living being itself. This mode is not a structure of experience but a structure of reality; not an a priori form of knowledge but a material, objective a priori. Positionality is the fundamental mode of organic life, and plant, animal, and human are its declinations.

Positionality is produced by an active struggle with the environment (and not by adaptation). Like Scheler, Plessner thinks that all organic beings are limited by their environment. Their limit (*Grenze*) is not only the border of the entity—it delimits the being actively. The living being shows its vitality by resisting the absolute limitation presented by its environment. One could say that it presses itself against its limit, transcends it, and thus remains, nay, becomes itself. Life is a contradiction or, as Hegel would say, a maintaining of oneself in contradiction. By demarcating itself, the living literally takes place: it does not only live *in* an environment, it lives *of* it (and despite of it). This is how living beings are "centered on themselves," that is, positional. A positional being has a center where it affirms itself; it is both stable and unstable, a being characterized by plasticity, for it has a form that can also be transformed.[21] Plants are to some extent exceptions to this general principle because they do not have a proper center but they are entirely open, ecstatic toward their environment but incapable of resisting it. The animal's positionality is properly centered because its form is closed. Unlike the plant, the animal is not simply open toward its environment, but it resists its environment actively and the environment resists it in turn, such that the organism becomes the site of interaction between its center and its environment. The animal's form reflects the resistance of the environment and its own resistance; it is affected by its demarcations. This is the origin of its "self," which is not yet an *I*.

The human being's positionality is eccentric.[22] Like the animal, the human is positional and lives from its center. Like Scheler, Plessner thinks human specificity comes forth when the functional circle is broken. Eccentricity means that the human is a reflexive being who can examine its own position as if from the outside. The human is not limited to one center, whether physical (time, place) or cultural (a given way of life), but it is constitutively eccentric: "Therefore, because of his form of existence, he [the human being] is by nature artificial. As an excentric being without equilibrium, standing out of place and time in nothingness, constitutively homeless, he must 'become something' and create his own equilibrium."[23] Animals are centric because they can live outside of their own center but they always fall back into it. On the contrary, human existence is definitively eccentric because it is defined by the chasm between exteriority and interiority.

Whereas in Scheler's *The Human Place in the Cosmos* human eccentricity leads to ideation (*Ideierung*),[24] in Plessner's *Levels of Organic Life and the Human* it culminates in technics. Technics is the main modality of all three anthropological laws formulated by Plessner: natural artificiality (*natürliche Künstlichkeit*), mediated immediacy (*vermittelte Unmittelbarkeit*), and utopic place (*utopischer Standort*). Plessner thinks that because the human being is without essence—*nichtig und exzentrisch*—its essence is its artificiality and technicity. The human being is constitutively homeless and unstable, naked and defective, and therefore it needs supplementation by technics: it must build the artificial ground of life that it does not naturally have.²⁵ Precisely because the human being is groundless, it can regard the things of the world as well as itself as simple objects to master; it can create tools and instruments and finally build an entire cultural world. All products of the human's technical activity attest to their maker's wits. The products also last longer than their maker and constitute the heritage of a culture.

The second anthropological law, mediated immediacy, says that the relation between the human and the world is the human being's *expressive* behavior that is at the same time the direct contact between organism and world and the indication of their distance. Unlike the animal, whose world relation is mediated by the senses, the human world-relation is mediated by the expressivity visible, for example, in music, image, dance, and language. Plessner points to the "necessity of expression as such, which is prior to individual modes of expression, with the essential connection between the excentric form of positionality and expressivity as the mode of life of the human."²⁶ Plessner does not think language as representation but as technical mediation, and this is why there are necessarily many languages, none of which is the "right" one.²⁷ "The expressivity of the human thus makes him a being who even in the case of continuously sustained intention continues to push for ever new realizations and in this way leaves behind a history."²⁸ The world is manifest in cultural products; the human never looks directly into the open, like the animal of Rilke's eighth *Duino Elegy*, but expresses its relation to the world in culturally conditioned ways. The human world is never given, it has to be constructed technically and culturally. Being not at home anywhere, the human being can live anywhere, but in every site it is of course tied to the available elements of world-construction. Technics is not a tool or an instrument: it is the entire spectrum of mediations between its organism and the environing world.

Finally, the third anthropological law, utopic place. "Excentrically positioned, he stands where he stands and at the same time not where he stands."²⁹ Its form changes because it constantly differs from the form that it already has and sketches a new form. The human being's relation to other human beings is not immediate either, but it is mediated by roles, masks, social techniques, and cultural conventions. Its community is nothing like a natural group, it is a cultural world that it constantly creates anew with technical tools, expressive means, and social acts: the utopic place. Now, if the human being is an originarily technical being, can't it also be the object of technical formatting? After all, the idea of human plasticity underlay the plans for the creation of the New Man that inspired Nazis, Stalinists, and, more discreetly, liberal eugenics policies from the 1930s onward. These ideologies dreamed of creating a new human type or *Gestalt* (the worker, the Aryan, etc.) by means of education, propaganda, and even eugenist medicine.[30] Assuming that humankind is naturally divided into races, these ideologies also presumed that the human genetic inheritance and mindset could be developed by means of voluntary biopolitical action. The anthropological idea of originary human technicity explains why this is possible. But it is also important to note that Plessner's idea of a utopic place is explicitly directed against this kind of anthropotechnics, where the human is only the raw material for state politics. Fleeing anti-Semitism in 1933, Plessner had already criticized (in 1924) the radical ideologies that aimed at reducing social existence to biological animal instincts. In *Die verspätete Nation* (banned in 1933 and republished in 1959) he criticizes totalitarian states that abolish the frontiers between private and public life. In opposition to biological racism, Plessner calls the human being the *Homo absconditus*, the groundless human. Its freedom consists in the power to undo any theoretical definition that claims to fix its essence. The eccentric human is not a perfect specimen of a political community; it is a rootless, exiled, homeless person.[31] On the one hand, the life of every individual is irreplaceable, as human mortality attests. But at the same time, "the human *is* the shared world" in which everybody can be substituted for and replaced by anybody else.[32] Social life combines these two aspects of existence when it both protects intimacy and mediates contact between people. This is possible when social life respects indirectness and artificiality "since respect for the other for the sake of the originary community of the shared world demands distance and concealment."[33] People wear masks in front of one another, not in order to deceive but in order to protect discretion: they make diplomacy possible. This is not Plessner's dystopia, but on the contrary it is the utopic place in which cohabitation with different people is possible.

Arnold Gehlen's *Man in the Age of Technology* has the same philosophical core as Scheler's and Plessner's theories, but he develops it in another direction by providing a different analysis of modernity. Echoing the ideas of his predecessors, he thinks (like Herder) that the human being is essentially a being of lack (*Mängelwesen*) and the technical supplements that try to remedy to this lack therefore belong essentially to the human being: "Clearly technique, in this highly general sense, is part and parcel of man's very essence. . . . Technique constitutes, as does man himself, *nature artificielle*."[34] Tools and other technical equipment so to speak replace the human being's lacking instincts and organs. The biologist Adolf Portmann had suggested that the human organism is "unfinished" at birth: the human child is born prematurely and cannot survive without being taken care of, not only by its parents but by an entire culture. For Gehlen, the human being's primary human interest lies in "a semi-instinctual need for stability in the environment."[35] The natural human condition is highly unstable, and it must construct stable structures by its own activity. This happens through technics, including both material technologies and the social technics of culture. Gehlen, however, understands such cultural technics very differently from Plessner. While Plessner outlined a utopic place where strangers play complex diplomatic games, Gehlen defends the construction of a stable artificial world, a homogenous culture whose institutions protect a definite life form against human and natural instability.[36]

Technics is as old as humankind: "Technique, from its beginnings, operates from motives that possess the force of unconscious, vital drives."[37] New technologies occasion historical change, the most important of which for Gehlen are the Neolithic Revolution and the modern invention of nuclear energy, which represent absolute cultural thresholds.[38] Today we live within an unprecedented industrial technology, which has changed human life and society significantly. "Today it is vital to understand the functional connection between natural science, technique, and the industrial system."[39] Even modern technology cannot be understood as applied science and it has also generated a mass culture that leads to platitudes and stupidity.

Gehlen's criticism of modern technological society echoes Heidegger's (and many others: there was a strong and widespread feeling of alienation in the mid-twentieth century), but they arrived at a similar politics from two different philosophical directions. Let us now look in a more general manner at how Heidegger responds to the overall program of Philosophical anthropology.

Martin Heidegger: Inhabiting the World of Technics

Martin Heidegger's magnum opus *Being and Time* (1927) is directed not only against the classical metaphysical theory of subjectivity established on Descartes's *ego cogito ego sum* but also against modern anthropological, psychological, and biological theories of the human being. According to the groundlaying definitions of section nine of *Being and Time*, "The essence of Dasein lies in its existence," such that "when we designate this entity with the term 'Dasein,' we are expressing not its 'what' (as if it were a table, house or tree) but its Being."[40] Dasein's being is "in each case mine [*je meines*]," and in function of this "mineness [*Jemeinigkeit*]," its two modes of being are *Eigentlichkeit* and *Uneigentlichkeit*, literally "being-own" and "not-being-own," generally translated into English as "authenticity" and "inauthenticity." Heidegger distinguishes Dasein from the anthropological subject by affirming that Dasein is an answer to the question "Who?"(authentic "I" or inauthentic "they"), whereas the anthropological subject is an answer to the question "What?" (a being present-at-hand). In section ten, Heidegger insists further on the specificity of his question concerning the being of Dasein compared with traditional anthropological and biological interpretations of the human being, which had understood the human subject as a substance and were completely blind to the question of this being's *being*. Even though in his own time the Philosophy of life and Philosophical anthropology, developed by Wilhelm Dilthey, Henri Bergson, Edmund Husserl, and Max Scheler, had overcome the understanding of the human being as a substance and interpreted it as a unity of experience, this does not yet grasp the fundamental ontological question.[41] In the 1929–1930 lecture course published as *The Fundamental Concepts of Metaphysics*, Heidegger adds that Scheler's later work, in which he constructs a philosophical anthropology on the basis of the concept of life, "is a fundamental error" because, according to Heidegger, the basic notion of "life" still remains unquestioned.[42]

According to Heidegger, the project of building anthropology and psychology into the framework of general biology can aim at an ontological interpretation of life but cannot amount to a fundamental ontology because life and being are in fact fundamentally separate. "Life, in its own right, is a kind of Being; but essentially, it is accessible only in Dasein. The ontology of life is accomplished by way of a privative interpretation; it determines what must be the case if there can be anything like mere-aliveness [*nur-noch-leben*]. Life is not a mere Being-present-at-hand, nor is it Dasein. In turn, Dasein is never to be defined ontologically by regarding it as life (in an ontologically indefinite manner) plus something else."[43] Heidegger would probably have seen the bio-philosophy formulated in Plessner's *Levels of Organic Life and the Human* as an ontology of life that was incapable of replacing the ontology of Dasein but should instead be defined privatively in relation to it. As *Levels of Organic Life and the Human* was written at the same time as *Being and Time*, it is understandable that Heidegger does not comment on Plessner's work here—but he and his followers did not pay attention to Plessner's book later either.⁴⁴ Plessner comments on this silence in the 1965 introduction to the second edition of *Levels of Organic Life and the Human*, as well as in *Diesseits der Utopie*,⁴⁵ where he points out that Heidegger opposes his concept of human existence to the anthropologists' interpretation of the human being as a living being, and he responds that this makes Heidegger incapable of accounting for human corporeity and sociality.

If the basic concepts of Plessner's philosophical anthropology and Heidegger's fundamental ontology thus diverge, their approaches to human existence nonetheless have several parallels. Of course, neither thinks of existence as a substance but rather as a capacity articulated in terms of modalities (positionality, mineness). Furthermore, neither approaches human existence from the point of view of rationality but from the point of view of ordinary practical life—as Heidegger puts it in section eleven, Dasein must be interpreted in its *everydayness*. Most importantly for us, both think of human existence in function of an environment, *Umwelt*, and both analyze this in terms of *technics*. They articulate their insights differently because Plessner takes a third-person point of view while Heidegger investigates from a first-person point of view, although the philosophical tasks themselves have several similarities.

In section twelve, Heidegger states that Dasein is being-in-the-world. This does not mean that Dasein is situated inside the world as in a container. On the contrary, it is not a separated being and world is not a limited place, but they constitute one another just as the inhabitant and the habitation are constituted by the single event of inhabiting (wohnen).⁴⁶ Dasein's worldly existence consists in understanding the world not as an object of its theoretical contemplation but as the site of its practical concerns. Dasein does not "see" itself theoretically because it is dispersed in its worldly occupations and Dasein does not "see" the world as such either because the world itself only becomes apparent in terms of its concern. Another name for Dasein's closest world is *Umwelt*, environment. Heidegger had found the term in biology as well (in *Being and Time* he quotes von Baer, who was also an important figure for Plessner, but in *The Fundamental Concepts of Metaphysics* he refers to Uexküll in particular), but he claims biologists use the term without understanding it and that fundamental ontology reveals its fundamental sense as Dasein's world.⁴⁷

In section fifteen, Heidegger describes the being of entities encountered in the environment. "The Being of those entities which we encounter as closest to us can be exhibited phenomenologically if we take as our clue our everyday being-in-the-world, which we also call our 'dealings' [*Umgang*] in the world and with entities within-the-world."⁴⁸ Most of the time, the world is by no means present as such but is the inapparent horizon of our dealings. Heidegger shows how being-in-the-world gradually becomes perceptible starting from a thing (*Ding*). A thing is encountered in being-in-the-world not as a theoretical object but as a practical one, a *pragma*. Heidegger's word for it is *Zeug*, which is an informal word for any stuff, gear, or material; Macquarrie and Robinson's translation of *Zeug* as "equipment" is thus a bit too formal, but we will make do with it here. Strictly speaking, "There is no such thing as an equipment. To the Being of any equipment there always belongs a totality of equipment."⁴⁹ Equipment's mode of being is readiness-to-hand (*Zuhandenheit*).⁵⁰ It is not visible to theoretical sight (*Sicht*), yet using it practically does not mean using it blindly, for this use has its own kind of sight: circumspection (*Umsicht*). Readiness-to-hand has the paradoxical character of having to withdraw from view in order to be available for circumspection and therefore authentically ready-to-hand. Technical equipment must hide itself in order to be the medium that it is meant to be. In everyday life we are principally concerned with the work to be done and not with the equipment with which we carry it out. Hence, somewhat paradoxically, only work allows us to discover the equipment, the totality of equipment to which it belongs, other entities ready-to-hand (tools and resources), and other Dasein—and at the same time work "works" only if these remain inapparent as such. When we are concernfully absorbed in whatever work we are undertaking, the world lies close to us (the workshop, the public world, the environing nature)—but it does not come forth as the phenomenon of world.

In section sixteen Heidegger gives his famous description of the interruption of work when equipment becomes in one way or another conspicuous, obtrusive, or obstinate—this can happen, for example, when equipment breaks down. In such moments, readiness-to-hand withdraws but does not disappear altogether, and equipment becomes present-at-hand (*Zuhanden*) as such. This event is where the thematization of beings begins: equipment is no longer just given to circumspection but it is "seen"; it is seen as equipment, and starting from its context it can be seen as the totality of references (e.g., as the world of work, as a workshop with different equipment and people). The world is not a totality of equipment, but it is the implicit, trusted environment in which equipment can be encountered, generally as ready-to-hand, and sometimes as present-at-hand. When equipment becomes present-at-hand, the implicit world shows itself. But it is only much later in the book, in the analysis of anguish, that the world comes forth as such, as the explicit question of being-in-the-world as a whole.

This overview of Heidegger's famous account should suffice to show that his basic analysis of being-in-the-world takes technics as its starting point. From there, the world itself comes into view as a technical world. More precisely, technics is associated with Dasein's inauthenticity, whereas later texts will associate its authenticity with poetry and philosophy. The modes of inauthenticity and authenticity thus draw upon the two modes of techne, technics and art. Both are also modes of language, the language of inauthenticity being instrumental, like any equipment, and therefore an inapparent part of concernful dealings with the world, while the language of authenticity is poetic, present *as* language, and capable of indicating the question of being.

Let us now see how Heidegger links technics essentially to human existence and not to animal life in the lecture course from the winter semester 1929–1930 published as *The Fundamental Concepts of Metaphysics: World, Finitude, Solitude* (*Die Grundbegriffe der Metaphysik: Welt, Endlichkeit, Einsamkeit*). In this course, Heidegger examines the same question as in *Being and Time* but from a different viewpoint. Instead of a fundamental ontology he now uses a comparative approach that differentiates between human, animal, and stone with three theses: (1) The stone is worldless (*weltlos*), (2) the animal is poor in world (*weltarm*), and (3) the human being is world-forming (*weltbildend*).[51] Heidegger thus seems intrigued by the comparative approach that was common at the time, and this lecture is his own attempt at illustrating the essence of the human being by contrasting it with the animal (and the stone, instead of the plant). This was also the general approach of the Philosophical anthropologists. However, Heidegger does not discuss any of their works explicitly (Plessner is never mentioned and Scheler's late philosophy is discarded), but it is obvious that their philosophy would be rejected as an attempt to interpret human existence on the basis of life because according to Heidegger all such approaches fail to interrogate the being of life as such. Let us see how he claims to go deeper than any biological or biologically inspired philosophical theories of the human being by studying the terms of the comparison.

Firstly, like Philosophical anthropologists, Heidegger refers to Uexküll, whose idea of *Umwelt* is analogous to his own idea of being-in-the-world.[52] But instead of simply using the term *Umwelt* as in *Being and Time*, he problematizes it and analyses the world itself as a phenomenon and as a horizon of phenomenality. In these comparative considerations, the question "What is world?" provides an entryway to the question of Dasein's finitude and solitude and thereby to the fundamental question of metaphysics.[53] For Heidegger, the world is not just the space-time opened by a living being's perceptions: the world is the given totality of significations. In *Being and Time*, world was thought from Dasein's point of view as the implicit horizon of its everyday life that could also become explicit in exceptional situations (like anguish). In *The Fundamental Concepts of Metaphysics*, on the other hand, Dasein is thought of from the point of view of world; more generally, the analysis of world allows access to the different inhabitants of the world, namely stone, animal, and human.

Heidegger criticizes Uexküll for presupposing the animal is a simple substantial *vorhanden* entity, whereas one should really investigate its mode of being in the world. "Since J. von Uexküll we have all become accustomed to talking about the environmental world of the animal [*Umwelt*]. Our thesis, on the other hand, asserts that the animal is poor in world [*das Tier ist weltarm*]."[54] Poverty-in-world is one mode of having-a-world. In *Being and Time*, Heidegger had described Dasein's relation to the world with the rather elusive term *being-in* (*insein*), but in *The Fundamental Concepts of Metaphysics*, he expands on the existent's relation to in terms of *having*. The animal is "poor in world" because it has world in the mode of not-having it, and the human being has world in the mode of world-forming (*weltbilden*).[55] The preposition "in" becomes thus extended into the verb "having." In *Being and Time*, Dasein *inhabits* the world (*wohnen*), whereas in *The Fundamental Concepts of Metaphysics* it *forms* it (*bilden*); in a later essay (*Bauen wohnen denken*) Heidegger will further develop the theme of "inhabiting the world by building it" by describing human existence in terms of *building, inhabiting,* and *thinking* (*bauen, wohnen, denken*). Heidegger thinks of different existents as different *modalizations*. In the 1929–1930 lecture course he differentiates between the animal and the human in terms of their different modes of having the world. One could also note here that Plessner does not think of life in terms of substantial readiness-to-hand either but in terms of a modality, that of *positionality* that is conjugated into the animal's centric one and the human being's eccentric one. The modalities of having-a-world and positioning-oneself-in-a-world are not the *same*, but they represent similar philosophical approaches to existence.

Secondly, Heidegger's fundamental point of reference in his comparative work on animality and humanity is not biology but Aristotle. He does draw on Uexküll by, for example, adopting his criticism of Darwinism and much of his analysis of animality by rewriting of Uexküll's concepts. But as soon as Heidegger poses the question of the metaphysical essence of the animal, he turns to Aristotle's *De anima* (which he taught many times in these years, e.g., in 1921, 1929–1930, and 1931[56]). He is certainly not the only German philosopher to do so—*De anima*'s influence was already strong in Hegel's *Philosophy of Nature*, a work whose implicit influence can be seen up to Philosophical anthropology. However, Heidegger develops an original phenomenological reading included in his own thinking of animality and humanity in *The Fundamental Concepts of Metaphysics*. In this lecture course, the explication of the human being's specific way of "having a world"—world-forming—is finally effectuated through a reading of Aristotle's word *logos* that, according to Aristotle's well-known definition, is what separates the human from the animal.

Heidegger thus transforms Aristotle's *"zoon logon ekhon"* into the "human is world-forming." This is of interests to us because the word *bilden*—to form but also to produce, create, sculpt, and mold as well as teach and educate—refers to the semantic fields of both technics analyzed in *Being and Time* with the notion of equipment (*Zeug*) and to the imagination (*Einbildungskraft*) examined by Heidegger in *Kant und das Problem der Metaphysik* (1929). According to Heidegger, any kind of equipment—*Zeug, Werkzeug, Maschine*—can only exist because it belongs to the world (*weltzugehörig*), because it is produced (*erzeugt*) by humans, and this is possible only on the basis of world-forming (*Weltbildung*).[57] Heidegger takes great pains to define the notion of world-forming as the opposite of the voluntarist notion of producing. The human being is not world-forming in the sense that it would fabricate the world to its liking. On the contrary, "World-forming happens, and on the ground of this, the human being can exist. Human as human is world-forming: this does not mean the human being such as it walks along the street, but Da-sein in the human being is world-forming."[58] World-forming happens, *geschieht*, as if by itself, when the world *is* the horizon of a given existence. The world is not a totality of entities like the totality of tools in a workshop, but the world is its own *way of being* accessible: its openness and manifestness, for example the availability of tools that are are used in a workshop without even thinking of it. However, world is of course not a workshop, it is the entire horizon of existence. "World is appearing of being as such in its entirety [*Offenbarkeit des Seienden als solchen im Ganzen*]."[59] In its appearing, the world has a structure or a form. Heidegger's central point here is that this world-form is not an ideal structure made up of eternal truths but a historical configuration that is destined to the historical Dasein. "World is formed [*bildet sich*], and world is what it is only as such forming [*Bildung*]. Who forms the world? According to the thesis, the human being."[60] And yet the human being does not fabricate the world, but world-forming "happens"; it is the world's way of being. At the same time, the world's form is not independent of Dasein but formed by it. The world is formed not by any wilful anthropos but by the Dasein in the human (who is a very impersonal figure here). Dasein is the one who produces the world (*her-stellen*), puts it into an image (*ein Bild geben*), and contains it.[61] In order to describe this strange reciprocal production, Heidegger also uses the term *Entwurf*, which means "draft," "architect's plan," or "construction plan." The fundamental character of the happening of the world is *Entwurf*: world is world; it "worlds" when it is drafted by Dasein and Dasein drafts the world only when world sketches itself and gives itself to it as such a sketch.[62] When world thus drafts itself, it also drafts Dasein in turn and transforms it.[63]

In *The Fundamental Concepts of Metaphysics* this "forming" is presented in and through logos. Heidegger's reading of Aristotle's logos is too long and complex to be discussed here, but I just want to draw attention to three striking features. Firstly, logos is not human speech or reason: it is the fundamental structure of the world, the synthesis and *dihairesis* that makes of it a totality and that Heidegger described using the quasi-technical terms *Bildung* and *Entwurf*. By speaking of world, he designates a finite place instead of the infinite totality of being, and by choosing the "technical" terms *Bildung* and *Entwurf*, he designates a historically changing transcendence instead of its ideal infinite dimension. Second, Heidegger interprets *logos* as "*als*." Logos is what permits *seeing something as something* such that (1) the *als-Struktur* brings together as a totality[64] and (2) it lets beings as beings and ultimately the ontological difference be seen.[65] This is a highly intellectualist interpretation of the human being's *differentia specifica*. Heidegger repeats Aristotle's *De anima* in it insofar as the animal's world-relation is defined by perception and the human being's by *noesis*, but he interprets noesis as the fundamental structure of transcendence as world. Third, while logos is properly speaking the structure of the world, it only becomes apparent through human logos (speech). The reciprocal structure of world-forming is thus reflected in the logos as well. As Heidegger says more concisely in his lectures *On Aristotle's Metaphysics Theta* (1931), he does not read logos in the sense of a logical proposition (*Urteil*) but in the sense of a production (*Herstellung*) of works. The analysis of equipment in *Being and Time* must also be read in this sense, rather than, for example, as a criticism of Marxism.[66] Logos as production has several forms that are only indicated here, but these are further developed in later texts. What is remarkable, however, is that through these forms logos becomes thought in terms of techne. The simplest forms, the prototypes of the world-forming logos, are equipment (*Zeug*) and the instrumental use of language described in *Being and Time*. The most developed forms are poetry (*Dichtung*) and art (*Kunst*). Technics and art are the two faces of techne. From everyday speech to philosophical language, human logos is thus also thought in terms of technicity.

Moreover, if Philosophical anthropology interprets human existence as a modification of a general theory of life, Heidegger rejects all interpretations of human existence as life. For him, life is the animal's way of being, whereas the human being's way of being is existence. This is the fundamental difference between Heidegger and the Philosophical anthropologists, and it also allows us to see their respective strengths.

Heidegger's definition of human Dasein in terms of existence reflects the leading question of his philosophy, the question of being. *The Fundamental Concepts of Metaphysics* does not really pose the question of being but, according to Heidegger, prepares for it. The purpose of the study of the structure of logos is "to *prepare our entering into the occurrence of the prevailing of the world [vorzubereiten das Eingehen in das Geschehen des Waltens der Welt],*"[67] that is, of understanding the specific mode of being of world. Only afterward can the question of being be posed.[68] In order to be attentive to this, Dasein must be in a fundamental attunement (*Grundstimmung*) to metaphysics, and for this to happen the Dasein in humans must awaken from its absorption in everydayness and be transformed (*verwandelt*) by proximity to the question.[69] Heidegger examines Dasein as existence in order to prepare for the philosophical question of being. Existence, and its main modalities authenticity and inauthenticity, depict Dasein as a metaphysical being. Given that the question of being coincides with the question of time, it is no wonder that the sense of Dasein's existence is its temporality.

According to Heidegger, this question cannot be brought forward when the human being is analyzed in terms of life. But the Philosophical anthropologists and, after and independently of them, many French phenomenologists starting from Merleau-Ponty, will emphasize that such an approach too has its price, namely Heidegger's relative inability to understand bodily existence, sociality, and spatiality. These axes of human existence are rooted in life, which Heidegger ultimately does not give an account of.[70] Heidegger's choosing of being over life is also reflected in his understanding of technics. Heidegger certainly does give technics an unprecedented philosophical role. In *Being and Time*, the paradigm of the thing is a technical object (*Zeug*) and the most important example of world is a workshop. In *The Fundamental Question of Metaphysics*, the transcendental horizon of world is described in quasi-technical terms (*Entwurf*, and later *Ge-stell*). However, technical objects and horizons are ultimately not thought in terms of technical activity itself but in terms of the question of being that they frame. In contrast to this, Plessner explains technics as an activity that rises from life itself: it belongs to the living being's efforts to secure its position and, in the human being's case, to invent a world of coexistence with other natural and human beings.

In an important later essay, "The Question Concerning Technology" (1953), Heidegger reaffirms that he thinks of technics in function of truth: "The essence of technology is by no means anything technological."[71] Already in an essay from 1939, "Vom Wesen und Begriff der *Physis*," Heidegger showed how Aristotle's explication of natural beings, *physei onta*, depends on comparing them with technical beings, *technai onta*[72]—but this means inversely that technical beings are aligned with the theoretical perspective in understanding the truth of natural beings. In "The Question Concerning Technology," Heidegger focuses on technical beings themselves, the essence of which will be a form of truth. One of his examples is a silver bowl, a technically produced artifact that is one of the examples Aristotle uses in the *Physics* in his account of the four causes—but in this the very account comes to reflect the mode of being of *technai onta*. *Causatio* is thus thought of as production, and *Hervorbringen* (*poiesis*) and *physis* itself comes to be thought of as a kind of production. "Production brings from concealedness to unconcealedness."[73] Hence, "Technology is therefore no mere means. Technology is a way of revealing [*Entbergen*]. . . . It is the realm of revealing, i.e., of truth."[74]

In his presentation of the ancient idea of technics, Heidegger refers to the interpretation of techne as a form of knowledge, know-how, that we already examined in chapter 2. The originality of his analysis lies in his underscoring that know-how presupposes a knowledge of the world, for example, an understanding of what can be done with different materials. As we saw earlier, techne does what physis makes possible—and it does what physis alone cannot do. Heidegger thinks that modern technics is essentially different from the Greek. Both are undoubtedly forms of revealing. But modern technics is based on the truth of the natural sciences—and, inversely, the natural sciences are dependent on technical instruments. According to Heidegger, modern technology reveals (*Entbergen*), but it reveals differently from the Greek *techne* for it does not reveal nature as the harmonious whole of the *kosmos* but as a domain of forces. It does not reveal by producing and bringing forth (*Hervorbringen*) in the manner of poiesis, but in the manner of provoking and challenging (*Herausfordern*). It not only accompanies natural forces (like an old windmill or bridge), but it seizes upon nature's energies with force and even violence. A famous example of this difference is the production of energy: an ancient windmill uses wind's energy by accompanying air's movement but without suppressing it, whereas modern nuclear power extracts the energy from the uranium atom by breaking it. Although Heidegger does not speak of this, the toxicity of nuclear waste reflects the brutality of fission. Furthermore, if ancient poiesis is adapted to nature's own potentialities, modern technology expands controllable processes as far as possible: "Regulating and securing even become the chief characteristics of the challenging revealing [*Steuerung und Sicherung werden sogar die Hauptzüge des herausfordernden Entbergens*]."[75] *Steuerung*, regulating, translates the Greek *kybernein*, and this is ultimately what cybernetics does: it programs the movements of the machine not to accompany the hazards of the material but to impose upon it predetermined operations (which aim to overpower material contingencies). Modern technology does not see nature as autonomous physis endowed with multiple potentialities but as a simple "standing-reserve" (*Bestand*) or stock of its own needs.[76] Although modern technology is obviously invented and produced by human beings, it is by no means understood and controlled by them. On the contrary, human beings have no choice but to adapt themselves to modern technology; they are formatted and produced by its system. Or to put it in terms of *The Fundamental Questions of Metaphysics*, the human being is a world-former (*Weltbilder*) who is itself formed (*bildet*) by the world, not that world is a conscious subject that molds humans like the mythological creator-god molds the first humans from clay, but world is a transcendental form that leaves, unnoticed, its mark on the human beings who find themselves in this world, thus gradually transforming them.

"Modern technology as ordering-revealing is, then, no merely human doing."[77] According to Heidegger's famous thesis, "The essence of modern technology shows itself in what we call Enframing [*Ge-stell*]. . . . Enframing is the gathering together that belongs to the setting-upon which sets upon man and puts him in position to reveal the real, in the mode of ordering, as standing-reserve [*Das Ge-stell ist das Versammelnde jenes Stellens, das den Menschen stellt, das Wirkliche in der Weise des Bestellens als Bestand zu Entbergen*]."[78] The essence of modern technology is *Ge-stell*, which assigns both humans and nature as being nothing more than standing reserve for the operations of technology. *Ge-stell* is nothing technological: it is the modern form of the *Bildung* described in *The Fundamental Questions of Metaphysics*, the transcendental horizon that makes everything appear in the light of the logic of technics. It is the form that truth takes in the epoch of technics: force, resource, utility.

Heidegger thinks that truth is epochal. This has nothing to do with a "relativism" that allows humans to choose their truth on a whim. One does not choose the transcendental horizon of one's existence. However, truth is not always the same. It changes with time, very slowly but irresistibly, and each time it is "sent" to humans, that is, it befalls them like an inescapable destiny. If in modern times the fundamental structure of truth is *Ge-stell*, the destiny of modernity is, according to Heidegger, indeed technical rather than, for example, scientific, capitalistic, or religious. Not only do we use nature and ourselves as the cogs and resources of technics, but we also see ourselves and nature as technical entities that can be used, manipulated, explained, repaired, and enhanced.[79] *Ge-stell* itself is neither good nor bad: it is how things are disposed now. But *Ge-stell* can lead to both good and bad ways of inhabiting the world. Heidegger was particularly sensitive to its dangerous consequences, which were then new and therefore striking (but these insights remain valid today, even though on a systemic level people seem to have become so familiar with them that they go unnoticed). As the human being "exalts himself to the posture of lord of the earth,"[80] "the decline of the truth of beings occurs necessarily, and indeed as the completion of metaphysics. The decline occurs through the collapse of the world characterized by metaphysics, and at the same time through the desolation of the earth stemming from metaphysics. Collapse and desolation find their adequate occurrence in the fact that metaphysical man, the animal rationale, gets fixed as the laboring animal."[81] With this development, the epoch of technics produces the epoch of nihilism, which combines the desolation of the earth with "objectified nature, the business of culture, manufactured politics and the gloss of ideals overlying everything," justified as the completion of metaphysics but really the general rule of "machination" (*Machenschaft*).[82]

However, it must be noted that there is more to it than this. Technics is not necessarily a prison, it can also become an incitement to creation. Just as the essence of techne is ambiguous and bears the possibility of both technological closure and artistic openness,[83] the *Ge-stell* of the technical era too may have led to the nihilistic era of machination, but it could also contain the possibility of another beginning, provided that humans learn to welcome it. It is true that Heidegger says next to nothing about the positive potentiality of the technical *Ge-stell* itself. He seemingly sees it as so rigid that it is incapable of turning to matters other than machination. "Saving" existence can be prepared for, though, by making *Ge-stell* more flexible, by deconstructing and dismantling it. Heidegger thinks this can be done especially by deconstructing the past tradition in order to clear ground for something totally other to come, the tradition being the history of philosophy in particular but also to some extent poetry (this need not be dismantled but heeded more attentively). Hence technical *Ge-stell* is the destiny of our era, but our future depends on our capacity for making its rigid frameworks more fluid. Heidegger expects this from art and poetry especially; these are the reverse sides of techne.

Technics and Nothingness

The Philosophical anthropologists and Heidegger both think of human existence as being in an environment that is not only given but also actively "built" by the human being. The human being makes the environment that in turn makes it. This does not happen through a conscious decision but the historical world is already given when a child is born, it grows into it before coming to be able to modify it. As Heidegger explicitly says: world-building happens, the transitivity of "building" is first, and world and human become what they are through its happening. As different as they may be, Plessner, Gehlen, and Heidegger all look into this happening of world-making and say, in the last instance, that world-making is technical in the broadest possible sense of the word. Technics mediates between the human and the world: it frames not only technology but the entire culture, society, and everything that counts as truth; it educates the human and it articulates the world and the human together. This is why the analysis of technics is the key to understanding human being-in-the-world.

As we have already seen, although the Philosophical anthropologists and Heidegger share this general view of existence, they are also mutually opposed insofar as the former analyze the human being as life and the latter analyzes the human being as existence. Heidegger claims that the interpretation of human existence as life leads to an inability to open up ultimate ontological questions, while Plessner affirms that Heidegger's contempt for life blinds him to bodily and social existence. These criticisms are not as irrefutable as the adversaries would have us believe because in the last instance none of them are fair. If we look at Plessner, for example, his interpretation of human existence in terms of life does not amount to examining it in ontic biological terms. For him life is an interpretation of being itself that also conditions such ontic theories. His theory of life is an interpretation of being as a potentiality; it is declined as different modalities of positioning; the outcome is a theory of being as a plastic capacity for transformation without model. These are theories of being in the same sense as those of Hegel's *Science of Logic*. They start from the evidence that there is being rather than not, and given this, the philosopher's task is to understand how being determines itself. For Heidegger, this is not at all satisfactory. He thinks that the question of being must be grounded in the clarification of Dasein, for whom being is a question, and then it can be brought to words through patient description, with the help of de(con)structive readings of past philosophy, of the way in which there is time and there is being (*es gibt Zeit und es gibt Sein*). Both Plessner and Heidegger pose real questions of being, but only Heidegger poses it in terms of Dasein, who poses the question.

If Heidegger's question of being is more demanding than Plessner's, its concrete reach is narrower. Heidegger's Dasein is hardly anything other than a function of the question of being. He has little to say of the everyday existence that is the starting point of his analysis because, over and beyond the fine analysis of equipment, *Being and Time* ultimately rules out questions of bodily existence, nature, and other people. Moreover, this dismissal goes hand in hand with a certain forgetfulness of spatiality. Contrary to what Heidegger suggests, these are no mere ontic aspects of existence but they contribute to the very sense of existence, as many French phenomenologists from Merleau-Ponty to Nancy have pointed out. Although Christian Sommer, for example, has shown that Heidegger has more to say on the living body than these critics claim, and although Heidegger's insistence on the "abyss" between animal life and human existence[84] could actually explain the human being's relation to its own body and to other people in an enlightening way, it nonetheless remains true that unlike the Philosophical anthropologists, Heidegger is mainly interested in the question of being and not in sociological questions.

These are well-known debates in which the opposing positions are clearly demarcated. But what about our question of technics? How does the consideration of technics as world-formation reflect these positions and does it complexify them? How do these theories contribute to our general question concerning technical humanity?

Plessner describes the philosophical origin of technics as follows (but note that Plessner uses the term *artificial* rather than *technical*, the two words pointing at two different phases of the same process: technics is the activity of production and the know-how that goes with it, whereas artificiality qualifies the result of this production, which could not exist without technical intervention):

> The excentric form of life and the need for completion constitute one and the same state of affairs. Need, however, should not be understood here in a subjective or psychological sense. This need is presupposed in all needs, in every urge, every drive, every tendency, every volition of the human. In this neediness or nakedness lies the motive of all specifically human activity—that is, activity using artificial means that is directed toward the unreal. In it lies the ultimate ground for the tool and for that which it serves—that is, culture.[85]

Here technical-artificial activity is described as a response to human neediness and nakedness. It is called forth by a need that is the very essence of human life. The productivity that leads to technics "is ultimately due not to drives, to the will, to repression, but to the excentric structure of his life, to the form of his existence itself."[86]

> The expression of his nature that corresponds to the human's essential dividedness, nakedness, and existential neediness is artificiality. Given with excentricity, artificiality is the detour to a second native country where the human finds a home and absolute rootedness. Positioned out of place and time in

nothingness, the excentric form of life creates its own ground. Only to the extent that it creates this ground does it have it, is it carried by it. Artificiality in action, thought, and dreams is the internal medium by which the human as living natural being is in accord with himself. With the forced interruption by fabricated connecting links, the life circle of the human, into which he, as an organism of needs and drives standing on its own, is irrevocably forged, raises itself into a sphere superimposed on nature, and comes full-circle there in freedom. The human, then, lives only insofar as he leads a life.[87]

Technics is an *expression* of human eccentricity. This does not mean that it is the voluntary realization of an idea invented by an individual but that human expressivity takes place together with what could be called the expressivity of the world. "The human can only invent to the extent that he discovers. He can only make what "already" exists. . . . His productivity is only the occasion for his invention to become an event and to take form." "The secret of creativity, of hitting upon an idea, lies in the fortunate touch, in an encounter between the human and things. . . . The creative touch is an achievement of expression."[88]

In short, Plessner thinks human artifice, or technics in the broad sense of the word, as an expression of the eccentric nature of the human being. It is not a specific conscious voluntary action, however, but a general expressivity visible in all human activity and corresponding to the world's own expressivity or availability. Although Heidegger also thinks that technics comes to be in an encounter between Dasein and world, he shuns the term expressivity and insists much more strongly on technics as a world horizon. For him, technics is an epochal situation, a "sending of destiny [*Geschick*]" that conditions human know-how or "dexterity [*Geschicklichkeit*]" mostly in an unconscious manner.

One could say that in Philosophical anthropology, technics is thought in the framework of an *autopoietic* project (although this term was not yet in use at that time). The autopoiesis is not that of a human individual but of a human collectivity that produces an artificial place in order to maintain itself against the ultimate nothingness of existence (following this general principle, the collectivity can then be thought of as the democratic diplomatic community of exiles, as in Plessner, or as the conservative community of a firm institution, as described by Gehlen). The aim of technics is to maintain this utopic place (Plessner) or institution

(Gehlen). By maintaining its place, the autopoietic community cares for itself by assuring its integrity and consistency and by trying to hide the threatening nothingness. It protects itself from whatever tries to undo it. This sense of technics comes to the fore especially in Gehlen's *Man in the Age of Technology*. In Plessner's *Levels of Organic Life and the Human*, the sense of artifice is more nuanced, since in this book it is a means of expression, and mediation between strangers and the community is open, changing, and democratic.

In political temperament, Heidegger is fairly close to Gehlen in that he too seems to imagine human Dasein automatically in terms of a finite historical community that appears closed to strange elements (however, one must note that it is not defined against the enemy, as in Carl Schmitt, but against the distant historical past represented by ancient Greece). But philosophically, Dasein's interrogation of being is quite opposed to closure. The very sense of Dasein is transcendence: it aims at puncturing the sphere of habitual life in order to reveal its ground and unground. Dasein's aim is not to accept being such as it is but to ask its sense; it does not seek to hide the nothingness behind an autopoietic closure but on the contrary to learn to question the givenness of the world and discover the nothingness on which it rests. Nothingness must be found and existence must be chosen within it. There is, after all, an existentialist aspect to Heidegger. The reassuring form of the given world is only Dasein's inauthenticity. The philosopher's aim is to discover the possibility of its authenticity by breaking through worldly evidence, by learning to ask why there is being rather than nothing: Why am I here, after all? Technics is the way in which world is given in the first place, but the self-evidence of technics must break down in philosophical interrogation of the world. As in "The Question Concerning Technology," its presuppositions must be brought to light.

Both the Philosophical anthropologists and Heidegger think of the self as a negativity, but they understand this negativity in radically different ways. In *Being and Time*, nothingness is revealed in anguish and in being-toward-death as the discovery of the possibility of my not-being-there that prepares the discovery of the nothingness of being itself. In the dense terms of "What is Metaphysics?": "Anxiety reveals the nothing. We 'hover' in anxiety.... The nothing reveals itself in anxiety—but not as a being.... The nothing itself nihilates."[89] This fundamental ontological nothingness is neither a nonbeing nor a negative determination of being like in logics or dialectics. It is the possibility of thinking my nonbeing in the midst of my existence, and the nonbeing of being at the heart of being itself. Nothingness is the suspension of being that happens in being itself so that it becomes possible for the Dasein to question being. The nothingness of being must be distinguished from the kind of negativity that characterizes the epoch of technics. The epoch of technics is also marked by a certain negativity—the nihilism of calculation and the instrumental use of language that contribute to the general negativity of "machination" (*Machenschaft*) described in Heidegger's readings of Nietzsche and Jünger— but it is a different negativity than the nothingness of being because instead of giving being to think, it hides being from thinking. The negativity of nihilism annihilates thinking instead of making it possible.

The Philosophical anthropologists do not think of negativity in terms of non/being, but, on the contrary, they think of the negativity of human existence toward the possibility of its determination. Plessner describes existence as neediness and nudity, and Gehlen affirms that the human is a being of lack, *Mängelwesen*, but both these forms of negativity are directed toward a positivity that can satisfy the lack. The difference is that Heidegger aims to discover a nothingness that is normally hidden (especially by the negativity of technics) rather than overcoming this nothingness by an act of positing and taking artificial refuge from it. For him, the aim is not to discover a determination of man but to make it visible in order to de(con)struct it, to go through its horizon so that something else can occur. This is why his aim is to make visible the present epoch as the epoch of technics, to make visible its intimate error and nullity, and thus to clear the ground for the possibility of the event of something else. If, as we just saw, for Plessner nothingness is a dynamic source of expression, for Heidegger, the revelation of nothingness stops expressive activity, brings it to a halt, and aims to take a stance of listening. Hence, for the Philosophical anthropologists, originary technicity becomes the definition of humanity. The point is not to impose a technical figure upon man but to see technical activity itself as its essence. For Heidegger, on the contrary, there is no definition of man at all, for the aim is to cease defining him and to think of existence as pure temporality.

Both the Philosophical anthropologists and Heidegger think of the world as a home, *Heim*. When Plessner thinks that the human is originarily homeless (*heimlos*) and Heidegger thinks that existence is uncanny (*unheimlich*), both use a privative form of the word *home*, but they refer to its privation in different ways. Plessner imagines a home that is lost, away, and Heidegger imagines a home that is too close, to the point where familiarity turns into disquieting strangeness. This is why in Plessner's case the lacking "home," especially in the sense of a social domain of normative familiarity, must be rebuilt, whereas in Heidegger's case the "home," both its normative order and, more profoundly, its epistemic norms, must be dismantled.

In sum, all these authors think that the core of humanity is nothingness, and not in terms of any kind of figure of humanity. The Philosophical anthropologists understand this nothingness as indeterminacy. Indetermination then calls for determination; at its best, determination takes the form of a culture, a provisionally stable artificial construction. Heidegger, on the other hand, claims that the epoch of technics imposes determinations on human beings (figures like Jünger's worker) and that this technically produced disguising of the human chokes the possibility of questioning. Very likely Heidegger would extend this critique to Plessner's idea of the human being's position in the world. Isn't a position (*Stellung*) the result of posing (*Ge-stellen*), a technical product of ideological technics? He understands the nothingness of human existence as the enigma of existence that cannot be explained but that nonetheless calls for technical, but also poetic, artistic, and philosophical responses. Each time, technics appears as a secretion of nothingness. Technics too is this nothingness. It is groundless and unreasonable, and this is why it will always be dismissed by history when new forms of technics come forth, promising better shelters from nothingness.

Philosophical anthropologists show how the intimate nothingness of the human being pushes it to construct artificial worlds ever anew. Heidegger shows how this world is also invested by an intimate nothingness, which makes it bound to wither when its intimate meaninglessness comes to the fore. Heidegger shows this in detail in his readings of poetry. But in principle one could also extend these considerations to his analysis of the epoch of technics. Dasein calls forth new forms of techne because techne does not obey a straightforward rationality but other, more obscure laws of invention that it cannot be deduced from present forms of techne. Dasein can, however, clear the ground for their arrival. Heidegger thus lets us see why technics has a history. Because of its intimate nothingness, Dasein reaches toward the world, and because of the intimate nothingness of the technically framed world, the world petrifies, withers, and is destroyed.

This is all very well. But after all, neither the Philosophical anthropologists nor Heidegger see very far into technics itself; they are more interested in the human being. What about technics itself and its intimate ambiguity?

Chapter 4

De/constructing Humanity

We have seen how technics becomes for the first time a determining feature of the human being in the works of Philosophical anthropologists on the one hand and in Heidegger on the other. Although Plessner and Heidegger think of the human situation in the world very differently, both think of technics not as a simple means but as the entire horizon of human existence: it is the fundamental character of the human world, built by humans and forming humans in turn. But we also saw that what really interests these early twentieth-century thinkers is nonetheless humanity itself, not technics. Rejecting all references to an ideal human form, they continue the idea, already prefigured in Herder, Rousseau, and Nietzsche, of an originary plasticity, artificiality, and technicity of the human being. Both the Philosophical anthropologists and Heidegger think that what accounts for this technical capacity is a still more originary *nothingness* that determines human existence. Plessner described it as a neediness and a nakedness and the artifical human world as its expression, while contrary to this Heidegger discovered nothingness by inquiring into the transcendental conditions of the factical world that is (also) given in terms of technics. Technics opens the way to nihilism because it reveals the fundamental nothingness of human existence.

Thinking of human existence as nothingness is a powerful counterargument to classical humanism and above all to the classical subject of philosophy present to itself as a totality of positive forms of knowing, cognizing, and sensing. However, thinking of human existence as nothingness does not mean completely overthrowing the idea of human subjectivity. Some sort of a human subject remains a focal point of all theoretical and practical activity, though it thinks of itself as an enigma instead of as an evidence. The revelation of its nothingness was very important, not the least because it accounts for the possibility of change, without which neither technicity nor historicity can be explained. But this, which thus resembles the existentialist view of the void at the core of the human being, is also a paralyzing vision, a glance at the face of the Medusa where there is ultimately nothing to be seen, therefore leading to pure blindness without insight.[1] To some extent, though, it is possible to sound out what cannot be seen: when the face sinks into darkness, it is still possible to hear distant sounds and voices, as when one orients oneself in the dark using one's ears instead of eyes. Hearkening to unseen sounds is the way of psychoanalysis and of some other forms of indirect interpretation, such as in Philippe Lacoue-Labarthe's description of the subject as an echo chamber filled with more or less identifiable voices from the past.[2] But even if one can to some extent sound out the subject who thus appears as a dark cavern filled with echoes, one cannot drag its contents to light nor touch firm ground: the post-Heideggerian subject, explored by Lacoue-Labarthe and also by Maurice Blanchot, remains shadowy and nocturnal. As important as it has been to find indirect ways of making it appear, over time the repeated discovery of its negativity leads nowhere because the very definition of the human being as negativity makes it impossible to shed any light on it. In order to see technics itself, and not only technics as another projection of the void, and in order to see the effects of technics on the human being, we need to find another approach.

One such approach was developed in the 1960s and 1970s in France by a whole generation of authors sometimes grouped under the ill-fitting label "poststructuralism." In the following I will discuss the contrasting views of only Michel Foucault and Jacques Derrida. These authors turned against both positivism, which sees the human being as just an object, and phenomenology, insofar as it presupposes a pure, authentic *I* present to itself. Even if this view of phenomenology fits Husserl better than Heidegger,[3] the poststructuralists also criticized Heidegger's Dasein, albeit its negativity. In the same way they criticized Alexandre Kojève's and Georges Bataille's anthropological interpretations of Hegel, which had emphasized the negativity of man. Incidentally, Kojève's and Bataille's respective conceptions of human existence share more than one feature with the philosophical anthropologists, but they adopt the first-person point of view instead of speaking of humankind as a collective. The "poststructuralists" turned against these "subjectivisms" with the help of structuralism, which interpreted human beings in function of impersonal structures. But they also criticized structuralism insofar as it thinks human nature in terms of timeless structures, whereas poststructuralists saw human life in terms of changing historical constellations. Such constellations are also determined by technics. This is why the poststructuralists can open up ways of thinking not so much of the humans who produce technics (as Philosophical anthropologists did) but of the ways in which technics produces humans (Heidegger glimpsed this but regarded it only as the worst kind of technological alienation).

Poststructuralists managed to think the human being neither as the full self-presence of the philosophical subject nor as the nothingness that resulted from its "death" because they did not think of it as the *origin* (of thinking and acting) but as the *effect* or the *result* (of impersonal signifying and practical processes). Instead of the anthropological subject and the philosophical subject, Foucault and Derrida consider human subjectivity as an effect of *language*. This is of great interest to the present problematic because they also understand language as a paradigmatic technics, thereby inviting us to think subjectivation in function of technics. Their works have often been interpreted as representative of a "linguistic turn of philosophy" that many (but not I) would like to see as something past. However, today they tend to be interpreted as the first representatives of a "technical turn of philosophy," as interpreting language itself as a technics that we use and that uses us, instead of thinking language as the instrument for the subject's self-expression and communication. But the technological turn is not just another name for the linguistic turn. As Arthur Bradley says,[4] what is now called the technological turn of philosophy was inspired by developments both in biology, where "life" became interpreted as "technical" (François Jacob's *Logique du Vivant*, 1970), and cybernetics, which posed the question of "information" to animal and machine alike (Norbert Wiener's *Cybnernetics or Control and Communication in the Animal and the Machine*, 1948). In philosophy, this led to the consideration of what Derrida calls the "originary technicity of life itself" in the works of Derrida and Foucault but also in Jacques Lacan, Jean-François Lyotard, Gilles Deleuze, and Félix Guattari. As David Wills points out, the model of technology with which these authors worked was less and less epistemological and more and more biotechnological, thinking life in terms of a (quasi-)program and technics in terms of (quasi-)life.[5] We can say that the "bio-technical" interpretation of technics presented in the last part of chapter 2 here finds its full philosophical interpretation.

The decentering of human subjectivity by Foucault and Derrida led to positions such as the critique of the "metaphysics of subjectivity" and declarations about the "end of man." But one should not take these expressions too literally. Foucault famously writes on the last page of *The Order of Things*:

> As the archaeology of our thought easily shows, man is an invention of recent date. And one perhaps nearing its end.
>
> If those arrangements were to disappear as they appeared, if some event of which we can at the moment do no more than sense the possibility—without knowing either what its form will be or what it promises—were to cause them to crumble, as the ground of Classical thought did, at the end of the eighteenth century, then one can certainly wager that man would be erased, like a face drawn in sand at the edge of the sea.[6]

But, as for instance Alan Milchman and Alan Rosenberg show, Foucault's "death of man" is "not the death of the human being but the death of a determinant historico-cultural form or modality of the subject."[7] In "The Ends of Man," Jacques Derrida quotes this passage in the epigraph[8] without discussing Foucault explicitly in his text. Without "concerning himself with any given author's name,"[9] he discusses the anti-humanist and anti-anthropologist wave that followed the Sartrean era in France and insists that his contemporaries should have reread Hegel, Husserl, and Heidegger (like Derrida himself does) instead of simply rejecting them, for this rereading would have shown that the (anthropological) end of man is but a consequence of the philosophical question of the end/destination of man already laid out in Hegel, Husserl, and Heidegger's theorizations of human finitude.

> Man is that which is in relation to his end, in the fundamentally equivocal sense of the word. Always.[10]

> The end of man is the thinking of being, man is the end of the thinking of being, the end of man is the end of the thinking of being. Man, since always, is his proper end, that is, the end of his proper. Being, since always, is its proper end, that is, the end of its proper.[11]

In other words, the questions of human subjectivity, being, and finitude go together. In the epoch of the "end of man," the human being appears precisely as a being who questions its own end and who understands itself as such a questioning. But this being said, Derrida admits that there are more strategies than one for overcoming traditional humanism: the deconstruction of the philosophical tradition, which risks remaining in the same closure with it, and the brusque change of terrain, which risks remaining on the old terrain without noticing it.[12] These two strategies surely make us think of Heidegger vs. Nietzsche—as well as of Derrida vs. Foucault—but Derrida does not attribute them to anyone in such a straightforward manner.

Neither Foucault nor Derrida (nor any other poststructuralist) would contest human reality nor the importance of the study of human experience in philosophy. Both think in terms of human finitude and historicity. However, they set out to study humanity otherwise: not as the origin of all experience but as an effect of texts, discourses, and finally more general technical *dispositifs*. This is why by drawing on them we can focus more precisely on technics itself and show in what sense humanity itself must be thought as a "technical construction" or a "deconstruction of humanity by technics." These are preliminary approximations spelled out in quotation marks because describing human existence as a technical construction and deconstruction does not mean (as a hasty reading has made some people think) that the human being could shape itself at will, on the one hand becoming what it wants and on the other being just a relativist void as such, ready to be filled with just about anything.[13] Such hopes and fears must be ruled out at once: the poststructuralists too think that humans are always already thrown into a situation that makes them something and this is why they cannot become just anything.

Both Foucault and Derrida contribute decisively to what I have previously called the "bio-technical" interpretation of technics: Foucault by discovering the domain of biopolitics, which is fundamentally life as formatted by different governmental technologies, and Derrida by revealing the connections between life, writing, and tele-technology. This opens up new perspectives to technics because the paradigm is no longer the artisanal (hammer) or industrial (power plant) technologies that aim at transforming nature but a technics of life and thinking that affect the human being itself: power and disciplinary techniques for Foucault and techniques of signification (writing, tele-technologies) for Derrida.

Foucault and Derrida explore different and even opposed aspects of this complex. Foucault contributes primarily to our question of "humanity" by showing how (individual and collective) subjectivation is marked by disciplinary techniques, whereas Derrida contributes primarily to our question of "technics" by showing how different inscriptions, writings, tele-technologies and so on function in the first place. Although the two authors claim to be allergic to each other, contemporary thought needs them both in order to draw a comprehensive picture of the co-occurrence of technicity and subjectivation. We shall attempt to understand this in the following sections.

Michel Foucault: Technics of Governing Human Life

Historians and philosophers of technology do not necessarily count Michel Foucault among their basic sources. Foucault has certainly revolutionized the ethical, social, and political sciences and the self-understanding of many human sciences, but he is not known for treatises on technology. Foucault is indeed a thinker of individual and collective subjectivation. However, he is also a thinker who does not refer subjectivation to a conscious sovereign will but to various disciplinary, governmental, and ethical techniques that produce behaviors and ways of thinking unbeknownst to the producing and produced individuals. Hence he contributes to the question of technological humanity by showing how human subjectivities are produced technically. He does not analyze technics for itself but always in function of this larger question, and he understands technics in the largest possible sense as any technique for producing the desired subjectivations that can also use material technical objects and systems. Whether his notion of technics is sufficient or not is a matter of discussion that will be broached later on, but his notion of subjectivity is nonetheless indispensable to us.

I have already referred to Foucault in the chapter 2 when speaking about the different senses of the Greek word *techne*. Foucault read the Greeks and the Romans not from a metaphysical but from a practical point of view and he studied techne not as the art of fabricating objects but as the art of life *techne tou biou*, as presented by Plato in the *Alcibiades* and later taken up by the Hellenistic philosophers. Beyond these historical points of reference, what are the philosophical implications of Foucault's studies of the technics of the self, especially in view of the "history of the present"[14] that was Foucault's ultimate aim?

Readers of Foucault used to disagree on whether Foucault's interest in self-technics at the end of his career represented a break with his earlier interest in power[15] or a continuation of a lifelong interest in the different ways in which subjectivation takes place. Taking Foucault's own late remarks literally,[16] I find the second hypothesis more convincing and above all more fruitful. This does not mean Foucault's earlier and later works on subjectivity would amount to the same thing but that the latter complement the former in an important manner, showing how concrete processes of subjectivation result both from the ambient power dispositifs and from individual aesthetical and ethical reactions to them, as we shall see. In what follows, my aim is not so much to contribute to Foucault scholarship but instead to show how Foucault contributes to a philosophy of technics. In his works, the term is always related to the question of subjectivation, which itself results from different social technologies and self-techniques that can, but need not, use material technical objects and dispositifs. This is why I will first recall what Foucault understands by subjectivation, then show how he thinks of technics, and finally discuss how technics contributes to subjectivation.

SUBJECTION AND SUBJECTIVATION

Although Foucault reveals the role of many kinds of techniques in the constitution of human existence, he does not establish a theory of "technological humanity" because he puts the very notion of "humanity" in question. He rejects both the anthropological approach to the human being and the romantic conception of the subject.[17] However, he keeps the word "subject" but displaces the way in which it is thought. In an interview from 1982, he goes so far as to say that "the goal of my work . . . has not been to analyse the phenomena of power, nor to elaborate the foundations of such an analysis. My objective, instead, has been to create a history of the different modes by which, in our culture, human beings are made subjects."[18] In classical philosophy from Descartes to Husserl, the subject is a substance or, more importantly, the self-assured origin of cognition and action. Foucault turns against this understanding of the subject and seeks to think the subject as a sort of a construction—not as a finished product but as a constant process of transformation:

> I do indeed believe that there is no sovereign, founding subject, a universal form of subject to be found everywhere. I am very

skeptical of this view of the subject and very hostile to it. I believe, on the contrary, that the subject is constructed through practices of subjection, or, in a more autonomous way, through practices of liberation, of liberty, as in Antiquity, on the basis, of course, of a number of rules, styles, inventions to be found in the cultural environment.[19]

Foucault also analyzes subjectivity in terms of *self* (*soi*), whose grammatical form helps to avoid the reifying connotations of nouns such as "subject," but also of substantified pronouns used in philosophy such as "the I" (*le Je*) and "the self" (*le soi*). He thus draws attention to self as a *reflective movement* of becoming that characterizes subjectivity. Self is the singular person that is at the same time already subjectified and still in the process of subjectivation through ethical action in relation to oneself and others: self is the reflection between what it is and what it will do, and these are the two poles of its subjectivation.

Subjectivation is a term Foucault uses in his last works to describe a becoming-subject. Its main two aspects are, first, the *assujettissement* ("subjection" or, according to Nikolas Rose's translation, "subjectification"[20]) to a power that Foucault had studied in his earlier texts especially and, second, *subjectivation* as a process in which the subject fashions itself, ultimately by becoming the subject of free action and true discourse (that the Greeks called *parrhesia*).[21] Neither subjection nor subjectivation happen to a given, completed subjectivity. For Foucault, subjectivity as a substance or as a subject capable of identifying itself definitely does not exist, but the subject is in a constant process of becoming.

On the one hand, then, the subject is *subjected* to different power mechanisms and techniques: violences, disciplines, educations, freely chosen practices. These mechanisms can leave their mark on the subject's body, but more importantly, they grow a "soul" in it that ends up spontaneously acting in the way it was *made* to act. This is how, to put it crudely, the asylum gives the identity of "madman" to its patients and the prison gives the identity of "delinquent" to its inmates. As we shall later see, it has often been noted that Foucault's description of subjection reworks the Aristotelian conception of the formation of *hyle* by *morphe*.[22] However, the Foucauldian subject is not the passive object of form-giving, but a living being whose life fundamentally consists in forces and capacities. This is why even in the cruelest detention and disciplinary systems described in *History of Madness* and *Discipline and Punish* Foucault can find acts of resistance and flight. There is resistance in every subject, something that opposes submission, whether by openly revolting or by secretly withdrawing from reach. Judith Butler asks if this resistance finds its possibility in the very powers that seek to materialize themselves in both institutions and their inmates or in an unspeakable outside. Insofar as power is located in discourses, resistance takes its opportunity in the discovery of their incoherences, errors, and manipulations, and insofar as power acts by repressing and ignoring, resistance depends on the possibility of expressing unsaid things.[23] But the resistance can also be analyzed in terms of materiality, if materiality is interpreted differently, not as a materialization of a form (formative power) but in terms of yet unformed matter (power-less matter), which could well be the *materiality of the immaterial*, as Judith Revel titles her insightful article.[24] In any case, Foucaldian "materiality" does not mean the "body" of the classical "mind-body dichotomy." In terms of such classical thought, it should be said that materiality can be both bodily and spiritual: it is bodily in the sense that a body is an irreplaceable point of attachment in the world, but this body is a psycho-physical body of experiences that is already a result of the workings of different powers and it is constantly in the process of further reinterpreting itself. For Foucault, the materiality of the subject is not an independent domain outside of the reach of powers. It is not some thing withdrawn beyond experience, a kind of a *Ding an sich*, but materiality is enmeshed in life itself. Foucault's notion of power is an *action on action*.[25] This is why, as Revel points out, power is only capable of managing, directing, and exploiting people's existing actions, but it is not capable of inventing and creating new actions. According to Revel, new actions that are out of the reach of existing powers can be brought forth by the power of invention, which only exists as a *potentia*, which is the immateriality of material life itself. The resistance that a subject can set against its subjection comes from this *potentia*, from the immaterial creativity of life itself.

On the other hand, the subject results from different processes of subjectivation in which it actively produces and creates itself. As we already saw in the context of the *techne tou biou* in antiquity and as we shall see more precisely below, some processes of subjectivation are instigated by self-techniques in which the subject consciously fashions itself: it exerts power on itself in order to empower itself. But in the end, subjectivation in the full sense of the word is the entire process in which the subject creates itself aesthetically and affirms itself ethically. Such processes cannot be reduced to instrumental self-techniques for shaping one's body and lifestyle, for subjectivation really results from the entire ethical action in the community and especially from the courage that Foucault analyzes under the antique term *parrhesia*. This is the courage shown in political speech when the weak, despite the risks, speak out frankly against the strong.

What makes Foucault so interesting is his way of presenting these processes of subjectivation not as the actions of subjects but as processes that result in subjects. Subjection does not happen between two preexisting subjects, such as when a sovereign subject imposes its will and uses its power on an inferior subject that it transforms into an object. Subjection really happens to both parties when impersonal power structures assign human beings particular places in a society in which certain kinds of acting, thinking, and sensing are possible for some persons but not for others (of course, the subject who finds itself in a dominant position and the subject who finds itself in a dominated position do not suffer from subjection in the same way). Similarly, active subjectivation is an aesthetic and ethical action of somebody on itself and on others but it is not an autopoietic act in which a subject creates itself purely out of itself, rather a process in which the subject submits itself to disciplines and practices that already exist as shared, social practices and eventually interprets them in a creative manner. Now, what are the technics that contribute to these different aspects of subjectivation?

Technics of Subjection

Foucault's entire work can be read as a description of the functioning of what we have called "bio-technics" applied to human life. His work is bio-technical in the sense that he studies techniques and technologies whose very object is (human) life (*bios*) and because he studies technics that in their functioning imitate life (e.g., a disciplinary system in which an individual is immersed frames the individual's life and determines what counts as ordinary life). Foucault does not use the term *bio-technics*, and the kindred term *biopower* comes to the fore of his work only in the *History of Sexuality* but under different terms, particularly that of "technology of power." Foucault's earlier works also describe the same phenomenon, as we shall see in this section.

Let us first see what technologies of subjection Foucault investigates in his early works, where he focuses on the different forms of power that sick, deviant, criminal, and ultimately all kinds of people are subjected to.

These ideas take form notably in the *History of Madness* (*L'histoire de la folie à l'âge classique—Folie et déraison*, 1961), *The Order of Things* (*Les mots et les choses—une archéologie des sciences humaines*, 1966), and *Discipline and Punish* (*Surveiller et punir—Naissance de la prison*, 1975). In these texts, power works through many different systems; more precisely, it does not preexist these systems but exists as medical or juridical institutions, architectural arrangements, industrial organizations, scientific discourses, social conventions, political orders, and so on and finally as the intricate interlacing of many such systems along abstract assemblages and dispositifs, as we shall see. Some forms of power use specific social techniques, others rely on concrete technological equipment, and most are discursive and epistemic constructions above all else. But fundamentally, all of them can be interpreted as more or less abstract *technics of subjectivation*.

In the *History of Madness*, Foucault describes a history of Western internment that had up to then been scarcely heeded. In broad outlines, he first recalls how in the Middle Ages, lepers were excluded from the community and interned in leper hospitals. Then he shows how, as leprosy disappeared at the end of the Middle Ages, the spaces used to confined the lepers were left abandoned for some time, until some of them became reused and other spaces of confinement were created during the Classical and Modern periods in order to intern other types of unwanted people. Foucault pays particular attention to the edict of 1656, which founded the Hôpital Général in Paris in which paupers, criminals, unemployed people, and madmen were interned, not because their faults were contagious but because their physical and moral flaws made them unproductive. Still another milestone is the 1790 *Declaration of the Rights of Man* that forbids the arbitrary imprisonment of people, except for criminals and madmen, who were judged to be dangerous for the society. With the advent of positivism, madmen were separated from criminals such that doctors, for instance Pinel and Tuke, could try to heal them. From then on, madmen have been confined in asylums where they are studied and treated but also fundamentally judged (as objects of the specialists' gaze in which they are but powerless objects). In this history of madness, the most important treatment technique is exclusion, internment, and confinement. These procedures require physical installations (asylums), social rules (juridical or medical decisions), and discursive justifications (theories of madness), all of which constitute the technics for dealing with madness.

Foucault's ultimate aim is less in historiography than in its philosophical and political consequences. He shows how certain modes of human existence are produced by the institutional frameworks in which individuals are enclosed such that, to put it (too) bluntly, madness is produced by asylums rather than (and as well as) healed by them (i.e., madness is produced as a problem, as a deviation, as a reality—of course, the asylum does not produce the entire phenomenon of madness but it produces its interpretations, which, especially in a case like madness, are an integral part of the ailment). Thus, as Bannett sums up, the *History of Madness* is not a historical study of psychiatric disorders but "a history of the process by which the community alienated certain values and certain types of thought, certain forms of behaviour and certain types of people by excluding them and making them 'other,' and then characterized them as 'alienated'—the mentally ill, the asocial elements, the disaffected, the outsider, the other."[26] Or as Foucault puts it:

> Lepers were not excluded to prevent contagion, any more than in 1657, 1 per cent of the population of Paris was confined merely to deliver the city from the "asocial." The gesture had a different dimension: it did not isolate strangers who had previously remained invisible, who until then had been ignored by force of habit. It altered the familiar cityscape by giving them new faces, strange, bizarre silhouettes that nobody recognised. Strangers were found in places where their presence had never previously been suspected: the process punctured the fabric of society, and undid the familiar. Through this gesture, something inside man was placed outside of himself, and pushed over the edge of our horizon. It is this gesture of confinement, in short, which created alienation.
>
> It follows from this that to rewrite the history of that banishment is to draw an archaeology of alienation. What is to be determined is not the pathological or police category that was targeted, which would be to suppose that alienation pre-existed exclusion, but to understand instead how the gesture was accomplished, i.e., the operations which together, in their equilibrium, composed its totality, and the diverse horizons from which those who suffered the same exclusion originated, to investigate how men of the classical age experienced themselves at the moment when familiar faces began to become strange, and lose their resemblance with that image.[27]

This is how the technics of alienation produce, concretely, the alienated subjectivity of the interned and, theoretically, a general theory of subjectivity in which normality is defined against madness.

Discipline and Punish is an account of the *Birth of the Prison* (as its subtitle goes) where Foucault broadens the horizon to the treatment of criminals. Using the striking images of Damiens, spectacularly tortured to death in 1757 for attempted regicide, and the Panopticon, a supposedly more humane and rational modern prison theorized by Jeremy Bentham,[28] he inquires into the meaning of the change in practices of judgment and punishment in Europe after public execution was abandoned in the eighteenth and nineteenth century reforms of the juridical system and more humane punishments became the standard until our days. Modern prisons were not meant to exclude criminals but to correct them, and for this, it was necessary to understand the criminals so as to better imagine new kinds of techniques for their betterment. Together with this "science," an entire "political technology of bodies"[29] was born. It functioned as a discipline imposed on bodies, but still more importantly, the procedures of punishment, correction, and surveillance formed the subject's "soul," which is the best of the punitive techniques because it makes the human being believe in its own punishment and realize it in its life: "The soul is the effect and instrument of a political anatomy, the soul is the prison of the body."[30] Such a discipline is an excellent example of political technologies that create both subjection and subjectivation and in so doing make people into the individuals that they are.[31] By and by, Foucault shows how a new kind of a disciplinary society is born: "Panopticism is the general principle of a new 'political anatomy,' whose object and end are not the relations of sovereignty but the relations of discipline."[32] Panoptism appeared to be not only more humane and democratic than the ancien régime[33] but also more efficient and productive. Political disciplinary technology gradually became widespread in the entire social body: hospitals, schools, factories, armies, and finally society as a whole became a disciplinary society in which we believe that we think and act freely but actually do what we are made to think and do by various powers. And although it is possible to say that today we have already moved from disciplinary society to a control society,[34] many aspects of our lives still reflect the disciplinary society.

If the most important technics of subjection described in the *History of Madness* were the confinement that hardly produces anything other than unproductive alterity, the disciplinary technics described in *Discipline and*

Punish are more varied. Since the nineteenth century, ever more meticulous surveillance and regulation not only of space but also of time, gestures, and forces[35] make productive individuals by fashioning new kinds of bodies and through them, new kinds of "souls" that animate them. These souls are entirely focused on effective production. The chapter "Le panoptisme" is a good example of the way in which Foucault thinks of disciplines as "technologies" and "mechanisms" that can certainly use concrete technological devices but, more importantly, that are abstract technics themselves: " 'Discipline' may be identified neither with an institution nor with an apparatus; it is a type of power, a modality for its exercise, comprising a whole set of instruments, techniques, procedures, levels of application, targets; it is a 'pysics' or an 'anatomy' of power, a technology."[36]

No doubt, disciplinary technologies, like political technologies in general, are for Foucault instantiations of power. Power is not the expression of a subject's will exerted on another subject, but power has its own, specific mode of being, precisely that of a mechanism or a machinery that connects an action to another action.[37] Power works through technics, and technics are power technics. As Foucault explains in the introduction to *Discipline and Punish*, political technologies "cannot be localized in specific institutions or state apparatuses": although institutions can use them on subjects, their own mechanisms and effects are situated on a very different level that Foucault calls a *microphysics of power*.[38] According to him, political technologies are multiform instrumentations, sets of diffuse and disparate tools and procedures, bits and pieces that do not make up coherent systematic methodologies and discourses. Political technologies wield power, but it is not a power possessed by someone, it is power as an available strategy, maneuver, tactics, and technics that can be used to connect apparatuses and material elements together in always provisional ways. They do not form a uniform whole and this is why their use implies innumerable collisions, adjustments, and crises. Furthermore, "We should admit rather that power produces knowledge . . . that power and knowledge directly imply one another; that there is no power relation without the correlative constitution of a field of knowledge, nor any knowledge that does not presuppose and constitute at the same time power relations."[39] As we shall later see, power and knowledge are reflected in one another.

In an extraordinary account of *Discipline and Punish* in his book *Foucault*,[40] Gilles Deleuze explains the philosophical consistency of Foucaldian power technologies:

De/constructing Humanity 123

The thing called power is characterized by immanence of field without transcendental unification, continuity of line without global centralization, and continuity of parts without distinct totalization: it is a social space. . . . Power has no essence, it is simply operational. It is not an attribute but a relation: the power-relation is the set of possible relations between forces which passes through the dominating forces no less than through the dominating.[41]

Deleuze chooses one of Foucault's words, "diagram,"[42] to characterize the dimension of the microphysical power. "Diagram," he says, "is a map, a cartography, that is coextensive with the whole social field. It is an abstract machine."[43] It is, he continues, a continual intersocial evolution that happens as an unstable, fluid, everchanging spatiotemporal multiplicity:

It is neither the subject of history nor does it survey history. It makes history by unmaking preceding realities and significations, constituting hundreds of points of emergence or creativity, unexpected conjunctions or improbable continuums. It doubles history with a sense of continual evolution. . . . What is a diagram? It is a display of the relations between forces which constitute power. . . . We have seen that the relations between forces, or power relations, were microphysical, strategic, multipunctual and diffuse, that they determined particular features and constituted pure functions. The diagram or the abstract machine is the map of relations between forces.[44]

In short, Foucault's notion of power, which is always a relation between powers, can best be described as the diagram of the coupling of such powers. This diagram is an "abstract machine": a technology or a mechanism (as Foucault says) or a machine (as Deleuze says). Insofar as power needs to be described in terms of technics, technics has now become an indispensable term for philosophy. Obviously, Foucault's concept of technics is very different from the term in its traditionally accepted sense. He does not think of technics as a tool or an instrument, which would be extensions of a subject's intentions: subjects can use such technical equipment but technics cannot be reduced to these uses. He does not think of technics as a realization of scientific ideas either: although technological progress reflects scientific progress and vice versa, singular technological systems do not incarnate the comprehensive overall rationalities of sciences but are just provisional and unstable local solutions to the problem of coupling certain actions with others. Such solutions reflect and create knowledge dispositifs as well as power dispositifs, such that both function in this context as abstract machines, not as scientific systems (we shall discuss this in more detail below). In other words, technics does not incarnate an idea; it is just a diagram, a map of a particular constellation of forces. Deleuze emphasizes that what Foucault calls a machine is a social technology before being a technical one, a human technology before a material one.[45] In the same sense, Deleuze and Guattari's *Anti-Œdipus* is a remarkable study of social machines, different types of psychical and social situations articulated as machines. Foucault's abstract machines too, which operate by connecting actions to actions, are mostly social techniques, but they also include material technologies as well as epistemic machines.

In order to understand how epistemic situations function as "abstract machines," it is useful to cast a quick glance at *The Order of Things*. After the *History of Madness*, Foucault also described another technics of subjection/subjectivation to that of simple physical coercion, namely specialist knowledge that includes, for instance, court judgments interning individuals and doctors' studies and diagnostics deciding about the right therapeutics but also philosophical speculations about the relations between madness and reason, which are then reflected back upon the people who consider themselves to be normal. Foucault's objective is not to write a history of such statements; instead, he seeks to reveal the epistemic situations that make these statements possible in the first place. Such epistemic situations are not known to the (individual and collective) subjects who utter the specialist statements, but they provide the framework from which these statements draw their sense and power. Foucault focuses more precisely on the epistemic conditions of subjection/subjectivation in *The Order of Things*, where he shows how the anthropological figure of the human being became the indispensable but unthought center of what gradually came to be called the human sciences. When it comes to epistemic and discursive power constellations, Foucault does not speak of "discursive or epistemic machines" (Deleuze might do so), but he introduces a new term, *episteme*. He develops this term in *The Order of Things* where he studies the history of Western human sciences in the Renaissance, Classical, and Modern periods. If the *History of Madness* had traced the otherness excluded from normality, *The Order of Things* carries out an archeology of the sameness that constitutes this normality; if *Discipline and Punish* will subsequently study the political technologies of power, *The Order of Things* examines the forms of knowledge that justify particular uses of power.

To give a general idea of episteme, it suffices to recall the main types of episteme presented in *The Order of Things*. First, Foucault distinguishes between the figures of thought that structure knowlege in different periods: Renaissance science relied on resemblance, looking for analogies between all things of the world; the science of the Classical period sought for identity and difference, establishing tables and classifications that rely on representation; and ever since the beginning of nineteenth century, modern science has been interested in functions that come to the fore especially in the new life sciences and history. Secondly, Foucault is particularly interested in the modern epistemological field that splintered into different disciplinary directions that hardly communicate with one another but that have a common, albeit unconscious, ground: anthropological human subjectivity. Modern thinking divided it into its empirical and transcendental aspects. As an empirical object of scientific study, the human being appeared as a speaking, working, and living being, while transcendental subjectivity is seen as a finite instance whose cogito is forever thrown back upon its unthought and unthinkable grounds. According to Foucault, the internal conflicts in this conception of subjectivity lead the modern episteme to an impasse that can only be overcome by ceasing to see the anthropological figure of the human being as the ultimate condition of knowledge. This is what Foucault means by the "death of man": not the death of the anthropological man but the end of an episteme that is entirely centered on the human being (indeed, according to Foucault it would be better to center questions of knowledge on the idea of language).

The only thing I want to take from *The Order of Things* is the notion of *episteme*. This is a complex notion and one that Foucault tends to use rather than define in this book. Bannet summarizes it well as a group of formal principles that "determine what objects can be identified by the community, how they can be marked, and in what way they can be ordered; they make certain perceptions, certain statements, certain forms of knowledge possible, others impossible; and by regulating diverse aspects of the community's mental activity, they stamp these diverse activities with a certain fundamental Sameness."[46] An episteme is not a science, it is the condition of sciences and other intellectual operations that select which objects and operations are possible in their domain. One might say that an episteme is a kind of ontology presupposed by these forms of knowledge but that does not coincide with everything that there is, nor with everything that is true, but only with what can be and can be true at this particular historical moment. However, even if the notion of episteme explains the operation of *The Order of Things* (and of *Archeology of Knowledge*), it ultimately does not suffice to meet Foucault's needs, and this is why he later replaces it with *dispositif*. He explains this well in an interview entitled "*Le Jeu de Michel Foucault*" ("The Confession of the Flesh,") where he criticizes the notion of episteme for being limited to questions of knowlege only, while what he really wants to speak about is the entire sphere of existence, including both knowledge and power relations. Hence the need of a more general term than episteme, dispositif[47]:

> What I'm trying to pick out with this term [dispositif] is, firstly, a thoroughly heterogenous ensemble consisting of discourses, institutions, architectural forms, regulatory decisions, laws, administrative measures, scientific statements, philosophical, moral and philanthropic propositions—in short, the said as much as the unsaid. Such are the elements of the apparatus [dispositif]. The apparatus itself is the system of relations that can be established between these elements.[48]

The figure of thought that started out as just an abstract power technology, then evolved into an episteme, thus finally took the form of a dispositif, that is, a general constellation of material, power, and discursive conditions of existence in a given historical situation. I find it instructive to compare the notion of dispositif with Heidegger's notion of *Ge-stell*.[49] Both function as the transcendental horizons of a historical situation that is figured through a framework of technics (*techne*) instead of the framework of universal reason (*logos*) that constituted the horizon of the subject of philosophy, especially in Kant, Hegel, and Husserl. In comparison with the classical logos, one could characterize *Ge-stell* and dispositif as being artificial horizons; however, they are not artificial in the sense of being productions (*poieumata*) of some subject (God, man), but only in the sense that being neither ideal (*mathema*) nor natural (*physis*) they frame the human world as a construction. Of course, neither *Ge-stell* nor dispositif are concrete technological installations. They are abstract frameworks within which concrete technical installations are possible: one is a "non-technical ground of technics" and the other is an "abstract machine." Both dispositif and *Ge-stell* are transcendental horizons because they dispose of people and of things by indicating what is and what is not, and what are the possibilities and the impossibilities of these beings. However, the parallel also has its limits. Foucault's dispositif is above all a social machine that says how people are produced, while Heidegger's *Ge-stell* is above all an ontological framework. Furthermore, even though Heidegger's *Ge-stell* marks the tearing (*Riss*) of a world and the separation (*diapherein*) of its inhabitants, it remains a unity, as indicated by the unifying prefix *Ge-*. On the contrary, Foucault's abstract machines are partial and overlapping tactics and strategies that function within everchanging multiplicities that cannot be unified. Hence, the horizon that to Heidegger appeared regrettably ruptured seems joyously multiple to Foucault.

SELF-TECHNICS

Whereas the different forms of subjection to disciplinary technics described above could to some extent be applied to other than human "bio-technical" processes (e.g., domestication of animals and even plants), the particularity of human "bio-technics" is the fact that humans can also use them consciously on themselves or on each other. Foucault describes such processes under the term of self-techniques and technologies of the self.

As already noted, Foucault's last period poses problems to commentators because of a turn toward questions of subjectivity that some interpret as a turning away from the earlier problematics of power and others as an inquiry into aspects of life that the earlier works already implied without explicitly discussing them. Of course, the last word in these debates is bound to remain unsaid insofar as Foucault's project was not completed but interrupted by his untimely death. However, he did complete plenty of works on technics of subjectivation that will be of concern to us in the following. We have the first formulations of Foucault's renewed interrogation of subjectivity in the first part of the *History of Sexuality*, *The Will to Knowledge* (*Volonté de savoir*, 1976), followed eight years later by the next two parts, *The Use of Pleasure* and *The Care of the Self* (*L'usage des plaisirs* and *Souci de soi*, 1984), while the fourth part, *The Confessions of the Flesh* (*Les aveux de la chair*) was only published posthumously in 2018. From 1980 to 1984 he lectured at the Collège de France on ancient Greek *techne tou biou* (self-techniques) and *parrhesia* (free speech), thus exploring Greek ideas of subjectivation where the government of self and government of others go hand in hand. In all these works, Foucault delves deep into ancient Greek and Roman cultures and brings forth fascinating readings of these periods. However, he did not have the time to fully develop the consequences of these works to encompass modern subjectivity except on a very general level, for instance in his readings of Kant's *What is Enlightenment?*, which culminates in a description of the task of philosophy for us: "I shall thus characterize the philosophical ethos appropriate to the critical ontology of ourselves as a historico-practical test of the limits we may go beyond, and thus as work carried out by ourselves upon ourselves as free beings."[50]

I think that by far the best presentation of Foucault's late ideas of subjectivation can be found in the last chapter of Deleuze's *Foucault* in which, as it was his particular gift, Deleuze brings forth the philosophical core of Foucault's thinking, in this case, his late thinking of subjectivation. This is why, after a short reminder of the import of ancient philosophy to the Foucaldian question of subjectivation, I will conclude this section with Deleuze's interpretation. Maybe Deleuze goes further than Foucault, but whether or not his interpretation of Foucault is faithful to his friend's intentions does not really matter when the main objective is to articulate a modern theory of subjectivation.

We have seen how Foucault interprets Classical and Modern conceptions of subjectivation above all as procedures of subjection in which subjects are fashioned by the dispositifs of power and knowledge under which they live. We have also seen how he finds in antiquity another approach to human formation, technics of self, the ultimate form of which is free speech. To be sure, *techne tou biou* and *parrhesia* only concern free men in Greek democratic cities and in the late Roman empire. But despite their limited scope, these studies of ancient technics of self lay the ground for Foucault's idea of an *aesthetics of existence* and of an *ethopoiesis*, in which subjectivation is an emancipatory process instead of merely a subjection to imposed disciplines.[51]

In the introduction to *The Use of Pleasure*, Foucault tells what he means by self-technics:

> After all, this was the proper task of a history of thought, as against a history of behaviors or representations: to define the conditions in which human beings "problematize" what they are, what they do, and the world in which they live.
>
> But in raising this very general question, and in directing it to Greek and Greco-Roman culture, it occurred to me that this problematization was linked to a group of practices that

have been of unquestionable importance in our societies: I am referring to what might be called the "arts of existence." What I mean by the phrase are those intentional and voluntary actions by which men not only set themselves rules of conduct, but also seek to transform themselves, to change themselves in their singular being, and to make their life into an *oeuvre* that carries certain aesthetic values and meets certain stylistic criteria. These "arts of existence," these "techniques of the self," no doubt lost some of their importance and autonomy when they were assimilated into the exercise of priestly power in early Christianity, and later, into educative, medical, and psychological types of practices. Still, I thought that the long history of these aesthetics of existence and these technologies of the self remained to be done, or resumed.[52]

The techniques of the self are an important element of the Greek and Greco-Roman ways of asking what human beings are and what they should do. They are the arts of existence by which men ask how they should conduct themselves, transform themselves, and go as far as "[making] their life into an *oeuvre* that carries certain aesthetic values and meets certain stylistic criteria." Following Aristotle, who describes the production (*poiein*) of technical beings (*technai onta*)—"oeuvres"—with the help of the theory of the four causes, Foucault's description of the making of one's own life into an oeuvre is generally interpreted in terms of Aristotle's four causes.[53] Paul Patton summarizes this production of oneself very well[54]: the material cause of technics of self is the ethical substance, not the body as a physical thing, but the sensing, desiring, and enjoying part of life that the Greeks interpreted as "aphrodisia or the acts linked to pleasure or desire," Christianity as the flesh, and the modern world as sexuality.[55] The formal cause of technics of self is the mode of subjectivation according to which individuals relate to themselves: the Greek aesthetics of existence, the Christian subject of divine law, or the modern subject of the universal law imposed by reason. The efficient cause that produces the subject is the ethical work of the self on self, the concrete *techne tou biou*. And the final cause is the particular ethical conduct that one would like to adopt, the type of self that one aspires to become.

In this articulation, our question of technics pertains in particular to the efficient cause, the specific means through which the self is being reworked. As we saw in the chapter 2, the means of the ethical work of self on self can be simple physical practices like diets, exercise and regular habitudes, or intellectual exercises conducted alone or under the supervision of a mentor. But what is the essence of such means? In *The Use of Pleasure*, Foucault suggests that the fundamental structure of technics of self is *use*, *chresis*, as opposed to possession. The work of self on self aims at learning how to use not just one's body but one's pleasures (*chresis afrodision*).[56] The substance that is thus being elaborated is *aphrodisia*, the works of pleasure,[57] of which the Greeks prefer the temperate, such that these constitute a kind of a golden mean of pleasures where life is neither barren and pleasureless nor dissipated. Foucault emphasizes that unlike the later Christian sexual morals, the Greek conception of the good use of pleasures does not set down forbidden and the permitted acts but aims at the proper regulation of any pleasures in view of a full and satisfying life in general. The most important technics—"know-how" or "art"[58]—of regulating pleasures mentioned by Foucault are restraint, the right moment, and the appropriate role in sexual intercourse. The essence of such technics is the *use*, *chresis*, of self by self. Use differs from possession, which exerts a total control over a passive object that can even be suppressed because use is a power used on another power such that the used power remains the power that it is, but it is conducted and directed in the desired way. *Chresis*, use, thus appears as the positive side of power: it is a power that acts on another power, but instead of submitting violently it just conducts skillfully (but can one draw a strict line between these two modes of power?). In sum, the material cause of Greek self-technics is the mobile domain of pleasures (*aphrodisia*) and its formal cause is the ideal of aesthetic existence. Their efficient cause is the self's battle over himself that pursues the right use of the self. The final cause of the use of pleasures is the individual's freedom that is indissociable from truth.[59]

Foucault emphasizes that despite the ideal of self-control, the Greeks never thought that a person could fashion himself alone. One needs an educator, a mentor, a friend who advises and supervises the formation of self. The mentor cannot give orders to the disciple, either, for he does not possess the disciple's life and has no right to use it either. But discussions and correspondence with the mentor help the self in his battle with himself by helping him clarify his idea of good life, which is the aim of this process. Other people are not only among the means of self-fashioning, they are also its principal aim. Greek self-techniques were above all meant to help in the education of rulers to govern others well: one has to govern oneself in order to govern others. The use of pleasures has an aim, that of educating a prince not to be a plaything of his desires but able to keep them in check and to therefore govern according to reason, not according to his whims.

The man who can govern his pleasures is not a rigid image of an ideal type. His pleasures may have any form they will, but the main thing is that they are regulated such that they ruin neither personal happiness nor virtuous political life. However, according to Foucault, a good life is not only a life of reasonable pleasures. It is above all a political life that is satisfying under certain conditions, as Foucault analyzes in his long studies of parrhesia. Parrhesia means free-spokenness and free speech: "Parresia is a virtue, duty, and technique which should be found in the person who spiritually directs others and helps them to constitute their relationship to self."[60] Parrhesia means free speech, but it is not what is meant today by freedom of expression, which is a *right* belonging to any juridical person as such; instead it is a constitutent of *ethos* and indexed to the person's belief rather than to factual truth. Parrhesia is the capacity to speak out freely the truth because it is the best for the community:

> It is then a discourse spoken from above, but which leaves others the freedom to speak, and allows freedom to those who have to obey, or leaves them free at least insofar as they will only obey if they can be persuaded. . . . Parrhesia consists in making use of logos in the polis—logos in the sense of true, reasonable discourse, discourse which persuades, and discourse which may confront other discourse and will triumph only through the weight of its truth and the effectiveness of its persuasion—parrhesia—consists in making use of this true, reasonable, agonistic discourse, this discourse of debate, in the field of the polis. And, once again, neither the effective exercise of tyrannical power nor the simple status of citizen can give this parrhesia.[61]

Parrhesia belongs to a democracy in which it is the discourse of those who impose their superiority on others, not by their birth or status but only because by telling the truth can they come to direct the consciences of others. Because the discourse of truth takes place in democracy, it is spoken in front of others and also against others in situations of rivalry and confrontation, and this is why it can even involve risking one's life when speaking out the truth.[62] On the other hand, parrhesia is not necessarily the free speech of one among equals in an open deliberation, but it can also be the free speech that the mentor or the counselor addresses to the prince. In both cases, parrhesia can be dangerous, especially if it aims to defend the weak against the injustices of the strong.

For Foucault, parrhesia is a part—a *summum*—of the care for the self. Of course, by speaking freely and truthfully a self does not shape itself in the same ways as when it regulates appetites and sexual pleasures. Parrhesia is more than the care for oneself because it is care for oneself only because it is care for the city. As we saw, parrhesia is a technics that consists in a right "use of words/reason (logos)," *chresthai logô*.[63] But it is more than the ordinary art of words—manipulative rhetorical skill—because parrhesia is expected to be a simple discourse whose persuasive force comes from its truth, not from rhetorical effects. The parrhesic discourse that strives to tell the truth for the good of the city is not an instrumental technics of the self like diets and sexual disciplines; it shapes the person in a higher sense, for it forms a style of existence, character, and destiny. It is more than an aesthetics of existence, it is an ethopoiesis where the person shapes its properly ethical character. If the use of pleasures contributes to subjectivation by subjecting the self to the direction of self, parrhesia contributes to subjectivation by making a praiseworthy individual, someone who has taken a place in the political community before others, whatever the cost. Thus Foucault's analyzes of *techne tou biou* and *parrhesia* account for the two sides of subjectivation that are both necessary to make a full person: subjection to a rule and subjectivation by free acts.

Sexuality and the Fold

In *The Will to Knowledge*, Foucault draws a general outline of the very different situation of subjectivation that followed antiquity. Now, it must of course be noted that Foucault's theory of subjectivation is surprising, or at least very different from the traditional philosophical theory of the subject, because he does not study the subject in terms of its self-consciousness or rationality but in terms of its sexuality. According to Foucault's key thesis in *The Will to Knowledge*, while antiquity gave to *aphrodisia* a positive but not dominant role in the fashioning of subjectivity, later European (Christian and Modern) thinking did not repress sexuality, as it is often thought, but on the contrary spoke of it more and more as a secret that must be confessed such that, by and by, "the project of the science of the subject has gravitated, in ever narrowing circles, around the question of sex."[64] I am not interested in sexuality itself here nor in the precise history of its production but only in Foucault's claim that ever since early Christianity, European subjectivity has been thought of in terms of sexuality, which has itself been produced through various power technologies.[65]

In order to understand this, it is necessary to recall the main phases in the *History of Sexuality* as Foucault presents it. Foucault's object is not sexuality as a phenomenon of nature (an object of biology) but "sex as history, signification and discourse."[66] In other words, the subject's substance, its material cause, cannot be found in biology but in the desires and pleasures in which it first folds upon itself, and that our time interprets as its sexual life. According to Foucault, sexuality is a changing discursive formation. This does not mean, as contemporary conservatives think with loathing and as contemporary transhumanists think with jubilation, that one could choose and construct one's body and sexual identity at will. It means that sexuality is at every time a dispositif: it is a discursive, epistemic, and material "abstract machine" that determines certain modes of sexuality as thinkable and others as unthinkable, but it is not a coherent system of reason that some conscious sovereign will could have imposed on people.

Another way of describing sexuality as a dispositif is Foucault's affirmation that sexuality has been produced by different power technologies.[67] Once again, these technologies are not technological operations on a physical body, they are power technologies that produce a "soul" that controls and "imprisons" a body. One of the most important of these technics is confession (of the desires of the flesh), first used in Catholic Christianity. In the eighteenth century, the Church's power over sexuality diminished as a number of new sciences, notably pedagogy, medicine, and demography, took over the discursive power over sexuality. These were keen on distinguishing different forms of "healthy" and "perverse" sexuality. Since the nineteenth century, the new sciences of sexuality led more and more effectively to a politics whose aim was the elimination of unhealthy and unproductive forms of sexuality, not for the individual's sake but for the sake of the population, whose health and heredity was supposedly improved by a number of racist and hygienist practices. According to Foucault's influential definition, this was the beginning of the era of biopolitics, which is a power whose highest function was perhaps no longer to kill, but to invest life through and through.⁶⁸

One would have to speak of bio-power to designate what brought life and its mechanisms into the realm of explicit calculations and made knowledge-power an agent of transformation of human life. . . . But what might be called a society's "threshold of modernity" has been reached when the life of the species is wagered on its own political strategies. For millennia, man remained what he was for Aristotle: a living animal with the additional capacity for a political existence; modern man is an animal whose politics places his existence as a living being in question.⁶⁹

The dispositif of sexuality thus uses many different techniques: Greek and Roman techniques of self-control, early Christianity's techniques of confession, modern science's interpretative and therapeutic discourses, and finally contemporary biopolitical measures that extend from social segregation (racial discrimination) and legislation (concerning sexual deviance, marriage, parenthood, etc.) to medical care (forced sterilizations, denial of abortions) and educational programs (child health counseling, sex education). Finally, the great contemporary biopolitical dispositif uses an unequal mix of all of these techniques as well as others, and this is constitutive for the formation of the experiential substance of contemporary subjectivity.

But if sexuality is mainly what we are subjected to, what are our possibilities for active subjectivation? Foucault indicated how the ancient aesthetics of existence had their modern counterparts in certain forms of Renaissance life or in Baudelairean dandysm, but these are rare and singular examples of people who help us see how the available powers can be composed in original ways but that do not necessarily show how the available powers could be truly contested. I think Deleuze's explication of Foucaldian subjectivation in the chapter "Foldings, or the Inside of Thought (Subjectivation)" in his book *Foucault* goes more directly to the core of the question.⁷⁰

Deleuze explains the Foucaldian subjectivation as folding: the subject is a fold, *pli*. One could say that the fold is the opposite of reflection: reflection starts from a subject, passes by a reflecting surface, and returns to the subject whose self-consciousness has been enriched by the detour, whereas the fold starts from the outside that folds upon itself such that the outside's reflection on itself produces the effect of the subject. The folded subject is the result, not the starting point of inquiry.

The starting point for the thinking of the fold is the outside, *le dehors*.[71] The outside is outside of any subject: not just the exteriority already constituted by power and knowledge but the totally strange region beyond any exteriority. As Deleuze points out, such an idea of the outside owes much to Heidegger's and Merleau-Ponty's late ontological considerations (there is being, there is time, there is the interlacing of the "flesh" of the world), but it also differs from them since for Foucault "there is no 'savage' experience" for "any experience is caught in relations of power."[72] The outside of the world framed by prevailing dispositifs of power-knowledge is the unthinkable that can only be pointed at in language, especially literary language. No wonder that it is precisely Blanchot who shows Foucault a still more radical idea of the outside no longer aligned with any expectation of the truth of being: "Foucault therefore discovers the element that comes from the outside: the force. Like Blanchot, Foucault will speak less of the Open than of the Outside."[73] Now, the fold is the inside of this outside:

> But is there an inside that lies deeper than any internal world, just as the outside is farther away than any external world? The outside is not a fixed limit but a moving matter animated by peristaltic movements, folds and foldings that together make up an inside: they are not something other than the outside, but precisely the inside of the outside. . . . The unthought is therefore not external to thought but lies at its very heart, as that impossibility of thinking which doubles or hollows out the outside.[74]

The outside is farther than exteriority and when it is folded, it becomes more intimate than the interior. This fold is our intimacy, it is the outside folded in us and we have folded it because something in the outside has become a problem, a reason to protest against some aspect of existing power-knowledge. Deleuze explains that the outside is not the void or the empty domain of death. This is why it is something other than Heidegger's (and the Philosophical anthropologists') view of human existence, whose core is the nothingness that is also mirrored in the nothingness that grounds being. Foucault's outside is a structure of powers and discourses and this is why it is a multiplicity filled with conflicts, coincidences, and chances. A fold starts to emerge when something in the world becomes problematic and requires new solutions: the unthought, the outside, are out there in the world, but because they become problematic they find themselves in us. As Butler said, too, the problems of the power-knowledge-situation in which we are already trapped are the source of creativity and freedom because they make us resist the situation and try to show its inconsistencies and make its problems visible and articulable: this is precisely how a situation is folded. Resistance is always resistance to existing powers and knowledges and this is why it is both rebellious and emancipatory, creative and criminal. But Deleuze thinks that we can also be touched by the outside as if by a force, like the uncontrollable alien forces Foucault discovers in Blanchot's texts. As Revel also said, novelty and invention are possible because of such forces that, when folded, may become forces for breaking existing power structures and introducing new powers into situations.

"The most distant point becomes interior, by being converted into the nearest: life within the folds. This is the central chamber, which one need no longer fear is empty since one fills it with oneself."[75] If Heidegger and the Philosophical anthropologists that we studied in the previous chapter only found impenetrable negativity at the core of human existence, Foucault finds a very different thing, not a frightening void but a full and potentially creative fold, the intimacy of the outside. Because the inside is not trapped in me but filled with the outside, it is always absolutely full of new forces and new chances. Of course, these new possibilities are not always nice. As the forces of resistance are contrary to existing powers, they are also contrary to laws, morals, common sense. Being resistant, they can also be delinquent and even criminal. Resistance is ambiguous and emancipatory. Also, as Deleuze said, Foucault does not believe in pure and savage experience, but as soon as something appears it is already interpreted, such that the outside is filled with articulated sights and sentences. This is why the forces of resistance lead to new articulations and new visibilities and not just to a pure virginal experience. The world is not just simply open, it is something that is said, seen. This is also why its criticism does not lead only to empty liberation but to the inventive production of something new. One never creates a new totality, one always makes partial, conjunctural, localized inventions: the world is transformed, not ended and then made again from scratch.

This is how Foucault's later thinking of subjectivation completes his older idea of subjectivation to power structures with an idea of creation. Let me quote Deleuze once more:

> The subject . . . is to be created on each occasion, like a focal point of resistance, on the basis of the folds which subjectivize knowledge and bend each power. . . . The struggle for a modern subjectivity passes through a resistance to the two present forms of subjection, the one consisting of individualizing ourselves on the basis of constraints of power, the other of attracting each individual to a known and recognized identity, fixed once and for all. The struggle for subjectivity presents itself, therefore, as the right to difference, variations, and metamorphosis.[76]

Foucaldian subjectivity is not a permanent universal structure. If it has to be created on each occasion, this creation is never ex nihilo but always in relation to a substance that is already formed and already fashioned by existing power-knowledge structures or by personal self-techniques. Subjectivity lives as long as it struggles against what it is already and by this struggle invents itself anew. Subjectivity results from many kinds of technics, but because of them, it can also invent new becomings and new technics.

Transition: Discipline and Technology

In the previous main chapter, we saw how both Heidegger and the Philosophical anthropologists learned to think of technics as the fundamental characteristic of the world. In return, human existence in the world of technics was defined as negativity. According to Foucault, technics characterizes the world as well, although technics is not for him a system of technological equipments but an abstract social machine. Contrary to Heidegger, Scheler, Plessner, and Gehlen, he does not think human existence in the nihilistic world of technics is determined by nothingness, but he thinks of it more positively as an existence fashioned by different mechanisms of power and knowledge. Foucault's thinking is above all a detailed description of different ways in which human beings are formed: they are subjected to mechanisms of power and knowledge, they are engaged in self-techniques, and their ethical action can make their life into an oeuvre. The human being's substance is the result of all these different technics and when it folds itself upon itself, it can also transform them into new arts and technics. In order to better explain these processes, Foucault describes in detail what different social techniques and self-techniques consist of and how they function. Foucault's human being is integrally a technical construction.

But has Foucault thought of the imprint of technics on human life carefully enough? In *Taking Care of Youth and Generations*, Bernard Stiegler claims that despite appearances, technics in the sense of technological equipment is really the unthought of Foucault. According to Stiegler, Foucault's interpretation of Kant's *What is Englightenment?* overlooks the role that Kant gives to reading and writing, although they are major elements of any self-technics, and more generally in all his work Foucault has overlooked the role of technics in the constitution of self and of the community, especially in the constitution of contemporary "noo-technics" (Stiegler's term for technics that contribute to *noesis*).[77] Stiegler's own project resembles Foucault's ontology of the present in the sense that Stiegler examines the contemporary modes of subjectivation by asking how existing technological systems contribute to them. Contrary to Foucault, however, and rather like Heidegger, he studies subjectivation from the point of view of material technologies rather than from the point of view of abstract social machines. Of course he cannot reproach Foucault for ignoring contemporary digital technologies, whose reach was still inimaginable when Foucault died in 1984, but he does reproach him for paying insufficient attention to the philosophical role of material technological equipment in general. Above all, Stiegler thinks that Foucault should have paid more attention to the structures of *pharmakon* and of *the supplement* first laid bare by Jacques Derrida.[78] We will come back to Stiegler later, but we shall first study the idea of technological supplementarity in the light of Jacques Derrida's thinking because Derrida is the first theoretician of writing, as well as of the structure of supplement and pharmakon instantiated by writing, as Stiegler says.

Now it is curious and somewhat provocative to say that Foucault obscured the question of technics—since he is after all the thinker par excellence of self- and social techniques. But it is true that the technics studied by Foucault are mostly social techniques—methods of acting, thinking, and directing behavior. He thinks of power in the sense of influence on human beings, not in the sense of energy to move machines. Stiegler, on the contrary, thinks precisely of the influence of material technological systems on the constitution of human life and mind. In Foucault, different kinds of equipment are sometimes means of power-knowledge technologies (*Panoptikon*), but then they are only considered in their instrumental role and not for themselves. Stiegler, on the contrary, studies their particular character and effect. In his book on education, Stiegler shows how material technologies can have an effect similar to the disciplinary techniques Foucault describes, only that they may be even more insidious: they contribute to subjectivation by subjecting people to their logic. Although in principle they can also become means of emancipative and creative subjectivation, at present these new technologies of spirit mainly propagate an incapacity to think, whereas they should be transformed into enabling equipment.

It is true that today Foucault's descriptions of self- and social techniques remain relevant, but they need to be revised in function of recent developments in digital technologies and so-called artificial intelligence. Contemporary existence is increasingly what Éric Sadin has called "algorithmic life" because it must adapt to what Antoinette Rouvroy and Guido Berns, drawing on Foucault, call "algorithmic governmentality."[79] Some aspects of the new algorithmic governance consist simply of services like commerce and marketing, which are increasingly online, or banking, where both private banking and trading are not only online but also automatized. Tele-education comprises not only online courses but also automatic assignments, gradings, and entrance exams. Media are available online and media content streams are suggested to different publics, content that algorithms moderate and even produce. Even politics is adapting to the digital space: electors are profiled and candidates create profiles to match the results of constant automatic polling. The lives of individuals are thus strongly formatted by algorithmic governance not only because individuals use digital services but also because they actively construct professional and private digital identities to match the virtual possibilities. While such virtual identities enlarge certain activities (shopping) they also narrow others (information), especially when they confine people within so-called filter bubbles in which search algorithms inform them about only the kinds of things that have previously interested them and leave out contrasting views. The algorithms do not just provide punctual services, they provide the entire place that people can occupy in a society: there are algorithms that determine whether people can have an insurance policy and for what price, what kind of health services they are entitled to, how much taxes they should pay, what kind of education they can get, and where they can be recruited. Certain governments also use algorithms to surveil and direct citizens' communication online, track them in the public space, and even determine what kind of a punishment they deserve if they commit an infraction. This is how algorithms contribute to individual and collective individuation. Their *statistical governance*, as Rouvroy and Berns call it, does not so much control the real but it structures what is *possible* and simultaneously suppresses divergent virtualities.[80] While some areas of statistical governance are determined by public powers,[81] much larger areas are within the domain of big technological companies who created the very architecture of the system. They function using the principles that Shoshana Zuboff calls "surveillance capitalism,"[82] which "aims to predict and modify human behavior as means to produce revenue and market control" and actually "thrives on unexpected and illegible mechanics of extraction and control that exile persons from their own behavior."[83] As Rouvroy and Berns point out, the world is not run by a self-conscious mega-AI that usurps God's role. Even though artificial intelligences increasingly run algorithms, AI is not a thinking mind but just a set of complex machine learning programs. The world is managed by innumerable large and small algorithmic systems that are more like nerve fragments of the contemporary social body. Everything and everybody is being monitored all the time, not by some*body* or some *consciousness* but by the innumerable impersonal automatic mechanisms that constitute what Dominique Quessada calls *sousveillance* ("subveillance" in contrast to traditional *surveillance*).[84]

These and many other new historians of the present are studying the way in which algorithmic governance formats psyche and society today; I do not need to repeat their work. What interests me in the following is the philosophical sense of the material technologies that occasion these transformations. This is what Derrida's work will help us understand better than Foucault's.

As Stiegler notes, Derrida was the first to study the role of writing and other technical means to the constitution of thinking, although he did not do it under the heading of technics, like Stiegler does. Derrida did not criticize Foucault for forgetting technology. His debate with Foucault was limited to a single exchange on Descartes after the publication of the *History of Madness* and this had the unfortunate consequence of definitively alienating the two philosophers. They did not really (try to) understand one another but each simply defended their own ground.[85]

The Derrida-Foucault debate began with Derrida's article "Cogito and the History of Madness" where he claims "Foucault wanted madness to be the *subject* of his book in every sense of the word: its theme and its first-person narrator, its author, madness speaking about itself. Foucault wanted to write a history of madness *itself*." "It is a question, therefore, of escaping the trap or objectivist naiveté that would consist in writing the history of untamed madness . . . utilizing the concepts that were the historical instruments of the capture of madness—the restrained and restraining language of reason."[86] Derrida shows, so to speak, why Foucault's project of speaking in the language of reason about what is without reason and excluded by reason is itself madness. Foucault replied to this hurtful criticism firstly by recognizing its validity (he omitted the original preface of *History of Madness* where he first claims to speak for madness in subsequent editions)[87] and secondly by counterattacking in a dry and ironic article later published as an appendix of *History of Madness*, "My body, this paper, this fire." While this article scolds the textualist Derrida for a careless reading of Descartes, "Réponse à Derrida"[88] expresses the real stakes of the debate, namely the reduction of 673 pages of historical analysis to three pages on the philosophical reading of Descartes in which the arrogant philosopher claims to locate the unconscious law that regulates four hundred years of medical, social, scientific, police, and political history.[89] For Foucault, Descartes is not the secret core of epochal episteme but on the contrary yet another of its symptoms. The debate, where the two smart Parisian scholars demonstrate their impressive analytical skills, seems to run over Descartes's dead body, but the real stakes are the much more vital questions of the philosophical method today. As Derrida says, Foucault wants to execute "a structural study of an historical ensemble—notions, institutions, juridical and police measures, scientific concepts—which holds captive a madness whose wild state can never in itself be restored."[90] Foucault unearths this repressive structure by archaeological and genealogical work through such historical ensembles. From Foucault's point of view, Derrida's work on philosophical fragments reflects philosophy's traditional contempt for history and other sciences, such that Derrida ends by capturing all thinking in the closure of philosophy absorbed in the "infinite commentary of its own texts without relating to any exteriority."[91] This is of course a nasty accusation to level at Derrida who had spent all his forces attacking traditional philosophy's "phallo-logo-centrism" and whose work consists in exploring the margins of philosophy and sounding the traces of the totally-other in its texture.[92] The other of philosophy is the central object of his philosophy, however he claims one cannot access it directly but only by carefully reworking, deforming, and reversing its texts.[93] Foucault claims to think the "singularity of the event" and asks how could a philosophy of the trace, following tradition and keeping tradition, be sensitive to the analysis of the event?[94] But this question misfires. Derrida, too, thinks the singularity of the event above all else: "The event, the singularity of the event, that's what différance is all about."[95] In "Cogito and the History of Madness," Derrida said that far from defending anything like a *philosophia perennis*, he on the contrary calls to "account for the very historicity of philosophy"[96] that is the condition of Foucault's project as well. From Derrida's point of view there is no historicity unless the singular event leaves a trace open to repetition—that can be very different from "tradition," for example when it is unconscious, repressed, introjected, incorporated, haunting, and so on. As we already saw when reading "The Ends of Man," every exchange between Foucault and Derrida boils down to the question of method: Can one meet the unexpected for which there is no language—the silence of madness, the singularity of the event, the pure exteriority (*le Dehors*)—and speak for it? Or must one follow the fissures and flaws of existing discourses in order to tease out its unconscious, unheard-of, unsaid margins from which alone anything like an event might beckon? Must one choose an archaeology of silence or a deconstruction of traces? Don't both, ultimately, work on texts? The aims of the two are not totally opposed—both aim at disturbing and shaking the classical metaphysical subject as well as the classical humanism that appears to follow it—but the approaches are different because one shakes it from the outside and the other from the inside.

Rather than arbitrating the difference between Foucault and Derrida, I am instead only interested in seeing how each of them can contribute to the question of technological humanity. While Foucault makes invaluable analyses of technics for individual and collective individuation, he treats material technical equipment mainly as means of these processes, not as problems in their own right. Derrida, on the contrary, draws attention to the material technics themselves and shows why they are much more than simple means. Contrary to Stiegler, he does not analyze concrete technical things but the "quasi-machinal" structure of writing,[97] and he does not analyze the matter of "materialism" but the "materiality"[98] that is as abstract as Foucault's or Deleuze's "abstract machines."[99] Derrida mentions numerous technical objects as metaphors for a more general "quasi-transcendental" technicity that he brings to the fore. But by far the most important of the techniques he is interested in is writing, which is not an ordinary tool but the exemplary technical aid for memory and especially for thinking. Both verb and noun, writing is a technique associated with thinking and memory rather than a material machine—and also the material machine, this text here, the noun *écriture* and not the verb *écrire*, whose materiality and machinality are central to Derrida. Maybe from Foucault's point of view, Derrida concentrates too exclusively on questions of thinking and even of metaphysics. But then on the other hand, from Derrida's point of view, Foucault concentrates too exclusively on social and political themes and leaves the way in which knowledge functions uninterrogated. From his perspective, Foucault does not pay sufficient attention to the unconscious aspect of knowing because he takes knowledge to be a positivity, whereas in Derrida's thinking everything is indirect, nothing is really positive, everything is hollowed out by the "negativity" attached to any piece of writing, any proposition, any encounter.

Derrida: *Technics* and *Humanity* as Terms

Jacques Derrida does not present a theory of technological humanity. His contribution to the development of this question is rather his way of deconstructing it, refusing the term *humanity*, displacing the term *technics*, and referring both to *originary technicity*. This book investigates the ubiquitous omnipresence of the question of "technological humanity" in contemporary culture, despite the instability, incoherence, and finally misleading character of this expression. Derrida lets us see the reasons for its instability and shows one promising way of dealing with it. In this section I will briefly show how he rules out the terms *humanity* and *technics*, and in the next sections I will concentrate on the structures that their deconstruction brings forth.

Like Foucault, Derrida refuses the term *humanity*, not to mention the term *humanism*. Both think, following Saussure, that "language [which only consists of differences] is not a function of the speaking subject" but that "the subject is . . . inscribed in language, is a 'function' of language."[100] As Derrida indicated in "The Ends of Man," if "the human" is something, it is a finitude (*finitude*) that questions the aims, ends, and limits (*fins*) that delimit it; the human is not a definite entity but the process of interrogating and transgressing its definitions, which ceaselessly turn into limitations. The human beings do not only reflect their own existence by means of language, but much more fundamentally, language, and especially writing, articulates (is and says) the limits of existence. For example, given that one obvious limit of finite existence is death, Derrida has often worked on the recurrent metaphors of writing as a *tomb* of signification[101] or of the (intention of) the author[102] and thereby drawn attention to the way in which writing doubles death; this work echoes the setting of Maurice Blachot's *The Space of Literature*, where the questions of death and of literature are the "same," although not identical.

Derrida does not have a specific theory of technics either. It is certainly striking that technical metaphors abound in his works, especially those pertaining to technics of writing, such as handwriting, typewriter, mystical writing pad, email, computer, television, Internet, and the entire tele-technological apparatus that is so characteristic of our era.[103] If these examples are regarded simply in function of the question of technics, it is clear that the sense of technics is palpably displaced when the paradigmatic technologies are not hammers or steam engines anymore but writing devices. Derrida's early discussions of writing have a deconstructive effect on classical interpretations of technics as tool and instrument because he does not think technics as tools, that is, as neutral means of human intentions, but as autonomous mechanisms that repeat whatever code they incarnate, sometimes following human aims but more often operating independently of them. This resembles Foucault and Deleuze's interest in repetitive mechanisms and machines; this also has some similarities with Gilbert Simondon's ways of analyzing the technical object in itself before studying its function to the human subject.[104] Instead of thinking technical objects in terms of instrumentality that uses objects as means of human intentions, Derrida shows how technical objects are ambiguous prostheses that condition intentionality itself (e.g., by thwarting, creating, or altering intentions).

However, the aim of Derrida's early texts is not a reformulation of a theory of technics but a radical rethinking of signification. In these works, writing as a general structure instantiated in these and other devices and procedures is the major *terminus technicus* for examining the (quasi)transcendental conditions of signification that Derrida was most interested in at that time. As Rodolphe Gasché has shown very early, Derrida does not build his thinking around a single master signifier (such as *différance*) but articulates an infrastructural chain of quasi-transcendentals.[105] "Now if we consider the chain in which *différance* lends itself to a certain number of nonsynonymous substitutions, according to the necessity of the context, why have recourse to the 'reserve,' to 'archi-writing,' to the 'archi-trace,' to 'spacing,' that is, to 'supplement,' or to the *pharmakon*, and soon to the hymen, to the margin-mark-march, etc.?"[106] With these displacements, Derrida turns the question of signification like a kaleidoscope and shows its implications for a number of philosophical problematics. It is nonetheless striking that all of these structures are regularly if not always described via technical and mechanical metaphors: trace, writing, supplement, *différance*, and the like are thought of as *technical and artificial supplements* and their functioning is explained as *machinic repetition*. These are not "real machines" that have been intentionally constructed by someone, they are structures of abstract, artificial (conventional) machinality, as will be discussed in what follows.

In his later texts Derrida consolidates his transcendental investigation of the conditions of signification, not by formulating an ontology but by deconstructing onto/theo/logical answers to questions of time and being by showing how the historical names of *"khora"* and "messianicity without messianism" give to thought an older dimension of time and being than any onto/theo/logy could. This work is accompanied by inquiries into the possibility of ethics, justice, and democracy especially in the context of contemporary globalization. These works refer constantly to contemporary technological reality and especially to its archival and tele-technological dispositifs. These are not simple facts of the contemporary world that are to be accounted for. In the hands of Derrida they become, not exactly concepts, nor merely examples, but *inevitable names* for thinking absence in the heart of presence, the thought of which is necessary both for the interpretation of being-with in the contemporary world and for the explication of *khora* and messianicity that frame experience.

For these reasons, Derrida does not present a positive theory of technological humanity. In what follows, I will show how he on the contrary deconstructs the classical conception of the human being—*animal rationale*—by showing how *rationale* is conditioned by the technicity of writing and by showing how *animality* is thought in terms of *life*, which modernity has thought of through technical metaphors ever since the Cartesian man-machines and right up to contemporary biology's description of DNA as program and code. Each time, Derrida takes a step back from a signifying and living positivity (reason, animality) toward its conditions (writing, life). Both writing and life exemplify an "originary technicity," which at the same time puts technics at the heart of human thinking and life and changes the current understanding of what technics is. As the importance of originary technicity for Derrida is well known today but its intimate structure is rarely examined for its own sake, I will try to shed light on it.

It is important to note that Derrida's final aim is by no means the production of a new theory of technological humanity but to work toward rethinking philosophy itself. Thinking of the *animal rationale* in terms of a life of writing and thinking, life and writing in terms of originary technicity does not lead to a positive hypothesis that could ground a new philosophical anthropology. On the contrary, it leads to the discovery of a *spectral* figure that does not exist as such—it is not the new figure of man—but that nonetheless comes forth in counter-relief as the *unconscious presupposition* of modern discourses on the *anthropos*. Derrida thus shows that the sense of animal rationale has already changed into an "originally technical anthropos" in the "text" of the contemporary world, for example in certain works in linguistics (de Saussure), biology (Jacob, Canguilhem), and anthropology (Lévi-Strauss, Leroi-Gourhan). Deconstruction thus not only undoes classical philosophical ways of thinking (*animal rationale*) but it also shows the figure that looms in its place, although it does not really exist (the spectral apparition of a textual living being). It makes "specters" appear and asks how to deal with them as specters and not as positive realities.

In what follows, I will show how Derrida's deconstruction of the classical figure of humanity reveals a figure of *prosthetic animality* that is a hidden nonintentional presupposition of contemporary (scientific) culture. It begins to emerge when Derrida's deconstruction of the instrumental conception of technics brings forth an idea of prostheticity. Thinking technics in terms of prostheticity also alters the idea of humanity, which appears as essentially invaded by its supplements. This is where Derrida's work prepares the way for later positive theories of human prostheticity and cyborg life.[107] After examining the general idea of prosthetics, I will first present writing as the principal prosthesis of the spirit and second, Derrida's theory of the life supplemented by such technics. Finally I will show how Derrida expands his theory of writing into a theory of tele-technology, which accompanies the discovery of spectrality at the heart of Dasein and of the thinking of *khora* that displaces the question of being. With these successive moves, I present Derrida's contribution to the question of technological humanity while at the same time emphasizing

De/constructing Humanity 147

that this is not a positive thesis about anything like a prosthetic animal but, on the contrary, a deconstruction that brings forth this figure as the spectre of such a position.

PROSTHETICS

The focus of modern philosophy (let us say, for the sake of simplicity, from Descartes to Husserl) has not been on the *anthropos* but on the thinking subject in its epistemic relation to the object of experience. Heidegger showed that prior to this theoretical relation that defines self-consciousness, Dasein is already related to things in its practical activity. Before the constitution of objectivity, things come forth as tools (*Zeug*), which are technical things that disappear in their use (*Zuhandenheit*) and emerge only occasionally (*Vorhandenheit*) through disturbances in use. What a present-at-hand tool reveals is not the truth of the natural world (*cosmos*) but the world as an epochal horizon, which is in our case markedly technical (*Ge-stell*). Foucault reinterpreted this idea of a technical *Ge-stell* in terms of a power dispositif that imprints its forms on human beings and really produces them as humans in the first place. Derrida agrees with Heidegger and Foucault's fundamental intuitions concerning a technical marking of human existence, but he shows in more detail how this happens by interpreting the subject-object-relation not only as Dasein's relation to its tool but as life's relation to its prostheses. Let us now consider what Derrida means by prosthesis and how it implies a deconstructive reinterpretation of the *anthropos* in terms of originary supplementarity.

The idea of the prosthesis results from a deconstruction of the idea of instrumentality that had dominated classical theories of technics. An instrument, for example a telescope, is a neutral means of technical activity: a prolongation of an organ, it allows an intention a greater reach without changing the intention itself. A *prosthesis*, on the contrary, is not a simple means but fundamentally a dimension of *subjectivation* that not only realizes an intention but makes it possible in the first place, delimiting, engendering, and transforming what can appear as possible intentions. By definition, a prosthesis patches up a lack in a living body.[108] It is an artificial addition to a living body, for instance a walking stick or a pair of glasses, but also a pacemaker, a heart implant, or a heart drug. A prosthesis is piece of lifeless matter that is indispensable to life; a dead supplement added to a living body such that its living force would be enhanced, or quite simply such that the body could stay alive. Because the prosthesis is indispensable to life, it also reveals the weakness and the helplessness of bare life: if your heart was good, you wouldn't need a pacemaker, a heart implant, or a heart drug, and more generally, if human life was self-sufficient, it would not need technics. In sum, a prosthesis is not simple equipment because it really inhabits its user and installs an ambiguous dead machine in the fragile living substance—such as insulin pumps, organ transplants, electronic stimulators, and pharmacological regulators located deep within the organism.

Today, the prosthesis is increasingly that of a living body and not only that of its mental capacities. Perhaps the most famous example of this is Jean-Luc Nancy's text *The Intruder* (2008, originally published in 2000 as *L'intrus*), in which Nancy shows how the same logic functions in a technically supplemented human body. Describing his personal experience of a heart implant, Nancy emphasizes that the prosthesis consisting of another person's heart is far from being a simple means to better health that the body incorporates. It is marked by the ambiguity of the prosthesis that we have just described. On the one hand, the implant supports the life of the body that would otherwise die. But on the other hand, instead of becoming one with the receiver's body, the implant remains alien to it. Its alienness is seen, for instance, in physical rejection reactions, in psychological estrangement, in technological and social dependency on the medical institution, and so forth. Instead of restoring health, the implant induces a technically produced chronic illness: the implant is an ambiguous intruder that protects the body from one kind of a death by introducing another kind of a death into it. This is how the prosthesis is and remains an external thing encrusted in the living being. Even the most perfect prosthesis is not one with the body but a stranger within the body proper: "My heart became my stranger: strange precisely because it was inside."[109] As Derrida might put it, it "haunts" the living being as a presence that is at the same time strange and intimate. It has the uncanniness (*Unheimlichkeit*) that constitutes the intimacy of the living being by rendering it a stranger to itself. A prosthesis is not just a contingent addition to the living being but the very condition of its life that both enhances and intoxicates it. The prosthesis constitutes life by alienating it from itself; it haunts life like artificial supplements that impress their strangeness on life.

In Derrida's analysis, all kinds of technics function in a prosthetic manner: psychological techniques, material technical objects, total technological systems—all of them end up supplementing human life as its prostheses. However, he is especially interested in the prostheses that support thinking, most notably writing, which he studies in all his early works. This idea comes forth particularly clearly in Derrida's influential interpretation of Plato's *Phaedrus* in "Plato's Pharmacy," in which he shows why writing is a pharmakon—a drug, which is of course also a prosthesis. In "Plato's Pharmacy," Derrida presents and reinterprets Plato's theory of writing as a supplement to memory. Memory is a human faculty; it is even the most indispensable faculty of human thinking. Because human thinking is conditioned by memory, it is rendered fragile by the ever present possibility of forgetfulness. In "Plato's Pharmacy," Derrida studies writing as the technical invention that is expected to remedy memory's forgetfulness.

As it is well known, in "Plato's Pharmacy" Derrida rereads the myth, told by Plato, of the invention of writing by the Egyptian god Theuth. Inventor of writing, Theuth is also the god of the dead, which is why writing is so easily associated with death. Of course the edifice of the tomb is a memorial sign and funerary inscriptions are among the oldest writings, but moreover, the myth really presents writing in general as the tomb of living speech and as the memorial that replaces living memory.[110] Theuth is also the inventor of numbers and calculus, hence writing has something of the mechanicity of calculus.[111] Theuth is the god of both science and magic; accordingly, his inventions always have the ambiguity of the pharmakon, poison-remedy: like drugs, writing and numbers both heal and intoxicate.[112] Plato's myth tells how Theuth's invention is rejected by Ammon, the king of gods, or by Thamus who represents him. Thamus condemns writing because he sees its ambiguity: although Theuth presents it as a remedy for forgetfulness, Thamus sees it is a poison to memory since a memory that thinks it can rely on text ceases to make the effort to memorize and remember. Writing is not good for living, knowing memory, *mneme*, it is only good for re-memoration, recollection, consignation, *hypomnesis*.[113]

Derrida's analysis of Plato's theory of writing centers on two terms. The first is *techne*: even more technical than rhetorics, *techné té logón*,[114] writing is an artificial external supplement, an "artefactum which is also an art,"[115] a product of technique that is itself a technique. Plato insists on the difference between writing as a technics of iterative repetition and the identical repetition of the idea:

> Nevertheless, between *mnémé* and *hypomnésis* . . . it is a question of repetition. Live memory repeats the presence of the eidos, and truth is also the possibility of repetition through recall. Truth unveils the eidos or the ontos on, in other words, that which can be imitated, reproduced, repeated in its identity. . . . Writing would indeed be the signifier's capacity to repeat itself by itself, mechanically, without a living soul to sustain or attend it in its repetition, that is to say, without a truth's presenting itself anywhere.[116]

Writing is only a dead mechanism because it repeats the signifier instead of resurrecting the signified, thereby admitting both the return of the same (*iter*) and its contingent alteration (*alter*) due to wearing, misunderstanding, context changes, and so forth. We shall examine its specific technicity in the next subsection, but for now it suffices to notice that the prosthesis of writing is both a remedy for forgetfulness and a poison that gives an illusion of the memory that it has in fact corrupted.

The second focus of Derrida's reading of *Phaedrus* is precisely the term *pharmakon*, which means remedy and poison at the same time. Writing is a pharmacological technique that intoxicates memory like a poison intoxicates a body[117]: it remedies forgetfulness while also luring memory to trust in an external support that itself leads to a weakening of the memory proper. Being the ambiguous supplement to living memory, a supplement bringing forgetfulness as well as memory, writing is a *prosthesis* that substitutes a "mnemotechnic device for live memory, the prosthesis for the organ"; "the perversion consists in replacing a limb by a thing . . . substituting the passive, mechanical 'by-heart' for the active reanimation of knowlege."[118]

For Derrida, the pharmakon of writing discovered in "Plato's Pharmacy" is the prototype of the prosthesis that at the same time supports the "life of the spirit" and reveals and even causes its weakness. Throughout all his early texts, Derrida shows that "writing" is indispensable to us humans—meaning some kind of writing, actually any trace of a signifying activity, for the term has such a totally general signification for Derrida that it can also be attributed to animals. We are dependent on signs because we need them in order to elaborate our thoughts, conserve them from forgetfulness, and share them with other people. Furthermore, we cannot be conscious of everything we think, not to mention what other people think, and most of the time we need to think without being conscious of it. Thinking happens through the pharmakon of signs, words that we alone have not invented, texts and archives that come to us from our past and from other people. In signs, writing haunts us, and we are what we are only thanks to this haunting by the pharmakon of writing. Writing is the prosthesis of language and language is the prosthesis of thought, from the simplest unconscious stirrings to ultimately philosophy itself.

The prosthesis of writing is a pharmakon. This means, on the one hand, that the life of thinking could not live without it and is truly inhabited by it. On the other hand, this means that the prosthesis incorporated by thinking is ambiguous, both beneficent and maleficent: it enables and disables thinking and makes it nontransparent and alien to itself. Language functions in us as a writing that "speaks" in us without our being able to control all that it does. It induces both memory and forgetfulness, both our words and the words of others, both correct retrieval of past ideas and their false understanding. We cannot control everything we say, and yet what we are is only accessible to us in our enunciations relying on just these signs. Of course, "errors" are not all that language does, on the contrary, we can very well use language and writing to express our intentions. But the *possibility* of such errors is nonetheless constitutive of all uses of language, something that the language users need to be conscious of.

What does this general prosthetics make of our original question of "the human being"? Let us now see how prostheticity transforms philosophical anthropology (in later chapters we will also see how it affects the idea of Dasein and the idea of life).

Of Grammatology explores the perspectives that the question of writing opens up in human sciences and in particular anthropology. Deconstructing the principal starting point of classical philosophical anthropology—the human is an animal endowed with logos—*Of Grammatology* examines logos as language, and language not as an extension of human intentions but as a general structure of traces described in terms of arche-writing. Derrida does not ask *what* language says but *how* it says, how it functions, what kind of a machine it is. For Derrida, the human is an existent that is prosthetically equipped by an archi-writing machinery.

In *Of Grammatology* and other early works, Derrida does not pay much attention to the second part of the definition of the human as *zoon logon ekhon*, namely *zoon*, animality, and when he does, he is not interested in animality as a biological substance. He is interested in language as a system of traces, whether human or animal, and a reduction to biology could not explain a system of signification. As we shall see later, in other works Derrida disturbs different frontiers drawn between animality and humanity and explains both in terms of life. For now, let us only examine *Of Grammatology*, in which humanity is not regarded as a *zoon* but as the kind of existence produced by language. As we saw, the way of being of language is that of a prosthesis grafted onto and incorporated into life. We will now see why this life is neither a natural given nor a pure and authentic state of existence to which the technical supplement would subsequently be added, but technicity and supplementarity make the human from the word go.

In *Of Grammatology*, Derrida explains humanity as supplementarity through readings of two authors in particular, André Leroi-Gourhan and Jean-Jacques Rousseau. Leroi-Gourhan is a paleoanthropologist who thought of technics not only as one attribute of the human being but as its very nature, ever since the Zinjanthropus first seized a flintstone in order to use it as a tool and, so to speak, as "a 'secretion' of the anthropoid's body and brain."[119] Technicity grows from the anthropoid body through a quasi-biological force such that "the Australanthropians . . . seem to have possessed their tools in much the same way as the animal has claws . . . as if their brains and their bodies had gradually exuded them . . . chopper and the biface seem to form part of the skeleton, to be literally 'incorporated' in the living organism."[120] For Leroi-Gourhan, human evolution has ceased happening in the anthropoid body because it has continued outside of the body in tools and language (and language is just another part of technicity). For him, these extensions of the human body and brain are essentially an exteriorization and a liberation of the human *memory*. The tool records an experience of the workings of the world (how such and such a stone breaks when hit, how such and such a piece of wood burns when stroked by a flintstone, how electricity circulates in computer chips); it does this so well that the memory can also be shared by other and later members of the community when they use the tool, even though they are not conscious of the world understanding inscribed in the tool and even though they do not share the original world experience deposited in the tool. As Derrida puts it, for Leroi-Gourhan, the technical object has the general structure of a *grammè*, which is both a memory inscription and a program of its functioning. Its program-character is instantiated on all levels from genetic codes and technical procedures to systems of writing and cybernetic systems. "It must of course be understood in the cybernetic sense, but cybernetics is itself intelligible only in terms of a history of the possibilities of the trace as the unity of a double movement of protention and retention. This movement goes far beyond the possibilities of the 'intentional consciousness.' It is an emergence that makes the *grammè* appear as such."[121] The *grammè* goes beyond intentional consciousness because, even if from an evolutionary perspective technical objects are extensions of the brain, this does not mean that from an individual or collective perspective they are projections of conscious intentions (just as being conscious does not mean being conscious of the brain by which one is conscious). Human beings and human communities do not need to be conscious of the functioning of the world if they know how to use a tool that incorporates this knowledge in its very structure. Technical principles do not belong to individuals but to technical objects themselves and through them to the communities that use them. They belong to what Gilbert Simondon calls the "associated milieu" projected by the technical object, that is, the milieu in which the object works and into which humans insert themselves in order to use it.

If Rousseau formulated the idea of language and technics as *supplements*, Derrida underlines the *originary* character of these supplements. There is no "state of nature" prior to the supplement, but it is on the contrary the supplement that projects the idea of a prior state of nature as a kind of a necessary fiction:

> Thus supplementarity makes possible all that constitutes the property of man: speech, society, passion, etc. But what is this property [*propre*] of man? On the one hand, it is that of which the possibility must be thought before man, and outside of him. Man allows himself to be announced to himself after the fact of supplementarity, which is thus not an attribute—accidental or essential—of man. For on the other hand, supplementarity, which *is nothing*, neither a presence nor an absence, is neither a substance nor an essence of man. . . . Therefore this property of man is not a property of man: it is the very dislocation of the proper in general. . . . Man *calls himself* man only by drawing limits excluding his other from the play of supplementarity: the purity of nature, of animality, primitivism, childhood, madness, divinity. . . . The history of man *calling himself* man is the articulation of *all* these limits among themselves.[122]

Supplementarity means, on the one hand, that the supplement has no proper sense in itself but it draws its sense from the thing that it supplements, the human. But on the other hand, the human is nothing in itself either: it only appears "after the fact of supplementarity" as its condition. Supplementarity is nothing and so is the property of the human. The human being is an "origin" of language and technics and yet it is nothing existing in itself but something projected and fictioned by the functioning of language and technics. Nothing given, it comes nonetheless forth as as a *desire* that pushes toward satisfaction, eventually, as a *lack* that wants to be supplemented.[123] A lack is nothing because it precisely lacks being, and it appears as a desire only from the point of view of satisfaction. Similarly the nothingness that the human being is comes forth as nothingness from the point of view of the supplement that seeks to remedy to it.

In Rousseau, the human is therefore "made" by its supplements. Rousseau's successors extend the same intuition into concrete historical situations and human existence as determined by specific texts, discourses, and technical dispositifs that make up a historical situation. The historical situation determines not only the *anthropos* but existence in a historical world. This is why ultimately Heidegger says that Dasein is nothing as such but qualified by its *Da*, by the historical world in which it finds itself and which is the opening through which sense is available to it. Foucault says that the human is marked by the prevailing power-knowledge structures in the world present to it. Although Derrida has referred in many of his lectures to current events, his aim is not the uncovering of the epochal features of current humanity but studying—in current events, too—the way in which sense functions in the human in general. This is why, like Gasché, I think that what is ultimately of interest to Derrida is the transcendental conditions under which any sense can operate in the first place. Transcendence appears as an epochal *logos* when it is a question of what the sense *is*; but transcendence appears as a *technics* when it is a question of the way in which any sense whatsoever makes sense to existence. Archi-writing is one name for such transcendental technics: it does not reveal the true logos but shows how iterative altering markings function, whatever their truth content. This is why Heidegger thinks of the logos in order to articulate the world as a site of truth, while Derrida thinks archi-writing marks existence without necessarily gathering a single world and that conditions both truth and falseness. Archi-writing does not aim at truth itself but the way in which any truth claim can be made in the first place.

For Derrida, the supplement of writing is a prosthesis, a repetitive iterative machine that is encapsulated in the "human." The human is in possession of writing that, in turn, "possesses" it; paradoxically, one possesses the other in both senses of the word "possession." This is where we can detect a difference between Derrida and Foucault. Both think that the ideas that make the world are imprinted on the human—ideas in a weak sense, "current ideas" that rarely if ever make a unique coherent totality but that pile up as a heap of disparate and often incoherent discourses—but, instead of simply imprinting a figure, they generate subjectivation, as if growing a soul that becomes animated by small signifying machines that continue functioning in it. Derrida emphasizes the alienness and even the uncanniness (*Unheimlichkeit*) of these small machines more than Foucault does. They do not educate the human, they are prostheses that do not exactly make the human but supplement it. They do not mingle into existence as if they were really one with it, they remain prostheses and parasites; they are alien, sometimes uncomfortable and even hurtful additions. Derrida pays particular attention to the uncanniness of the supplement by studying different ways in which alterity appears as alterity: it haunts the proper instead of making it homely. One cannot know what the "proper" would be—as we saw, it is nothing as such—but it makes itself felt as whatever reacts to the uncanniness of the prosthesis or withdraws from it. The nothingness that is supplemented and projected by the prosthesis is a forever inaccessible, savage, and abysmal region that is marked by strong desires: the desire to be and to be supplemented, the desire to flee the grip of the prosthesis and to not be. This is why in Derrida's terms the *human* cannot be figured to be a technical *construction* (as Foucault's human tends to be) but only as a surface of inscriptions that mark something that it is not. The noncoincidence of the technical supplement and what it is a supplement to is very important since this disparity explains why the human is never ready but always changing and underway. Even though one can find an analogical indetermination in Foucault, especially in his idea of resistance, Derrida's term *supplement* more clearly articulates the gap between technics and what resists it.

The idea of originary supplementarity thus deconstructs any idea of a prior, pure, given, human nature. If Derrida were made to answer to the question of human nature, his answer would not be unlike that of Plessner, for both think of an originary technicity that marks the human from the word *go* and makes it produce its "nature" ever anew. However, Derrida, unlike Plessner, does not mean to formulate a philosophical anthropology at all, for he rather shows, like Foucault, how philosophical anthropologies proceed from a misleading question that aims at erecting an essence that it simultaneously undoes. As we shall see, in the end Derrida is closer to the phenomenological way of examining existence.

So far, the figure of prosthetic humanity that looms in Derrida's early work dissipates a number of misleading ideas about humanity but it also results in a somewhat confusing mix of technics and life. In the end, this appearance of confusion is due to false expectations: Derrida does *not* build a philosophical anthropology, and making him answer anthropological demands can only get us so far. Let us now take leave of the anthropological dream and follow Derrida's more precise studies of the prosthetic object itself on the one hand, and of the life that it supplements on the other hand.

Writing, on *Khora*

In "Plato's Pharmacy," the prosthetic object, writing, comes forth through a deconstruction of logos—and indeed, no doubt the best known element of Derrida's early works is his deconstruction of logos. Logos is one of the two constituents of the classical definition of the human being as *zoon logon ekhon* and for classical philosophy certainly the more important of the two. Logos is both the reason for all (*en arkhe en ho logos, noesis noeseos*, the absolute) and the human language. *Zoon logon ekhon* is the being that has access to the universal reason and makes it exist in its language. Derrida destabilizes the idea of reason interpreted both as the absolute reason of being (as brought to its *summum* by Hegel) and as the self-transparency of consciousness (found in Husserl). This does not mean he simply refuses the very possibility of truth (as Derrida's first critics claimed), but that, like Nietzsche, he stresses the impossibility of making a clear distinction between truth and fiction and that, like Heidegger, he questioned the possibility of self-transparency. Thought was thereby brought to the terrain of finitude (and not relativism). In Derrida's first works the deconstruction of logos was effectuated by showing how logos (as absolute reason) was conditioned by language (which brings in the problem of reason's historicity) and how language was conditioned by writing (which brings the problem of language's materiality). Against the ancient accusation that speech is an inferior image of thought and writing an inferior image of speech, Derrida observes that thought can only exist in discourse and that discourse is always conditioned by "writing." Of course, this is not an empirical claim that every discourse is indeed written but the discovery of a quasi-transcendental structure that conditions signification.

Derrida focuses on the question of language and writing in his early works, *Of Grammatology, Dissemination, Writing and Difference*, and the *Margins of Philosophy*. In these works he shows that although writing has traditionally been interpreted as a contingent technical supplement to language proper (speech), it is instructive to consider it as the fundamental structure of all language as such: "Language is always already a writing."[124] Here is how, in *Of Grammatology*, he connects the problematics of language with the question of technics:

> With an irregular and essentially precarious success, this movement would apparently have tended, as towards its *telos*, to confine writing to a secondary and instrumental function: translator of a full speech that was fully *present* (present to itself, to its signified, to the other, the very condition of the theme of presence in general) technics in the service of language, *spokesman*, interpreter of an originary speech itself shielded from interpretation.
>
> Technics in the service of language: I am not invoking a general essence of technics which would be already familiar to us and would help us in *understanding* the narrow and historically determined concept of writing as an example. I believe on the contrary that a certain sort of question about the meaning and the origin of writing precedes, or at least merges with, a certain type of question about the meaning and origin of technics. That is why the notion of technique can never simply clarify the notion of writing.[125]

Instead of interpreting writing as a simple tool at the service of speech, Derrida interprets it as a transcendental structure (also called archi-writing) that conditions all language, speech included. At the same time, he shows how the interpretation of technics in terms of writing also changes the understanding of technics, no longer as a tool but as a specific "writing machine." This is how a certain idea of technics helps to formulate a novel interpretation of language, but the study of language also leads to a novel interpretation of technics: language and technics clarify one another without amounting to the same.

The study of the "meaning and origin of writing" on the one hand, of technics on the other hand, contains several gestures. First, Derrida locates and brings forth what classical theories of language actually say of writing—not much, really, because the theme of writing has usually been left at the margins where it was not examined attentively. Especially in *Of Grammatology*, Derrida shows how writing has been interpreted in classical philosophy and up to Saussure's linguistics as a contingent, technical, and artificial supplement to speech—"speech" standing for language proper. "The written signifier is always technical and representative. It has no constitutive meaning."[126] Ever since Aristotle, it has appeared evident that speech is an image of thought and writing is an image of speech, such that each time the imitation draws its truth from the model but in a diminished form. Being ontologically inferior to the original, the imitation has appeared not only artificial and inessential but also potentially weak, therefore false, and therefore dangerous. This is why it has appeared advisable to protect the original signification from the contamination induced by writing. Derrida finds the same logic everywhere from Plato to Saussure: writing is false and dangerous *because* it is just a tool, just a technical supplement.[127]

Second, Derrida does not refuse the interpretation of writing as artificial technical supplement but makes it explicit and asks what its technicity entails. He shows how the classical theoreticians of writing up to Saussure rely on an "instrumentalist and technicist concept of writing, inspired by a phonetic model which it does not conform to except through a teleological illusion. . . . This instrumentalism is implicit everywhere."[128] From its point of view, as also Rousseau says, "Writing serves only as supplement to speech," such that, as Derrida explains, "It is the addition of a technique, a sort of artificial and artful ruse to make speech present when it is actually absent."[129] A supplement provides a presence—but it is only the presence of a supplement or of a replacement that takes the place of the real presence (of the thing itself, of the signification, of the speaker), it is an "exterior addition," whose very exteriority carries the possibility of the "negativity of evil."[130]

Against this tenacious tradition, Derrida affirms that instrumentalism is not the last word of technics.[131] On the contrary, the fundamental character of language's technicity is its *machinic* character that is precisely illustrated by writing. "The originary and pre- or meta-phonetic writing that I am attempting to conceive of here leads to nothing less than an 'overtaking' of speech by the machine."[132] Writing is what can be repeated, like a machine repeats the code entrusted to it. Considered as such, writing is a machine that enables the rereading of the same inscriptions again and again by different persons and in different contexts. Writing also makes evident what differentiates machinic repetition from ideality. Derrida emphasizes that a machine does not reproduce the identical but it only reiterates the similar that brings the dissimilar with it. This is well illustrated by acts of reading where nothing guarantees that each reading of the same inscription revives exactly the *same signification* and the *same intention* as the one originally put in writing; it only reiterates the same *sign*, which can be interpreted very differently from one person to the next and from one context to another. Against the classical theories of language that have overlooked this distinction, Derrida shows that it is fundamental to sense. In "Signature Event Context" he explains:

> [Writing] must . . . remain legible despite the absolute disappearance of every determined addressee in general for it to function as writing, that is, for it to be legible. It must be repeatable—iterable—in the absolute absence of addressee or of the empirically determinable set of addressees. This iterability (*iter*, once again, comes from *itara*, *other*, in Sanskrit, and everything that follows may be read as the exploitation of the logic which links repetition to alterity), structures the mark of writing itself. . . . The possibility of repeating, and therefore of identifying, marks is implied in every code, making of it a communicable, transmittable, decipherable grid that is iterable for a third party, and thus for any possible user in general.[133]

It is because writing functions by *iterative* repetition that it is thought of in terms of a machine. No doubt, a machine is something that repeats the same code again and again. But it is a material thing that is exposed to wear and tear from one context to another. This adds variations to the repetitions, such that machinic repetitions are different from the re-instantiations of an idea that cannot not be perfectly identical (like a geometrical idea). The consequences of this are far-reaching. One effect of iterability is the way in which any concrete text, insofar as it can be read, functions like a machine. For example Hegel's text "functions as a writing machine in which a certain number of typed and systematically enmeshed propositions (one has to be able to recognize and isolate them) represent the 'conscious intention' of the author."[134] Thanks to such written propositions Hegel's text can be read and his logos can be rediscovered over and over again. But readings vary, the same text gives way to different and even conflicting interpretations and even to interpretations that could be valid even though they seem to go against Hegel's manifest intentions, like Derrida's deconstructive readings of Hegel that function in a manner akin to psychoanalyses of themes that seem to be unconsciously repressed by Hegel, such as the role of writing in philosophy. Hegel dealt with the question of writing under the terms *Vorstellung* and *Darstellung*, representation and presentation, which, according to Hegel, a philosophical reading can sublate (*aufheben*) into conceptual comprehension but that, according to Derrida, also keep working as writing, which can lead to a number of unwanted consequences, for example the consequences of the theme of writing both in Hegel's theory of language and in his very writing. Thus, according to Derrida, Hegel overlooks the machinic condition of the *Darstellung* of the system: "What Hegel, the *relevant* interpreter of the entire history of philosophy, *could never think* is a machine that would work."[135] Hegel cannot see what in a representation (*Vorstellung*, *Darstellung*) works *against* conceptual comprehension. This is not something that Derrida imposes on Hegel's text from the outside but that he reads in Hegel's very text, which itself ignores the possibility of such a reading.

Another more positive effect of iterability is the possibility of invention. The constitutive possibility of alteration induced by writing opens up not only a margin of error but also the margin of change and novelty. If every text is deconstructable, every text is both capable of critique and open to new interpretations. This is why the same logic opens the possibility of a certain thinking of art and technics, starting with the one Derrida finds in Rousseau, who thinks art (*techne*) as *mimesis*, which is also immediately a technique of imitation: "Imitation, therefore, is at the same time the life and death of art. Art and death, art and its death are comprised in the space of the *alteration* of the originary *iteration* (*iterum*, anew, does it not come from Sanskrit *itara*, other?); of repetition, reproduction, representation; or also in space the possibility of iteration and the exit of life placed outside of itself."[136]

Mimesis is at the same time the life and death of art because it is both imitation as sterile copying and imitation as poiesis, that is, creative invention and production, as Philippe Lacoue-Labarthe in particular has shown.[137] However, art in the proper sense of the word demands more, it demands invention that is expected to be more than just mimesis—or art is mimesis that overcomes itself as invention. As Derrida shows in "Psyché: Invention of the Other," one invents when one produces something new and unprecedented, especially in art and in technics. "On the one hand, people invent stories (fictional or fabulous), and on the other hand, they invent machines, technical devices or mechanisms, in the broadest sense of the word."[138] Invention *discovers*, not by creating ex nihilo, but nevertheless for the first time "it unveils what was already found there, or produces what, as *techne*, was not already found there but was still created . . . it gives rise to an event, tells a fictional story, and produces a machine by introducing a disparity or a gap into the customary use of discourse. . . . [Invention is] the event of a novelty that must surprise."[139] Invention is the production of a technical device or procedure that was not and that did not appear possible until the impossible happened: it was invented. The only thing that cannot be thus invented is the future to come, the "event of the entirely-other to come,"[140] which Derrida will conceptualize as messianicity without messianism.[141]

Third, what is thus illustrated by concrete writing is true of language and signification in general. In *Of Grammatology*, Derrida says, "If 'writing' signifies inscription and especially the durable institution of a sign (and that is the only irreducible kernel of the concept of writing), writing in general covers the entire field of linguistic signs."[142] Writing is therefore not an image or a symbol of speech associated with some determined systems of writing, but it is fundamentally a "*graphie* [unit of a possible graphic system] [that] implies the framework of the *instituted trace*, as the possibility common to all systems of signification."[143] Writing is not making concrete traces: the abstract possibility of trace is the root of concrete writing *and* of any signifying act whatsoever.[144] The trace is one of the most important articulations of Derrida's central philosophical task, that of contesting the transcendental signified postulated by metaphysics while opening another dimension that accounts for sense.[145] Trace lets the dimension of difference *as such* be thought, instead of referring difference dialectically to a signified that absorbs difference in a unity; trace marks the relationship with the other, but instead of postulating the other's presence, it marks the other's dissimulation and absence.[146] The trace is not the *trace of* some living presence, on the contrary, such a presence can only be thought from the trace. "The trace is not only the disappearance of origin—within the discourse that we sustain and according to the path that we follow it means that the origin did not even disappear, that it never was constituted except reciprocally by a nonorigin, the trace, which thus becomes the origin of the origin. . . . If all begins with a trace, there is above all no originary trace."[147] "The trace is in fact the absolute origin of sense in general. Which amounts to saying once again that there is no absolute origin of sense in general. The trace is the différance which opens appearance [*l'apparence*] and signification."[148]

In order to highlight that trace is not a concrete piece of writing but the general structure of sense, Derrida also speaks about *arche*-writing, which combines the general structure of *différance* with spacing and the "dead time" of temporalizing, which can never be present as such either.[149] Now although I cannot enter fully into the matter here, I want to point out that beyond their function of designating transcendence as pure differentiation, the notions of trace and of archi-writing also open the space, not of Derrida's ontology but of Derrida's early answer to ontology. He does not formulate an ontology in the classical sense of a doctrine of the good beyond beings because he, on the contrary, questions all onto-theological foundations. However, he thinks arche-writing as an *elemental* dimension thanks to which logos and existence can be thought but which is hardly thinkable itself. In order to present this elemental dimension to thinking, he recounts Plato's hypothesis of the *khora*, the strange "place" that is neither the heaven of "intelligible patterns" (*paradeigmatos*) nor the earthly world of things that are "only the imitation of the pattern, generated and visible," but the place in which forms and things are mixed, "the natural recipient of all impressions (*ekmageion*)," such that "the forms which enter into and go out of her are the likenesses of eternal realities (*ton onton aei mimemata*) modeled within her after their patterns (*typothenta*) in a wonderful and mysterious manner."[150] Derrida makes careful studies of *khora* such as Plato describes it in the *Timaeus* and draws attention to two specific features that are of particular interest to us here.

Firstly, Plato uses *typographic metaphors* when speaking of *khora* as the "third class," for "all these things 'require' (*Timaeus* 49a) that we define the origin of the world as trace, that is, a receptacle. It is a matrix, a womb, or receptacle that is never and nowhere offered up in the form of presence, or in the presence of form, since both of these already presuppose an inscription within the mother. Here . . . 'Plato's metaphors' are exclusively and irreducibly scriptural."[151]

Plato thinks of *khora* as space in which ideas are "inscribed" into things, not by simple impression but by a properly creative event. *Khora* is the matrix in which ideas turn into appearing things, "the impression-bearer," which does not mediate a simple reduplication but the profoundly metaphysical transformation between the eternal model and the finite thing. Of course, Derrida does not adopt the Platonic myth of the heaven of ideas: the ideas do not "really" exist somewhere, but the *operation* illustrated by the fable of the *khora* does, and this operation cannot be told without projecting the ideas. When contemplating visible things, we do not see ideas through them, but we still think of things as *traces* of ideas and the ideas themselves as origins of such traces, such that it is really the tracing that invites us to think of ideas in things.

Secondly, Derrida points out that *khora* is never present as such but it is an effect of a kind of a literary fiction designated to account for the possibility of being: "The discourse on the *khora*, as it is presented, does not proceed from the natural and legitimate *logos*, but rather from a hybrid, bastard or even corrupted reasoning."[152] As Naas underlines, *khora* is a unique narrative that Plato first invented and that was then commented on across the history of philosophy. *Khora* is a story of a unique place that cannot be reached or touched, that does not have an essence, that is not a subject: "She is nothing other than the sum or the process of what has just been inscribed 'on' her, on the subject of her, on her subject, right up against her subject, but she is not the subject or the present support of all these interpretations, even though, nevertheless, she is not reducible to them."[153] *Khora* is not a transcendent being that guarantees the truth of being, it is the imaginary space and the space of imagination in which ideas are thought to touch things such that *mimesis* and *methexis* can take place and things become thinkable. So to speak, it is not being, it is the dimension of the transcendental imagination of being that can neither be seen nor thought but that can be imagined and written about. Between the sensible things and the intelligible ideas, it is the originary *techne* that does not copy ideas from reality but allows their iterative reinvention in ever new things or allows the iterative rediscovery of ever new ideas in things.

Like archi-writing is not an originary form, *khora* is not a real space in which forms can be imprinted. *Khora* is the quasi-transcendental space in which the ever mobile originary technicity can take place. *Khora* lets transcendence itself be thought as originary technicity combining both iterative technics and imaginative art. By thinking through originary technicity Derrida can avoid several onto-theological postulates that have come to appear as dead ends, such as the explication of being through eternal structures (Platonic ideas, *causa sui*, or absolute spirit) and the explication of thinking through perfectly rational and conscious structures (the cogito or the transcendental ego). The originary technicity of the *khora* helps to articulate the opaque autonomous configurating of the experience that is neither the subject's nor the object's but that of their common emergence. It is not a positive thesis affirmed by Derrida but a kind of a necessary fiction inherited from the history of philosophy.

LIFE AS TEXT AND AUTOIMMUNIZATION

We have seen how the inquiry into technics has led to a reinterpretation of the classical philosophical subject-object-structure as prosthetics grafted onto the *anthropos* reinterpreted as originary supplementarity. We have studied writing as the prosthesis of memory and of thinking. Let us now see how Derrida deconstructs the subjective pole of the subject-object-structure. His undoing of the subject of philosophy (consciousness, transcendental ego, etc.) does not stop at a deconstruction of the subject of anthropology (human being) but aims at a more elementary thought of life that does not distinguish humans from other living beings but is on the contrary shared by all living beings. Derrida's deconstruction of the human-animal divide is not our main focus, but we can note that prostheticity is not proper to humankind but it belongs to all life. It is in Derrida's works on life that we find the first bases of what I have called "bio-technics": it is a thinking of life in terms of originary technicity that is valid for both human and nonhuman beings.

Derrida develops his most intense interpretation of biological life in *La vie la mort*, seminar given at the EHESS in 1975–1976 and only published in 2019. In this seminar he—following in the footsteps of Nietzsche's and Freud's encounters with Darwin and Heidegger's encounter with Uexküll (GA 1929–1930)—works on the fundamental concepts of the life sciences as formulated especially by the epistemologist Georges Canguilhem and the biologist François Jacob. He points out that while the biologists claim to reject heavy metaphysical concepts like "life" and speculative logics like "teleology," it is in reality difficult to find philosophers who have really used these concepts as clumsily as scientists claim. On the contrary, the biologists' concepts often remind him of Aristotle (*energeia*), Spinoza (*conatus*), Leibniz (*appetitus*), and especially Hegel (concept as life).[154] More importantly, while biologists believe that they formulate rigorous scientific concepts without postulating metaphysical essences, in reality they constantly presuppose an essence that Derrida sets about revealing. By unearthing and actually affirming such an essence of life—or rather of the living being (*le vivant*), as Jacob says—Derrida proceeds in an opposite manner to his readings of philosophy: while he deconstructs the essences postulated by philosophers, he *reconstructs* the essence of life that biologists deny.[155]

Derrida shows that the essence of life that Canguilhem and Jacob project is the "same" as the fundamental object of the human sciences that he studied in *Of Grammatology*: text. Methodologically it is important to see that by pointing out this "sameness," Derrida does not postulate a superior transcendent essence that both human and biological sciences would instantiate nor a model of rationality applicable everywhere, like the ones that semiotics and systems theory constructed at that time. He does not declare either "life" or "text" to be more originary than the other. He emphasizes that text is not the model of life and life is not the model of text, but both refer to one another in a circular movement in which the one helps in understanding the other but the one is also constantly differentiated from the other. Derrida emphasizes that it is the biologists themselves who resort to the textual model of biology: "What I will call the logic of the living, to use the title of Jacob's book, tends today, through the whole problematic of the *message*, the *code*, indeed the *genetic text*, to *decode* the living (le vivant)."[156] In François Jacob's *La logique du vivant*, the concept of *program* solves the problems attributed to finalism and teleology because it allows speaking of the internal finality of each individual without postulating an external fatality. What modern biology studies is heredity, and biologists claim that the study of heredity has at last become properly scientific when it has learned to speak of its object in terms of *code, information,* and *message*.[157] Thus the object of modern genetics is structured like a text.[158] The epistemologist Georges Canguilhem confirms this by affirming that "message, information, program, code, instruction, decoding—such are the new concepts of the knowledge of life,"[159] terms that have a solid "operative function" thanks to which they are truly concepts and not only metaphors. In his lecture, Derrida questions the biologists' assurance that they can maintain the strict distinction between a simple metaphor and an operatively functional concept. What is more important for us, however, is the firm choice of a linguistic vocabulary (instead of for example a physical or a chemical one) to ground biology as a scientific discipline.

Derrida has obviously no reason to contest the results of modern biology, but he is interested in the status of its concepts. He shows that contrary to what the biologists themselves believe, textuality determines the *essence* of biology. Furthermore, he shows that when Canguilhem and

Jacob describe the "message of life" in terms of writing and graphics,[160] they do not notice that interpreting textuality in terms of inscription and trace, instead of only information, makes a difference and actually leads to a deconstruction of certain fundamental distinctions in biology. One of the most important is Jacob's distinction between genetic memory and hereditary memory. According to Jacob, the genetic memory is an absolutely rigid program of heredity that can only reproduce the same, whereas the nervous memory of the brain is a supple memory of experiences that can transmit acquired characteristics: it can change.[161] Derrida shows that the biologist cannot maintain the distinction he draws in the first place. It turns out that the genetic memory does not reproduce itself in a similar way ad infinitum because contingent exterior events occasionally affect the process of reproduction. On the other hand, nervous memory too has a tendency to reproduce the same and to reject exterior events as contingencies. Thus both tend to maintain the same and both can also incorporate contingent exterior events—which does not mean that they would ultimately amount to the same but that their relation is more that of a differentiation than a direct opposition.[162]

Derrida suggests that this ambiguity can be explained if the genetic text is thought in terms of inscription and not only in terms of message. In *On Grammatology*, Derrida analyzes inscription as a supplement. In *La vie la mort*, he points out that Jacob often uses the term "supplement," most significantly when he describes the emergence of death and sexuality. While the biologist thinks that life consists in the reproduction of genetic programs in a way that is most manifest in simple bacteria that are said to reproduce themselves without sexuality and death, Derrida notices that in the biologist's own text both sexuality and death actually occur even on the bacterial level, albeit in a supplementary, auxiliary, accidental manner. In the case of the bacterium, this can result from a simple error in reproduction, but also in a more interesting way when a virus brings a fragment of the genetic program of another being into the bacterium. If the bacterium can integrate it into its own genetic program, this leads to a mutation, which can go on reproducing itself unless an environmental incompatibility prevents it. Derrida notices that from a biological point of view, the mixture of genetic programs is the very definition of sexuality, and the possibility of this kind of a mutation shows that the difference between sexuality and asexuality is ultimately not unambiguous. The same goes for death as, although the bacterium is said to divide itself instead of dying, the process of division can stop because of external circumstances.

Hence Derrida notes that both sexuality and death happen to the bacteria as accidents or supplements—in such a way that this supplementarity is ultimately not an accident striking the pure essence (reproduction) that then cannot take place, but it rather remains an originary possibility of the living.[163]

> The living being, insofar as it tends to reestablish the prior order or maintain the preexisting order, can thus never be a closed system, says Jacob: "It cannot stop absorbing food, ejecting waste-matter, or being constantly traversed by a current of matter and energy from outside. Without a constant flow of order, the organism disintegrates. Isolated, it dies. Every living being remains in a sense permanently plugged into . . . the general current which carries the universe towards disorder. It is a sort of local and transitory eddy which maintains organization and allows it to reproduce itself." All this might appear somewhat trivial, but I am quoting Jacob here only in order to underscore that this structural opening of every living system makes untenable those statements about bacteria not dying because death comes to them from the outside or about death in the proper sense of the term having to be inscribed in the organism, etc. It also makes untenable all the simple oppositions between inside and outside that subtend what the book says both about sexuality and mortality as accidents come from the outside that come to be inscribed within. Supplementarity is inscribed in the very definition of every system, every living or non-living system.[164]

By emphasizing the role of supplementarity, Derrida is also making a passing comment on the cybernetic theory first formulated by Norbert Wiener. Both the cybernetic system and the living system are defined through retroaction (feedback) in which the results of an action are reintroduced into the system in order to "oversee and redress the mechanism's tendency towards disorganisation,"[165] that is, its entropy. Instead of explaining this as a process of maintaining the same, Derrida stresses that the process lives off the heterogeneity introduced by the supplement, which functions as a pharmakon that enables self-maintaining, reproduction, and death.

Biology needs the textual model in order to explain reproduction— one could almost say that Derrida presents the ontico-ontological difference operative in biology as the difference between beings as textual things and being as reproduction. In biology, "An organism cannot be thought, as it were, in the present; it is not first of all the production of a present. It is first of all, in advance, what I will call an 'effect of reproduction.' It begins not with production but with reproduction. 'Reproduction,' says Jacob, 'represents [for the organism] both the beginning and the end, the cause and the aim.'"[166] The aim of the program—the essence of the living being—is to reproduce itself.[167] Reproduction goes together with selection: the living being must integrate novelty in its programs in order to ensure a better reproduction and dissemination, but excessive novelty results in nonviable monstrosities and dissemination at loss.[168] Derrida notes that since Marx, production has become a general master term. However, all forms of the production-of-something rely on a prior capacity for autoreproduction, which is the fundamental character of the living being. Autoreproduction is regularly explained by a comparison with the technical being (the cell is like a factory, the DNA is like a computer code, etc.) although at the same time these similarities are limited by the underlying difference between production and autoreproduction (the factory produces but does not reproduce itself, the program realizes a command but does not program itself[169]).

By linking the logic of supplementarity to a prior logic of reproducibility, Derrida shows the connection between the pharmakon and what has been called "originary technicity." In the contemporary culture, technicity is increasingly thought as a production following a program. We have seen that the idea of productivity also structures modern biology, which explains living beings through technical comparisons, especially comparisons with cybernetics. Like "writing," "technicity" is neither a transcendent idea nor a univocal form of rationality. More than a simple metaphor, it is an inevitable word for describing both living and textual beings that different sciences explain both in function of their similarity and in function of their dissimilarity. While Derrida's analysis of writing gave access to the differentiation presupposed by signification, his analysis of productivity and technicity gives access to the differentiation between real beings. It does not make sense to speak of a singular autopoietic being: life is conditioned by supplementarity that enables sexuality and death and can only take place in a plurality of living beings. Both living and technical reproduction can only take place in the plurality of what the biologists call a species, what the linguist calls a text, and what Derrida will finally analyze in terms of *Geschlecht*.

Derrida comes back to the analogy between biological and technical beings in a later text *Faith and Knowledge* (*Foi et savoir*, 1996). Within the confines of a short article he does not carry out a detailed reading of biological texts, but in reality his discussion of life in terms of autoimmunity implies a deconstructive move toward the theories of autopoiesis that had inspired both biology and cybernetics. In this article, life is not a distinctively biological term but a more general concept that includes "spiritual life." Neither biological nor spiritual life is an authentic plenitude that surges forth and propagates its potency, but on the contrary both appear as deferred origins that their prosthetic supplements make thinkable. In *Faith and Knowledge*, the prosthetic supplements take the form of technoscience and tele-technology, which complement Derrida's theory of writing. Let us see how *Faith and Knowledge* complements Derrida's interpretation of biological life and how it relates the apparently disparate questions of biological life and tele-technology.

A key concept introduced in *Faith and Knowledge* is *immunization/ autoimmunization*. To start with, autoimmunity is a biological term that designates a process in which a living thing, which normally maintains itself thanks to *immunitary* reactions that thwart alien influences (like a cell defends itself against a virus), develops also an *auto*immunitary reaction, "which consists for a living organism, as is well known and in short, of protecting itself against its self-protection by destroying its own immune system"[170] (this can happen in autoimmune diseases). Derrida generalizes this principle into a general logic of autoimmunization that "seems indispensable to us today for thinking the relations between faith and knowledge, religion and science, as well as duplicity of sources in general." The logic of autoimmunization is ambiguous and complex. Immunization first appears to be what a living organism does in order to protect what it is—"the indemnity of the body proper"—against foreign bodies. Derrida underlines, however, that in fact only the immunitary reaction *produces* the domain of the "unscatched" life, such that the "immune" does not actually precede the immunitary reaction but results from it. A living being is not first given (to itself) such that it would then immunize itself against foreign elements, but on the contrary only this twofold reaction of immunization—autoimmunization produces the difference between the domains of unscatched purity and dangerous alterity. In other words, the body proper and alterity are not two determined domains but they result from the constant balancing between immunization and autoimmunization. Immunitary reactions are always exposed to autoimmunitary reactions in a way that is not only beneficent to the regulation of the passage between the domains of proper and other, but the risky conflict between the two can also result in injury, death, and "radical evil."[171] This is how Derrida gives "life" to be thought, not as a given domain of pure life protected by more or less efficient immunitary reactions but as the effect of the activity of immunization—autoimmunization reactions.

This is because, as Derrida's reading of Jacob showed, life needs to be exposed to the exterior world. To put it bluntly, a living organism whose immunitary power would be so efficient that it would separate itself totally from all foreign influences would also cut itself off from its environment and choke itself to death (it could not reproduce itself without the supplements of nutriments, sexuality, and death). This is fundamentally why the immunitary reaction needs to be incomplete or blocked by autoimmunity.[172] However, autoimmunization is not a simple question of dosage, it is a conflict between opposing tendencies, one of which rejects everything foreign and the other attacking this protection in order to possess and use the foreign influence, which therefore appears in the full sense of both toxic and beneficent pharmakon. Derrida underlines the mechanical, non-intentional character of the entire process. In the case of biology (but also in the case of religion and technology), the foreign element appears to the organism not as a part of its life but as a dead supplement, an alien mechanism, a biological virus (or a tele-technological emission) that threatens to destabilize the organism's own functioning. But also the organism's reaction to the supplement is mechanic and automatic—it is not an intentional act but a spontaneous reaction of the organism: "The reaction to the machine is as automatic (and thus machinal) as life itself."[173]

> The relation between these two motions or these two sources is ineluctable and therefore automatic and mechanical between one which has the form of the machine (mechanization, automatization, machination or *mechane*) and the other, that of living spontaneity, of the *unscatched* property of life, that is to say, of another (claimed) self-determination. . . . Nothing in *common*, nothing immune, safe and sound, heilig and holy, nothing unscatched in the most autonomous living present without a risk of auto-immunity.[174]

Similar to how Derrida's reading of Jacob included a reference to cybernetics, his work on autoimmunization can be further clarified by comparison with the theories of autopoiesis presented first by Maturana and Varela and that were later developed into systems theory by Niklas Luhmann (who actually incorporates elements of deconstruction into systems theory).[175] Maturana and Varela concentrated on biological systems while Luhmann extends the idea of system into all kinds of material realities, especially psychological and social organizations. Following Gregory Bateson's definition, autopoietic systems are systems that recursively produce their own operations:

> Autopoietic systems are products of their own operations. They have properties such as dynamic stability and operational closure. They are not goal-oriented systems. They maintain their autopoietic organization of self-reproduction as long as it is possible to do so. Their problem is to find operations that can be connected to the present state of the system. In this sense they are what Heinz von Foerster calls nontrivial machines or historical machines. They use self-referential operations to refer to their present state to decide what to do next. They are unreliable machines, to be distinguished from trivial machines that use fixed programs to transform inputs into outputs. Autopoietic systems rely not on tight coupling but on loose coupling to move from one state to the next, and this makes it possible to evolve into different structural types according to random links between the system and its environment.[176]

Autopoietic systems are "machines" that function along an operational closure in which the system's operations apply recursively to the system itself such that, its environment permitting, it can go on operating in the same way or, facing environmental challenges, it must modify its operation in function of them, thereby "learning" from its situation.[177] In the last instance it is the recursive operation itself that produces the difference between the system and its environment. Unlike simple machines, autopoietic systems do not simply reproduce a code, but they relate to external information and adapt their operations to it. Information is not just something presented to the system by the environment (the environment as a whole is too rich and chaotic) but it results from selection such that, according to Gregory Bateson's definition, information is "a difference that makes a difference."[178]

We see that both deconstruction and systems theory apply a theoretical inspiration from biology to all kinds of psychological and social systems, including religion and law. Both theories of autopoiesis and autoimmunization describe the becoming of the organism in function of an external event that makes a difference and destabilizes the system such that the system is pushed to seek a new stability in a movement that both preserves and alters the system's functioning, giving it a history. However, while systems theory aims to explain how systems maintain themselves against difference, deconstruction wants to show how the discovery of a difference enables their deconstruction. If autopoiesis aims to protect the system, "auto-immunity is an aporia: the very thing that aims to protect us is the thing that destroys us."[179] Derrida would not use a word such as *autopoiesis* that suggests that self is an object of a production, its constituted starting point and result. In his account of autoimmunization, life reacts to both exterior factors and to itself such that both are not only reconfigured but truly result from this constant passage of immunitary boundaries. In an autopoietic system, what a living being properly is results from its past and from its reactions to present environmental factors, such that recursivity makes the system evolve toward future states. Contrary to this, Derrida emphasizes that the past inheritance too is an alien factor to which the system reacts. A being's past is both something that constitutes it and an alien inheritance that destroys it, although an active deconstructive relation to the past can reveal in it the chance of what is to come. Both systems theory and deconstruction pay attention to the external difference that triggers autopoietic recursion and reactions of immunization—autoimmunization, but for systems theory it is information absorbed by the system's recursive movement whereas for deconstruction it is a supplement or a prosthesis that may contribute to the living system or undo it. This is why, although theories of autopoiesis and *différance* converge because both regard living organisms in function of external difference, they ultimately relate to it in opposed ways. As Cary Wolfe puts it, autopoiesis and autoimmunization are complementary theories in that "Derrida and Luhmann approach many of the same questions and articulate many of the same formal dynamics of meaning (as self-reference, iterability, recursivity, and so on), but they do so from diametrically opposed directions."[180] Gunther Teubner notes that while Luhmann aims to undo the paradox presented by the environment, Derrida wishes to uncover antinomies and paradoxes, and in this sense the two approaches are, as Teubner puts it, "paranoiac" of one another so that Luhmann's idea of system is "Derrida's nightmare" and Derrida's idea of gift is "Luhmann's redemption."[181]

Life Death and *Faith and Knowledge* contain the "essence of life" that Derrida finds in contemporary biological texts. It is not something he invents from his own conceptual necessities. Instead, his deconstructive reading shows that it is already tacitly presupposed by the biologists even though they do not intend to rely on an essence. The main addressee of this reading may still be philosophy itself, for biology does not really care for philosophical essences. Similar to how the discovery of the logic of writing had a salutary effect on the philosophical idea of logos, the discovery of the biological essence of life has a salutary deconstructive effect on the philosophical idea of life: it makes the *animal rationale* tremble. Methodologically it is important to notice that the theories of writing and of life result from a deconstruction of contemporary sciences. The deconstruction does not postulate a new essence of the human being instead of the ancient one, as if a writing animal could replace the rational one. Deconstruction shows the impossibility of such an essence, but it also lets us see in counter-relief what technical life looks like in our time. The idea is not to say that this image is just an illusion. The idea is just to say that it cannot have the status of an *essence* but that it is nonetheless the *unavoidable image*—maybe the specter of an essence.

Some features of this unavoidable image are of interest to us here.

Firstly, life is not a positive given, like an expansive self-expressive force. It results from the action of external factors against which life reacts. Life is literally this immunization against the alien and autoimmunization against its own immunization such that external supplements could also enrich it.

Secondly, life is not a proto-subject defined by anything like a consciousness or even an intentionality. It is a mechanism that reacts to external factors—which also act on life mechanically. The relation between life and its external supplements happens as their limit. At this frontier, one element is constituted as the living being and the other one as its dead supplement (even if it may be alive for itself). Both appear as mechanical actions in relation to one another. The living mechanism absorbs or rejects the supplement, according to its program. The supplement appears to the living as a "text" that it can "read" (the supplement "makes sense," it can be used or must be avoided) or that it can reject as indecipherable, inappropriable, alien. When the supplementary code is absorbed, it continues to realize its program and starts to affect, infect, and contaminate its host—like a fragment of alien DNA brought into a bacterium by a virus that can add itself to the bacterial DNA or like a text read by a living mind can affect, nourish, infect, or contaminate this mind. Both the supplement and the living host act mechanically.

This is what prostheticity means: it is "like" the relation of a living host to the supplement (that counts as dead—to the host). This relation emerges as the deconstruction of the subject-object-relation in which the conscious subject can contemplate its object because the latter remains at an objective distance from where it presents itself to evidence and certainty. In a prosthetic relation there is no object but a supplement that is always too far or too close to be contemplated or even perceived: its mode of action is imperceptible infection. There is no subject either, only life that is supplemented and sometimes poisoned by its prostheses that are too close to be noticed anymore. This is not an intentional consciousness, this is the action of one mechanism ("writing") on another one ("life"). Life is therefore interpreted using a technical metaphor. But also, inversely, technics is reinterpreted as life: technics is a mechanism that follows a program, but this program also selects suitable supplements and can also be reprogrammed by them. "Originary technicity" is the technicity of life, the technicity of writing, and the technicity of prosthetic life, in which life supplements itself with new text fragments. Originary technicity is nothing other than the incessant circulation of these different versions of originary technicity.

This is how originary technicity functions on the level of life—which is always both nonhuman and human, biological and spiritual. Let us now see how life relates to its technical supplements in the case of human life specifically, where meaning and coexistence are at stake.

FAITH IN TELE-TECHNOLOGY

Especially in *Faith and Knowledge* and *Echographies of Television*, Derrida provides still another contribution to his potential theory of technics that we have examined in terms of prosthetics and writing: a study of *tele-technology*. By this he means on the one hand the epochal phenomenon created by the factual technological systems of the radio, the telephone, the television, the Internet (today we could add the entire digitalization of life): these tele-technologies now operate in all meaning-making (religion and science were studied in *Faith and Knowledge*, and politics was discussed in *Echographies of Television*) and extend to a global scale. But on the other hand, unlike Bernard Stiegler, who is his interlocutor in *Echographies of*

Television, Derrida emphasizes that tele-technicity is not unique to our epoch but a general feature of all technics of writing in any epoch. The so-called new technologies simply exacerbate certain features and bring them strikingly to the fore. The notion of tele-technology complements the notion of writing with considerations of a phenomenological variety. A tele-technological apparatus deconstructs the notion of presence that is particularly important for phenomenology. Tele-technology seems to reproduce a presence, but it also makes evident that this presence is fundamentally what Derrida calls an *artifactual construction* whose fundamental way of being is that of proximity betraying a distance.[182] The modern user of tele-technology does not really fall into the trap of false presence, but in order to use a tele-technological device one must agree to the illusion created by the machine and to play as if the presence was really present. The political problem arises from the difficulty of monitoring the conditions of the constitution of the artifactual tele-technological scene.

The reconceptualization of the phenomenon in terms of tele-technology goes together with a deconstruction of Dasein in terms of spectrality. Since *Specters of Marx*, spectrality designates existence—Derrida plays with describing is as a *spiritual life*—which is constituted in function of distant apparitions. They can be appearances of other, definitively distant Dasein (like Hamlet's dead father's ghost described in *Specters of Marx*), but their mode of appearing is tele-technological, in that they act from distance. Let us now see how Derrida describes tele-technology and how it constitutes "hauntological" existence.

What does Derrida mean by tele-technologies? In *Faith and Knowledge*, Derrida associated tele-technologies with the problem of autoimmunization: tele-technologies provided an interpretation of the difference that triggers recursion, not as just an environmental factor interpreted as information but as a technical prosthesis interpreted in terms of inscription. However, in this text Derrida interprets life in a more general sense, and indeed the term *tele-technology* qualifies the kind of life that can be called "spiritual." *Faith and Knowledge* was originally a conference on religion, and this is why Derrida refers faith first of all to the question of religion, especially to monotheisms, and among them mainly to Christianity that spreads out in a vast movement of "globalatinization (essentially Christian, to be sure)."[183] He refers knowledge to technosciences and these further to "tele-technologies," asking what today links religion with technics.[184] On the one hand, tele-technologies resulting from technoscience appear today as the prosthetic difference that triggers the immunitary reactions of religion: they are both the alien influence that the immunitary reaction of religion wants to reject—and the pharmakon that religion wants to appropriate to its own use.

> Religion today allies itself with tele-technoscience, to which it reacts with all its forces. It is, *on the one hand*, globalization; it produces, weds, exploits the capital and knowledge of tele-mediatization. . . . But, *on the other hand*, it reacts immediately, *simultaneously*, declaring war against that which gives it this new power only at the cost of dislodging it from all its proper places, *in truth from place itself, from the taking place of its truth*. It conducts a terrible war against that which protects it only by threatening it, according to this double and contradictory structure: immunitary and auto-immunitary. The relation between these two motions or these two sources is ineluctable, and therefore automatic and mechanical, between one which has the form of the machine (mechanization, automatization, machination or *mechane*), and the other, that of living spontaneity, of the *unscathed* property of life, that is to say, of another (claimed) self-determination.[185]

But on the other hand, Derrida draws attention to the common feature shared by religion and technoscience despite their apparent opposition, namely that both are fundamentally acts of faith. This is more evident in the case of religion, but this is also the case in technoscience insofar as its object is not given to certainty, such as supposedly occurs in classical science, but is constituted by technical means such that the certainty of results is conditioned by technical constitution of its object that has to be relevant and *credible*. "In this very place, knowledge and faith, technoscience ('capitalist' and fiduciary) and belief, credit, trustworthiness, the act of faith will always have made common cause, bound to one another by the band of their opposition."[186] Tele-technology (as media) is the pharmakon from which religion protects itself *and* by which it wants to be equipped, but tele-technology (as scientific instrument) is also the indispensable pharmakon of contemporary science.

Tele-technologies are surely a major epochal feature. In *Faith and Knowledge* tele-technology is associated with religious globalatinization, but in *Echographies of Television* tele-technology is shown to have a much more general signification, because it structures all (homely and political, intimate and public) forms of contemporary globalized life. As Derrida says in *Echographies of Television*, tele-technical equipment has penetrated all common and private spheres, going as far as reconstituting the very sense of presence, which is less and less "physical" and more and more reconstructed by artificial, "artifactual," spectralizing tele-technological means. Understanding and deconstructing this situation is certainly one of the major political challenges today for Derrida:

> What I would like to convey to this illusionless request [to have the right to reconstitute the conditions under which one is surrounded by teletechnologies at home] is the paradox of a task or a watchword: perhaps it is necessary to fight, today, *not against* teletechnologies, television, radio, e-mail or the Internet but, on the contrary, so that the development of these media will make more room for the norms that a number of citizens would be well within their rights to propose. . . . Who has the right of inspection over whom?[187]

Tele-technologies make obvious the distance at the heart of presence. In classical science, presence was the guarantee of objectivity and certainty. Tele-technologies cannot reach the evidence attributed to presence because tele-technological presence is always "artifactual," that is, constructed (and this is why it invites an uncovering and deconstructing of the technological, economical, political, religious, and in general ideological conditions of its construction). If "spiritual" life's relation to its prostheses thus cannot be certainty, it is, according to Derrida, *faith*. Here faith is no longer a religious attitude, but it is the act of trust that corresponds to the spectrality of whatever presents itself *in absence* and that is required by all thinking mediated by tele-technologies: religion, science (that appears therefore as technoscience, i.e., as science that needs technological mediation in the constitution of its objects), and the entire social and political sphere (that appears through and even happens as media). Whenever the object is not present (and all prosthetic presence is spectral and artifactual), life cannot be certain of it but it can only trust it, believe in it (and if necessary inquire into the grounds of its belief).

This faith, trust, or "credence" must be thought as a counterpart of the *promise* first thematized by Nietzsche in *Genealogy of Morals*, where he asks how human beings make themselves a memory. Nietzsche's book was very important for Foucault as well, who developed Nietzsche's idea that memory is made by imposing painful experiences on bodies, and it is inscribed on bodies and hearts by torture. Derrida pays attention to another feature of Nietzsche's text, namely his definition of the human being as a promising animal.[188] Promise is the human way of both opening and controlling the future. The future opened by a promise is not produced by any natural or logical necessity but only by the promise itself. The promise promises to be trustworthy, it promises that one can count on it—but as no necessity guarantees this for certain, one can only trust it and put one's faith in it.[189] Tele-technological writing is a promise, a promise of keeping the promise, a promise of iterability of the promise, that speaks to trust and faith. Hence, there is

> no discourse or address of the other without the possibility of an elementary promise. Perjury and broken promises require the same possibility. No promise, therefore, without the promise of a confirmation of the yes. This yes will have implied and will always imply the trustworthiness and fidelity of a faith. No faith, therefore, nor future without everything technical, automatic, machine-like supposed by iterability. In this sense, the technical is the possibility of faith, indeed its very chance.[190]

The one who makes a promise says, You can count on me. The tele-technological prosthesis includes such a promise, often of some kind of truth or value, but above all, of trustworthiness itself. By its very structure it is the promise of a technological control of a calculable, iterable future. The living organism to whom such a promise is made can in the last instance only trust the promise and give credit to it, as there is no definitive way of being assured of the tenability of the promise except by faith in it. The inheritance of a living organism too is a promise, a promise given in the past by those who have already passed—but a promise that, because it is a promise, binds the future in which the promise would be fulfilled. "Of a discourse to come—on the to-come and repetition. Axiom: no to-come without heritage and the possibility of repeating. No to-come without some sort of iterability, at least in the form of a covenant with oneself and confirmation of the originary yes. No to-come without some sort of messianic memory and promise, of a messianicity older than all religion, more originary than all messianism."[191]

The relation between autoimmunization and tele-technology must therefore be complemented with the logic of spectrality and messianicity.

Before that, let us just note what the analysis of tele-technology brings to the notion of life. Derrida's considerations on life show why he could not make a theory of "technical humanity" even though his theory is thoroughly a theory of originary technicity. He does not present an Enlightenment-inspired figure of the human being understood as a subject controlling its objects. Instead, he describes a form of life that is always already infected and contaminated by its prostheses, without which it could not live although they also introduce death into it. In *Life Death*, Derrida showed how the essence of life is the "same" as the essence of writing: supplementarity. In *Faith and Knowledge*, he showed how the essence of life is the "same" as the essence of the tele-technologies: a process of immunization/autoimmunization. Between the two texts, the thinking of life and the thinking of writing are enriched. The mechanism of supplementarity, which accounted for the machinic iteration of the prosthesis, turns out to be the "same" as the mechanism of immunization/autoimmunization that accounts for the living organism itself. Both are machinic and "nonconscious." The alien element that triggers the immune-autoimmune reaction is like a virus that cannot be perceived because it is always too far or too close to be seen: it infects and contaminates the living being and starts to multiply itself imperceptibly in its host. Tele-technological writing functions in a viral manner as well: it presents an artifactual stage that appears objective, but the technological conditions of this stage are always too far and too close to be perceived and this is why they act in us imperceptibly as well. With the double analysis of life and tele-technology, which are and are not the same, Derrida emphasizes the very way in which the prosthetic life functions. A prosthesis is not an object, and this is why it cannot be known and controlled as theoretical objects are. It is a graft that acts without being noticed, it infects and contaminates, it acts mechanically in the organism, which in turn reacts to it mechanically.

HAUNTING DASEIN

The anthropological question of the human, even when it is reformulated as the question of a form of existence de/formed by technics and resistant to it, is not what Derrida is aiming at. In his later texts he makes a different contribution to the question of human subjectivity, albeit without using the term. Most of the time he speaks simply of "life" and he studies it as existence with (human and sometimes animal) others in relations of mourning, friendship, ethics, and politics. Many of these studies could be read as comments on existential phenomenology as developed by Heidegger, Levinas, and Patocka and continued by unclassifiable authors such as Blanchot and Nancy. All of these authors are critical of anthropological interpretations of the human being and they develop the notion of existence such as it was first brought to philosophy through Heidegger's notion of Dasein. Derrida does *not* say that he would build a theory of Dasein as existence and being-with, on the contrary, he has written a series of very critical essays in which he shows why Heidegger's thinking of human existence, marked as it is by its *Geschlecht*, is so problematic.[192] However, his studies of existence—or rather "life"—make a lot of sense when they are read as new perspectives on the question of Dasein that the analytics of Dasein had merely opened up. In what follows, my focus remains the question of technics, so I will not explore Derrida's complex debts to Heidegger's Dasein—which would require a very long and multifaceted work—but I still want to point to the question of Dasein as an important source for Derrida's "spectrality" or "hauntology" that I will discuss in what follows.

In *Specters of Marx* (*Spectres de Marx*, 1993) Derrida develops a theory of spectrality that could indeed be read as a close deconstruction of Heidegger's notion of Dasein. Like Dasein, spectrality does not answer the anthropological question "*What* is the human being?" but the question "*Who* is it?" that delimits the domain of existential analytics.[193] Like Dasein, "life" studied here by Derrida cannot be grasped as a substance but only in terms of complex movements of temporalizing. The interpretation of existence as temporalizing is fundamentally inspired by Heidegger, but Derrida also turns against him and corrects his work in important ways, as Bernard Stiegler in particular has pointed out.[194] Derrida shows that Heidegger does not notice, or take seriously enough, the necessity of thinking of all ecstasies of temporality in terms of prostheses—which Derrida thinks less in terms of the natural time measured by clocks, which was rejected in *Being and Time*, than in terms of writing, which provides the material support for historical time. In *Specters of Marx* and in several texts published at the same period (*Echographies of Television, Faith and Knowledge*) Derrida examines the phenomenalization of writing in terms of tele-technologies. What interests us first here is the way in which one of the most important ways of being haunted is being touched by such techniques of distance.

Specters are generally people separated from us by time; what we are concerned with is their way of touching us by tele-technological means. Specters have been with Derrida since his early work. For example, phantoms are mentioned in "Plato's Pharmacy," where they accompany the *pharmakeus*, magician, who uses the ambiguous magical effects of the pharmakon; in *Given Time*, where they describe the impossible phenomenality of the gift that can only exist if, like a phantom, it has no present phenomenon[195]; and in *Of Hospitality*, where "the foreign guest appears like a ghost."[196] However, Derrida develops an entire theory of spectrality only in *Specters of Marx*, where spectrality is a form of phenomenality that allows the study of transcendental conditions of experience below intentional consciousness. It opens the sphere of a "hauntology" whose logic is "larger and more powerful than an ontology or a thinking of being" and that "harbors within itself . . . eschathology and teleology themselves. . . . After the end of history the spirit comes by *coming back* [*revenant*]."[197] Here I will only examine the subjectivity presupposed by "hauntology," which is not (an active) haunting subject but rather a (receptive) subject of haunting, a subject to whom specters appear. One could almost say that Derrida (like Stirner, whom he quotes) defines subjectivity by haunting spectrality: "Therefore 'I am' would mean 'I am haunted.'"[198] On top of these theoretical considerations, the declared primary aim of *Specters of Marx* is ethical, to "*learn to live finally*," to "learn to live with ghosts"[199] because "there is no Dasein of the specter, but there is no Dasein without the uncanniness, without the strange familiarity (*Unheimlichkeit*) of some specter."[200]

> It is necessary to speak *of the* ghost, indeed *to the* ghost and *with* it, from the moment that no ethics, no politics, whether revolutionary or not, seems possible and thinkable and *just* that does not recognise in its principle the respect for those others who are no longer or for those others who are not yet *there*, presently living, whether they are already dead or not yet born. No justice . . . seems possible or thinkable without the principle of some *responsibility*, beyond all living present . . . without this *non-contemporaneity with itself of the living present*.[201]

Spectrality articulates this non-contemporaneity with itself of the living present. The non-contemporaneity with itself of the present opens above all the question of historicity, that Hegel, Marx, Nietzsche, and Heidegger thought in terms of the teleological figure of the end of history,[202] to which Derrida opposes a whole crowd of specters that he first rounds up from Shakespeare's *Hamlet* and Marx and Engels's *The Communist Manifesto*,[203] and that introduce another thinking of historicity. Leaving aside the question of historicity proper, I will now just limit myself to its experiential core. Even so, the figure of the specter bundles together many senses.

First, like the gift, "the spectral *is not* . . . it is neither substance, nor essence, nor existence, *is never present as such.*"[204] This already follows from the most ordinary acceptation of the term "specter," the phantom of a dead person, that Derrida invokes as an exemplary phenomenon of a *non-phenomenal apparition*: the specter is the presence of somebody who is definitively absent because absolutely separated from us by death, either its death or our own, for our death is what ultimately separates us from those who are not yet born. More importantly for us, spectrality characterizes what Derrida has earlier called writing. Structurally, as Derrida said in "Signature Event Context," if we can read a text, it must be possible that its signification is lost and its author is dead. Spectrality is the phenomenological interpretation of this structure. Appearing without presence is the mode of being of writing (hence, of technics): it gives itself without proper Dasein, appearing as the specter of an absent thing (signification, author) from which it draws its sense. Spectrality explains the *tele-technological* character of all writing because it is a technics that allows the writer to touch us from a distance, from the inaccessible beyond inhabited only by specters without Dasein. The specter who haunts us beckons from afar, presenting itself without presence, pre/ab/senting itself in (psychic or material) writing that conjures up the spectral origin.

If "I am" means "I am haunted," spectrality delimits who I am. Most of my memory does not originate from my own conscious experiences but from others' experiences "written" in me throughout all my education and life. I am not a closed unity, I am the space of my own and alien experiences that have been inscribed in my memory, an archive of my own and strange records deposed in me, secretly delimiting my possibilities and impossibilities, suggesting who I am and what I must do. I am what I inherit—not just what is left for me but what I make of it.

> Inheritance is never a given, it is always a task. . . . To be . . . means to inherit.[205]

> There is no inheritance without a call for responsibility. An inheritance is always the reaffirmation of a debt, but a critical, selective, and filtering reaffirmation, which is why we distinguish several spirits.[206]

Spectrality brings forth the distance and the inaccessibility of the origin of the experiences that make us. Our past is not present to us, but we are haunted by it and we are (with) those who haunt us. This is why existence is always alien to itself, uncanny, *unheimlich*. Whatever is "inscribed" in our conscious or unconscious memory is written in strange words, languages, and characters by the dead. Furthermore, if the absent and even dead thing can haunt us in the first place, it is because it is constituted tele-technically. This is really an aporia:

> Without singularity, there is no inheritance. Inheritance institutes our own singularity on the basis of an other who precedes us and whose past remains irreducible. The other, the specter of the other regards us, concerns us: not in an accessory way, but within our own identity. From this point of view, technics is . . . a threat to inheritance. Now, at the same time, the opposite is also true: without the possibility of repetition, or reprise, of iterability, and therefore, without the phenomenon and the possibility of technics, there would not be inheritance either. There is no inheritance without technics. Inheritance therefore stands in a relation of tension to technics. A pure technics destroys inheritance, but without technics, there is no inheritance. This is why inheritance is such a problematic and ultimately aporetic thing.[207]

Second, spectrality describes phenomenality insofar as a phenomenon, even when it abstains from appearing here and now, is a phenomenon *to somebody*. Not just any archive makes us attentive, most are simply stocked in memory beyond attention, but whatever haunts us, somehow visits us,[208] speaks *to us*, calls us, requires something of us. Derrida's prime example of a specter is Hamlet's father's ghost who is a phantom par excellence who demands attention: it beckons, calls to terrible deeds, and finally drives Hamlet to madness. Derrida emphasizes the distance over which the specter affects us by calling it the "*visor effect*: we do not see who looks at us. . . . This spectral *someone other looks at us*, we feel ourselves being looked at by it, outside of any synchrony. . . . To feel ourselves seen by a look that it is impossible to cross, that is the visor effect on the basis of which we inherit from the law."[209] The ghost looking through a visor is not seen, it is a presence that is not really "there" but that beckons beyond absence so that the person who sees it feels moved, changed, pushed to action. Although its phenomenality is without presence, its effect is very real, even fatal and devastating as it is for Hamlet.

Third, life with specters is therefore a matter of ethics, politics, and justice to be done to specters. Because the specter calls for action it actually comes from the future that it wants someone to put right. Hamlet's father's ghost demands Hamlet avenge him and this is how his inheritance not only falls from the past but must be reaffirmed by choosing it.[210] However, beyond Hamlet who is fatally bound to his parents' past, Derrida looks for still another type of justice, "a justice that one day, a day belonging no longer to history, a quasi-messianic day, would finally be removed from the fatality of vengeance."[211] For Derrida, such a justice has everything to do with inherited texts (whether material or psychic inscriptions). One does justice to them when one reads them in an open and deconstructive manner, without simply repeating them as such but by reading them freely and critically. Justice also requires openness to the unexpected that this deconstruction can liberate but that can also come independently of any horizon of expectations set by past texts, for example when a stranger arrives and calls for "justice as incalculability of the gift and singularity of the an-economic ex-position to others."[212] Such a justice would discover the time to come not as the gathering to an end, which is the only kind of future that the thinkers of the end of history and even Heidegger can see. It would welcome a "desert-like," "chaotic," and "abyssal" messianism that arrives as justice rendered to the singularity of the other,[213] not by giving what we can to the other but by giving to the other what we do *not* have and what only belongs to the other. Ultimately, we are not to realize the past specters' will but to let those to come discover their will, life, sense. Ethics is thus less a matter of our emancipation but a matter of the emancipation of (specters) to come. This is why "one can never distinguish between the future-to-come and the coming-back of a specter."[214] What is to come comes also from the past, but it comes not in terms of our expectations but in its own, hitherto unknown terms.

Fourth, this is why Derrida's analysis of specters (of the past) is complemented by an analysis of messianicity without messianism. This important figure forms the climax of Derrida's thinking of temporality and historicity, and it also concludes the meditations on *khora* that were Derrida's earlier answer to ontology.

> What remains irreducible to any deconstruction, what remains
> as undeconstructible as the possibility itself of deconstruction

is, perhaps, a certain experience of the emancipatory promise; it is perhaps even the formality of a structural messianism, a messianism without religion, even a messianic without messianism, an idea of justice—which we distinguish from law and even from human rights—and an idea of democracy—which we distinguish from its current concept and from its determined predicates today.[215]

Messianicity is the dimension of the future which remains to come (*à venir*) in the demanding sense of the word: not future as the expected consequence of the past or the present but to come as the unexpected surprise. Such an unexpected event is not the eternal return of the self-same ghosts, nor the anticipated arrival of the religious figure of the Messiah whose advent puts an end to history. It is the possibility of the coming of the totally other—the event of the *impossible*, that is, of what now, in the light of present expectations, appears impossible. Only as messianicity without messianism is the future truly open, to come (*à venir*). The future of invention is also the coming of the impossible, but one does not invent the other, one welcomes it, and this is messianicity without messianism. This messianicity without sense of history is the temporal dimension of the materialism without substance of the *khora*: together they open experience to its distant, absent, non-phenomenal conditions, which on the one hand withdraw from experience and are nothing and on the other hand are a dimension out of which surprises may come.

> One may deem strange, strangely familiar and inhospitable at the same time (*unheimlich*, uncanny), this figure of absolute hospitality whose promise one would choose to entrust to an experience that is so impossible, so unsure in its indigence, to a quasi-"messianism" so anxious, fragile, and impoverished, to an always presupposed "messianism," to a quasi-transcendental "messianism" that also has such an obstinate interest in materialism without substance: a materialism of the *khora* for despairing "messianism."
>
> [Also] at stake, indissociably, is the differantial deployment of techne, of techno-science or tele-technology.[216]

The existential sense of messianicity without messianism can best be clarified in the context of Derrida's ethics and politics, where it opens the question of the relation to the totally other, to the other who calls for hospitality, for a "'yes' to the arrivant(e), the 'come' to the future that cannot be anticipated."²¹⁷ Derrida lays out the problem of hospitality in a lecture published as *Of Hospitality*, where he points at Plato's use of the figure of the Stranger in *Sophist* and Sophocles's treatment of the problem of hospitality in *Oedipus at Colonus* (the references to antiquity should not lead one astray: the question of hospitality is more topical than ever today). In *Oedipus at Colonus*, Oedipus, deposed and blinded, arrives to Athens and seeks hospitality. The king of Athens, Theseus, weighs up whether hospitality should be accorded: Should it be denied to this criminal who has committed several sacrileges or should it be accorded in the name of the divine duty of hospitality dictated by Zeus himself? From the Greek point of view, the stranger must be welcomed by asking where it comes from, who it is, and as questions continue, how its (ethical, juridical) case should be judged within the limits of this city. But this commandment is conditioned by a more general question of absolute hospitality toward the totally other, which is, as Levinas has underlined, a total welcome without question and condition.²¹⁸ Derrida shows how these two figures, the stranger and the other, condition one another, in ethics where the other must be heard but cannot be totally heard and as meant, and especially in justice, where it is at the same time impossible for the judge to solve the other's case in a way that is totally lawful (knowing everything that pertains to the deed and therefore able to calculate its lawful consequences), and nevertheless justice is done only if the judge solves the other's case even if the ultimate groundlessness of the judgment makes it a moment of folly.²¹⁹ While after Heidegger, a number of contemporary philosophers refer the question of temporality ultimately to the event, Derrida speaks about the future in terms of openness to the event of the coming of the totally other and in terms of a just welcome of whoever comes. Future also means dealing with past specters who interfere and ask for justice. Derrida's idea of historicity is displayed through the figures of ghosts, messiahs, and hosts of all kinds, and it spans between these dimensions: a spectral past that beckons from afar, a messianic future that calls for ethical work.

Now after this lengthy reminder of the different aspects of the questions of spectrality and hospitality, what really interests us is the way in which tele-technology conditions it. As Derrida said, "[Also] at stake, indissociably, is the differantial deployment of techne, of techno-science or tele-technology." What is technics, if "a spectral spiritualization is at work in any techne"?[220] And especially what are the media that are indissociably linked to their technological means? "The medium of the media themselves (news, the press, telecommunications, tele-techno-discursivity, techno-tele-iconicity, that which in general assures and determines the spacing of public space, the very possibility of the res publica and the phenomenality of the political) this element is neither living nor dead, present nor absent: it spectralizes. . . . It requires . . . hauntology."[221]

Despite appearances, tele-technologies do not refer here to specific technologies associated with the contemporary epoch (radio and television, telephone, Internet, etc.) but to a consideration of any technologies with regard to their "telepathic effect." As such, contemporary tele-technologies combine all aspects of technics that we have studied so far: they are fundamentally *writing*, they are by nature *pharmaka*, and they complement our lives *prosthetically*. Spectrality adds a phenomenological viewpoint of technics. First of all, the spectrality of tele-technologies refers obviously to the fact that those who appear to us tele-technologically call from afar, they are not present here and now, we cannot touch them, and most of the time we cannot even speak with them. Nonetheless, through diverse technics of writing, they "touch" and "move" us as though through a "visor." More importantly, Derrida draws attention to the technics themselves, not to people that they are supposed to connect with one another. He analyzes tele-technologies as *technics of contact* where the truth of the contact is a *separation*. Derrida describes the technicity of contact especially in *On Touching: Jean-Luc Nancy*,[222] where he shows how contact, which promises to provide maximal proximity, is really the opening of separation and distance: touching itself is a technics that makes both distance and proximity appear. This is true of all touching, but contemporary tele-technologies make this more obvious than ever. As Derrida shows in his essay "Artifactualities" included in *Echographies of Television*, tele-technologies like television and email create artificial conditions of space and place, they make up the "real time" that is not at all natural but completely artificial.[223] Although this artificial presence gives itself for living presence, it is really a set that is the specter of another dimension, that does not "really exist" (have Dasein) but that still projects as its origin the political and economic conditions under which we live. This is not "evil" as such, this is the condition of (political) community in its good and evil aspects. But this becomes preoccupying if it is so oppressive that we lose our means of deconstructing it, so dense that there is no space to welcome unexpected events. This is true of all technics, material or not, but with contemporary technologies this has become obvious and also more politically sensitive than ever. Today it is obvious that technological infrastructure is a political question.

For Derrida, human existence is temporalization. It is not just abstract temporality but concrete historicity in which distant ghosts touch us from afar, as a past that haunts us and a to come that surprises us. This touching of distance is always mediated by technics. It may be concretely a tele-technological device, an archive, or an image of a world to come, but whether it appears tangible or not, it is never "really" present. This is why our relation to it is not evidence but "faith" (into which we shall soon look in more detail). Faith is our relation to whatever is counter-timely and fundamentally futural. We cannot truly *know* it because it is not present, really present as such, like the phenomenon of a phenomenologist's dream; what gives itself as present is really a tele-technological text that refers elsewhere and touches from afar, and our relation to this "afar" is that of faith. We *believe* that the sign reports the distant thing correctly, but we cannot go and verify it, no more than as in Kant's first *Critique* we can go and see if experience really corresponds to the *Ding an sich*. This is not faith *in* something, it is faith as a dimension of temporalization.

As reached through faith, all technologies are fundamentally tele-technologies, that is, technics of distance. Writing was the first of the tele-technologies that we examined; if Derrida's studies of writing express the character of inscription and archiving, his studies of tele-technology express its particular phenomenality. Bernard Stiegler will develop this further, as we shall see in the last part of this book. Noting that arche-writing cannot be reduced to technicity,[224] he criticizes Derrida for not having paid enough attention to concrete technological devices that incarnate and even create these structures. But Derrida uses concrete technologies only as metaphors. What he really aims to show is how originary technicity constitutes the transcendental framework of experience. What touches from afar is not something to which our intention stretches without really reaching—it is the prosthesis that parasites and infects us whether we notice it or not.

We have seen that Derrida does not write an anthropology in the sense of a theory of what the human being is. The human has no fixed definition because its very definition is the originary technicity that undoes and reformulates every definition. Whatever form prevailing technicity imposes on the human is also immediately hollowed out by what withdraws from it, what is unconscious, inaccessible, resistant. Derrida instead studies the transcendental conditions of existence. Like Heidegger, he thinks of human existence as temporalization, but he criticizes Heidegger for still understanding the core of this existence as an authentic zone. Derrida always thinks of it from originary technicity of which any impression of authenticity is but a projection. Against Heidegger but with Levinas, Derrida also thinks of existence as being with the other, with others, *Mitsein* with others who were there before me. But unlike Levinas, he thinks of being-with in function of technics that operates the contact included in this "with": it connects only by distancing.

Rethinking existence as life haunted by its tele-technological prostheses, Derrida emphasizes the spectrality of past archives, given to faith. He shows how the future to come is prepared by the reiteration of the promise, but when the promise appears *as* a promise, it appears as the possibility of fulfilling or not fulfilling it and also of being surprised by something totally other. The latter is the abstract condition of "messianicity," which is not the promise of something or somebody but a more ancient "yes" of the very possibility of to come, to which the living can in the last instance only have faith that something still remains to come. This is not a particularly religious, cultural, or even human attitude: it is life itself. Such messianicity combines the apparently irreconcilable principles of the machine and of the event that "Typewriter Ribbon: Limited Ink (2)" tried to bring together by referring them to the *khora*-like materiality of what neither consists nor carries but simply withdraws from experience.[225]

Thinking of life as a relation of faith to tele-technical prostheses does not amount to choosing between religion or atheism or between conservative or progressive worldviews. It considers the fundamental constitution of life as it is not an existing *thing* but in a movement of temporalization between the inherited archives and the promise of a future to come. Such a movement does not originate in life itself but in the two sources from which, through its prostheses, life is given to itself, that Derrida names the messianicity and the *khora*. Beyond all questions concerning the sense of history, the messianic is openness to the surprise that interrupts any programmed course of things: "The messianic exposes itself to absolute surprise," which gives "the general structure of experience."[226] Beyond all questions about the sense of being, *khora* is a place "without age, without history and more ancient than all oppositions" that "remains absolutely impassible and heterogenous to all processes of historical revelation or anthropo-theological experience."[227] Saying that life can only relate to its tele-technological supplements with faith but not with certainty does not mean adopting a relativist or nihilistic attitude toward the possibility of knowledge. It shows the limits of any claim to certain knowledge and opens experience to the absolute heterogeneity of the absolute surprise.

After Foucault and Derrida: Technological Existence and the Question of Total Digitalization

In the previous chapters we have seen how Foucault and Derrida can contribute to our understanding of human existence under the technological condition. As both start with severe criticisms of traditional humanism, careless readers have sometimes believed that their philosophies lead to unlimited relativism and willful constructivism—and this would ultimately justify the transhumanist dreams of building a new posthuman reality at will. In order to dissipate such misunderstandings it is helpful to see how they actually see existence in the technological world—even though neither of them actually believes it to be desirable or even possible to formulate a positive theory of technological Dasein.

Unlike Heidegger and the Philosophical anthropologists examined in the previous chapter, Foucault and Derrida do not ground existence solely on the void of nothingness. We have seen how, in the case of Philosophical anthropology, nothingness both reflects the world's artificial quality and results in the construction of an artificial world and how in Heidegger's case the nullity of existence corresponds to the technical framework of the modern era but can also open up a poetic word that could break free from it. The world of technics, at least the world of modern technology, had a nihilist tinge in these works. Contrary to this, Foucault and Derrida give existence to be thought as an originary technicity that opens both existence and world as modifiable, plastic dimensions. Their thinking of technology is not marked by nihilism but by cautious curiosity. As such, technology is neither good nor evil but a pharmakon that brings both danger and promise.

The reversal brought about by these poststructuralist thinkers is based on a fundamental philosophical gesture that starts with the very *relation* of being-in-the-world and not with the *terms* that the relation brings together (the existent thought as a given, ultimately substantial thing and the world thought as a given place). The existent and the world are the *effects* of their reciprocal relation. Furthermore, this relation is now thought of in terms of *technicity*. If it was thought of in terms of ideation, the world would appear as given, and if it was thought of in terms of praxis, the existent would appear as the origin of action. On the contrary, when the relation between the world and the existent is technicity, nothing appears as originally given. The question is how the existent and the world come to be in the first place.

Derrida in particular has discovered concepts that are fit to describe the world of technics. The world of technics does not have the stability of a substance nor the permanence of an ideal world. It is by definition artificial and contingent. Instead of taking this contingency as a sign of nullity, Derrida helps us to see that this does not prevent it from being effective and significative. Philosophically, although the world of technics is not a stable substance, it can still be thought in terms of *khora*, and although it is not an ideal dimension, it can still the make sense as the specter of ideality. These are important discoveries because they help to counter the tendency of implicitly postulating a general ontology of code that could provide a common ground for biological entities, human brains, and computers. Derrida shows that all kinds of living and intellectual entities certainly *appear* in terms of codes, programs, and inscriptions today: so to say, codes and inscriptions are abstract, invisible phenomenality of entities. However, although this phenomenality seems to presuppose a certain materiality, this is not a substantial ground but just the hypothetical and actually fictive horizon that Derrida described in terms of the *khora*. The world is not "really" code, but it is imagined, today's scientists would say *modelized*, in terms of codes. Codes have become the a priori structure of our thinking, the transcendental horizon of our knowing—but they are abstract projections, almost fictive ways of understanding things, and they constitute only an artificial, historically changing quasi-transcendental horizon.

Derrida shows how life develops as a reaction to its own technicity. Foucault especially helps in understanding human beings' subjectivation processes in the world of technics. Born and thrown in historical contexts that they cannot choose, human beings are profoundly marked by the technical situation to which they find themselves exposed. However, they are not only submitted to it but can also become the fold in which resistance to and reflection on the situation becomes possible and the invention of new ways of being can take place.

This is how, from a purely philosophical perspective, thinking of technics has become an important way of thinking of existence after ontotheology. However, I think that on the basis of their work, one could take one more step beyond Foucault and Derrida. This step is toward a thinking of existence in terms of bio-technics. This term refers to the Foucauldian notion of biopower but extends it beyond human beings to all kinds of living beings. After all, the administration of life was first the power exerted on plants and animals in agriculture, sylviculture, and also in the culture that does not think about its traces in nonhuman nature, although it leaves them constantly. Ultimately this has created what is now called the *anthropocene*. The economy of the human society continues administering places in which human beings live, which one could call *ecology* in the most general sense of the term. Bio-technics therefore includes both human and nonhuman beings in the domain of technics. The term *bio-technics* also refers to Derrida's work in which both human and nonhuman life are thought in terms of originary technicity. Bio-technicity is the originary technicity of life, not only the possibility of thinking life after technical models but also life's possibility of reinventing itself technically. Technicity is not something that living beings do: instead, plasticity and originary technicity are fundamental *modes of being* of living beings.

Foucault and Derrida's works on technics also have important practical motivations and consequences, and they provide promising ways of dealing with new questions that they were hardly aware of. One feature of the contemporary world fits particularly well with Derrida's theories of writing and their extension into inquiries into tele-technology, namely, the omnipresent digitalization of all human life and even increasing aspects of nonhuman life. Digital devices are everywhere—computers, telephones, and wearables pervading the intimate space and large, sophisticated information systems managing increasing domains of the public space. Derrida's concepts help to comprehend the complex and often confusing effects of the reality innervated by digital tools and processes.[228] In particular, when we consider digital phenomena as *writing*, we see why their capacity to guarantee a faithful mechanical repetition of codes and messages does not amount to conserving the truth value originally intended. When we consider digital memories as gigantic *archives*, we see why they are not neutral but reflect a certain architecture and certain principles of selection that result from decisions that are both purely technical and also economical, ideological, and political. These decisions are not taken by us, nor necessarily by any human beings, and most of the time they are not known to us.[229] When we evaluate digital devices and archives as *prostheses* we see why they have become at the same time indispensable and alienating: digitality is the contemporary *pharmakon par excellence*.

When we think of digital phenomena as *spectral*, we understand our unease before their way of presenting things that are non-present and that yet touch from a distance. And when we think of them as *tele-technological* systems, we see, in addition to all this, how importantly and at the same time unobtrusively they frame our historical and political situation.

As Foucault has shown, the dispositif under which we live formats strongly what we think our situation and our capacities are. Accordingly, as Rouvroy and Berns have shown, Foucault helps us see that today digital dispositifs not only mediate but also determine the power-knowledge structure in which we live and the self-techniques that are at our disposal—what we know and what we think is in our power to do. However, Foucault does not speak directly of digital technologies, whereas Derrida has addressed them explicitly in his discussions of cybernetics and tele-technologies. Derrida has emphasized the way in which the tele-technological dispositif is *in* us while being at the same time *alien* to us: similarly, digitality is an intimate stranger. Tele-technological archives make our "Dasein" by constituting our inheritance and promising what is to come. Derrida uses the word *tele-technology* not to designate contemporary technologies in the narrow sense but to designate the way in which sense is given in general: technically, from afar. When tele-technology is understood in this way, it is possible to say that tele-technologies make our existence without being chosen and intended. Their mode of being is spectral and their origin withdraws beyond all possible experience. This is why their mode of being cannot but draw upon the fictional-fictioning powers of the *khora* and of the messianicity without messianism. They not only participate in but really make the materiality of the *khora* and the historicity of messianicity.

There is one element, however, that neither Foucault nor Derrida pay explicit attention to but that is characteristic of digital reality. This is, namely, that computers operate on the basis of an essential difference between the interface encountered by users and the programs that operate the interface without being visible, let alone comprehensible to all but very few users. There are good reasons to inquire into the structure of this digital infrastructure, for on the one hand it directs and formats what can appear on the interface and what users can do with it, but on the other hand it is not chosen by them but ordered by enormous technological, economical, educational, media, and political structures that are as a matter of fact the most powerful power-knowledge structures today (as shown by Rouvroy, Berns, and Zuboff). Furthermore, because of the very nature of computing systems, and especially of the new machine learning systems, the results of computational adminstration are not necessarily transparent to the administrations that order them in the first place, for they result from computational logics that are too quick and too alien to be followed by human beings. The technological infrastructure of the digital world is not the System, like the one forcefully rejected by situationists, it is just a heap of more or less successful technical solutions to various practical problems without much of a common plan. Furthermore, the visible side of digitality is conditioned not only by the invisible informational infrastructure but also by a material infrastructure consisting of raw materials, energy sources, systems of fabrication and distribution, waste and recycling systems, and so on. The material infrastructure implies its own, also very real, risks in the form of the exploitation of natural resources and the workforce, as well as pollution of environment and human welfare.[230] Making the invisible power-knowledge structures and material conditions of digital media visible and submitting them to critical evaluation is one of the big challenges of our time, and in the next chapter we will look more into this.

Now, many later authors have examined the digital world in the wake of Foucault and Derrida. One of the first to develop these impulses is N. Katherine Hayles, who lets us see how the mechanism operative in information theory and cybernetics is in may ways parallel to what Derrida has analyzed as writing: both are suplementary structures that make meaning possible without being exactly meaningful themselves. In a sense, Hayles follows an inspiration from Alan Turing's famous imitation game that compares thinking and computing, except what is at stake is no longer showing that a computing machine can "think" (whatever this means) but to enlighten the unthought basis of thinking by relating it to computing and *vice versa*. Hayles divides the history of cybernetics into three stages, the first of which was was centered on the principle of homeostasis; the second, on reflexivity; and the third, on virtuality.[231] She explains that while homeostasis means the system's capacity to maintain steady states in varying environments, reflexivity means its capacity to adapt itself to changing environments thanks to reflexive recursive movements. This is also why second-order cybernetics is concerned with questions of autopoiesis (Bateson, Maturana, Varela, von Foerster) and thereby joins with the systems theory that we had contrasted to Derrida's theory of autoimmunity. The third wave of cybernetics pays attention to the capacity to evolve, going as far as considering emergent new codes as new forms of life.

In *Unthought: The Power of the Cognitive Nonconscious*, Hayles affirms very clearly that most of our cognition is not conscious and does not even happen "in us" but in technical objects and the biological context with which we are concerned.[232] This "unthought cognition" operates our biology, psychical, and social life. Against the background of cybernetics, systems theory, and even Derrida's theory of writing, when it comes to information content it does not really make much sense to strictly distinguish between technical, cultural, and biological modes of the unthought since all of them function using analogical processes. Hayles examines by means that are close to systems theory the very problem that Foucault and Derrida have studied from the point of view of the finitude of experience: "It thinks" in us without our noticing it. This "thinking" happens through different mechanisms than our intentional thinking: whether one calls it cognizing, computing, machinating, or iterating, it is always a question of quasi-machinic modes of thinking that have other virtues than everyday human thinking. This machinic thinking is neither better nor worse than human consciousness, it is totally different from it. Human conscious thinking calculates and deduces much more slowly and makes many more mistakes than the machine, but on the other hand, human thinking is capable of breaking free from the limits of calculus; it is much more capable of inventing new ways and welcoming new events than any machine. As Plato said, the writing machine is capable of *hypomnesis* while the human memory is capable of *mneme*—and as Derrida added, the complete process of thinking cannot really choose between *hypomnesis* and *mneme* but needs them both. The two modes of thinking are not really rivals, but they need to find better ways of cooperating. However, as the unthought is—precisely—unthought and unconscious, bringing forth, evaluating, and changing this collective impersonal thinking is very difficult, as Hayles suggests. In the next chapter we will see how Agamben and Stiegler develop this problem much further.

This being said, the very latest in the information technology that works on the so-called artificial intelligence (machine learning) is regularly presented as equal and ultimately superior to human thinking, potentially in all its dimensions. In his book *Recursivity and Contingency*, Yuk Hui has explained the specificity of artificial intelligence insofar as it does not simply realize a program set once and for all but programs itself by applying to itself recursively to things that it has learned from its environment. Artificial intelligence in this sense is the realization of the principle of third-wave cybernetics, which Hayles also explained. Hui presents today's "artificial intelligence" as a new answer to Turing's question of a machine capable of rivalling human intelligence. Now, as we have seen, Hayles allows us to formulate one convincing counter-argument to the postulated identity between artificial intelligence and human intelligence: artificial intelligence presents *different* procedures of cognizing and different forms of embodiment than human consciousness, such that it does not supplant human consciousness but complements it as a new type of *unthought*. Derrida allows us to explain this further. In the last instance, information systems still function autopoietically, whereas life is characterized by an autoimmunity that not only adapts itself to new environmental factors but undoes itself and exposes itself to unforeseeable alterity both outside and inside of itself (including the alterity of death). The question is therefore not how (human, living, technical) systems constitute themselves but how they manage to resist to their systematic closure and open toward the outside, as both Foucault and Derrida have asked.

Chapter 5

Humanity and Inhumanity of Technical Communities

In the introduction to *The Inhuman*, a suspicion that animates Jean François Lyotard's thinking (in *The Inhuman* and elsewhere) crystallizes: the humanism imposed as a primordial cultural value is becoming hollowed out and human beings are becoming more and more inhuman both because they are crushed by the prevailing culture ("the system") and because they are, despite everything, still connected to a more savage part of themselves.

> What if human beings, in humanism's sense, were in the process of, constrained into, becoming inhuman (that's the first part)? And (the second part), what if what is "proper" to humankind were to be inhabited by the inhuman?
> Which would make two sorts of inhuman. It is indispensable to keep them dissociated. The inhumanity of the system which is currently being consolidated under the name of development (among others) must not be confused with the infinitely secret one of which the soul is hostage. . . . What else remains as "politics" except resistance to this inhuman? And what else is left to resist with but the debt which each soul has contracted with the miserable and admirable indetermination from which it was born and does not cease to be born?—which is to say, with the other inhuman?[1]

Maybe the "humanism" that Lyotard is criticizing thinks that it knows what is "proper" to the "human"; against such knowledge, our previous considerations on the role of technics in humanization have rather shown, firstly, that the "proper to humankind is its absence of a defining property," and secondly, that human existence takes innumerable forms because it is open to difference and to the surprise of the event. For both of these reasons we cannot really define "humanity" once and for all. But Lyotard says more essentially that we still recognize *inhumanity*, and living a just life demands above all resisting this (even in the form of the inhumanity of a system that poses as "humanist"). The capacity to resist, he thinks, draws from a totally different type of inhumanity that Lyotard finds especially in childhood, whereas Derrida led us to look for it in other secret domains such as animality, spectrality, and death. This is why Bernard Stiegler's expression in *Taking Care* is precise despite its ponderousness when he says that before the increasing risk of societies producing *being inhuman*, we ought to work toward *non-inhuman* societies.[2]

Our only concern in this book is that aspect of inhumanity that results from technology. Especially since the twentieth century, technics has often been given a pivotal role in humanizing, inhumanizing, and dehumanizing. The inhumanity of technics comes forth when technics is considered not as a neutral thing that can be made use of by humans with evil intentions ("It's not the gun that shoots but the human who holds the gun") but as a nonhuman thing in its own right that can turn into an inhuman reality that human beings need to resist (the gun). The dehumanizing effect of technics has deeper reasons than the harmfulness of some specific technological arrangements. The inhumanity of technics is not simply a technological flaw that objects with a better design could fix: it is an irreducible part of the ambivalent *pharmakon* effect of all technics as such. Of course, describing technics in terms of its humanity and inhumanity is very imprecise and actually misleading. It does not say anything about the quality of the technics in question but is simply a way of pointing at their alienating effect. One aspect of the alienating effect of technics depends on its character as supplement that we studied above with Derrida. Another aspect is the nature of any technics as a common thing: technics robs the individual of itself because it is by nature a collective formation belonging to what Heidegger described as the inauthentic existence characteristic of the "they" (*das Man*). In this chapter we shall concentrate on this collective aspect of technics and see how it is both an oppressive and an emancipatory force because it is a force that builds communities.

Technics was seen as an emancipatory force in the Enlightenment ideology, especially as one of the great outgrowths of humanism and also as a great occasion to promote technology as a means for humanization. Like education, technology was seen as a means to help people out of misery and ignorance—as it certainly was and still is. But Enlightenment philosophy was too intent on formulating the idea of humanity to pay attention to the exact role of technics in humanization: considered merely as an instrument of humanist ideas, technics was not evaluated as such. By an intriguing ruse of reason, the first industrialization that to some extent realized certain Enlightenment ideals also brought their reverse side to the fore, especially the exploitation and alienation of working people that led to a new kind of misery—a psychological, social, spiritual, and political misery in addition to simple material poverty. Marx and Engels already pointed out the role of industrial technology in this misery as did an increasing number of authors of all political sensibilities in the twentieth century: Adorno, Horkheimer, Heidegger, Jonas, Anders, Marcuse, and many others. They were in agreement concerning the dehumanizing effect of modern industrial technology on society, which technics helped to turn into a standardized mass society where individual differences were leveled down and abolished. To some extent, these authors differed with regard to the basis for this development—for some, technology was ultimately at the service of capitalism and for others, at the service of Soviet communism—but in the end all political ideologies appeared caught in the debilitating rhythm of the development of industrial technology. If many of the first important critics of industrialization were German, their French colleagues that we discussed in the previous chapters were generally more enthusiastic about technological progress—and at the same time sceptical of the ecological thought that was developing in Germany. However, the French philosophers of technology Gilbert Simondon and Bernard Stiegler have a much more nuanced picture of technology's promises and dangers, as well as of its ecological impact. The necessity for qualified judgments goes hand in hand with the need to evaluate different technological systems as specific cases instead of referring everything to a singular global judgment for or against technology. This is why many important philosophers of technology today—Bruno Latour, Don Ihde, Peter-Paul Verbeek—prefer situated and multifaceted interpretations of concrete technological assemblies and their associated milieus. Such interpretations show how technologies are rarely, if ever, simply humanizing or dehumanizing—the question is, rather, whether they transform their users and their contexts in enabling or disabling ways.

In what follows, we shall study technics' capacity to create communities. Technical communities are more inconspicuous than spiritual communities: while the latter spend much energy identifying themselves as such (think of cultural, linguistic, or religious communities), technical communities are often imperceptible, but they nonetheless push their users into parallel ways of life (think of those who use airplanes all over the world). Besides, unlike a spiritual community, a technical community is not just limited to the human species. Technical communities can very well assemble human, animal, and technical beings, for example in Deleuze and Guattari's description of the assemblage of man, horse, and stirrup[3] or in Latour's descriptions of actor networks and assemblages.

We have already touched upon several modern perspectives on the role of technics in community formation. The first one of them emphasizes that technics makes communities but that people find themselves alienated in technical communities. This is because technics are never individual but collective, people have to adapt to their logic in order to use them. Heidegger described this most precisely in *Being and Time*, where the technical utensil was the starting point for the analytic of inauthentic existence as "the they" (*das Man*) and later, for example, in "*Zum Seinsfrage*," "Overcoming Metaphysics," and "The Question concerning Technology," where the epoch of technics submitted human and natural beings alike to the inhuman logic of technology. As these analyses show, even though technical *objects* may belong to individuals, they possess only the object but not the technical *principle* itself, which constitutes the technicity of the object (to take a happier example than Heidegger, a violinist generally possesses its instrument but has not built it and has to learn its techniques instead of inventing them). The technical principle is for its part always collective and actually it is one of the main constituents of the collective domain (e.g., of a string orchestra and its public). The same holds of personal styles and techniques for using technical objects (e.g., a violinist's personal touch): in the last instance, they are dependent on supraindividual technical principles. Using a technical object thus robs me of myself because it submits me to this technical principle that is necessarily common—but at the same time it enables the development of individual styles and it brings people together (and makes a certain kind of music possible).

A second point of view on commonality of technics emphasizes that technics is not necessarily a means of alienation but it can also be a means of emancipation. Leroi-Gourhan has thus shown that although historians tend to equate technical communities with the ethnic communities in which the technical principles were invented, in the last instance technical communities do not need to coincide with ethnic communities because technics are easily learned and generalized and thus transferred from one community to another. Indeed, we know, for example, that although technical inventions (e.g., agriculture, the art of writing, or gunpowder) have their origin within specific civilizations, all civilizations have been able to adopt them as soon as they have appeared useful to them (and the secrets of fabrication have been unveiled). Sometimes new technical principles have been invented simultaneously in distinct places, such that apparently distinct communities are actually united by their technics. There are plenty of examples of technical inventions that have become generalized and that have spread quickly over community and state borders not only because they made life easier but above all because they have people better informed, better connected, and thereby freer. This has been the logic of generalization with respect to the printing press, television and radio, and today the Internet and other telecommunications.[4] These are telling examples of technical principles that have above all contributed to the liberation of people from their closed communities (i.e., family or religious and ethnic communities).

A third point of view regards contemporary technology and especially information technology from the point of view of globalization. Although technics has always tended to spread beyond the narrow limits of ethnic communities, today efficient technical inventions spread very quickly all over the world. One could even say that they are the main vehicle for and dimension of globalization. The most emblematic technologies of globalization are the information technologies and what Derrida called tele-technologies, which promise to reach everywhere around the world and connect virtually all people together: they promise both universal communication and panoptic surveillance. However, the space opened by worldwide digital technologies is not as worldwide as it seems. It displaces us to spaces that are far from our intimate physical place and that are shared by other similarly displaced persons.[5] But these places are neither the physical places of anybody else nor are they of a truly universal scope: they are just "displaces," places made only of displacements, of pure transitivity, that only enable the transit of information toward other displaces. The Internet is not a universal place, and certainly not a universal political community. It is a finite displace of a common alienation.[6] The worldwide scope of digital tools is therefore also an illusion. Although digital tools have the power to connect practically all places, they replace physical places by abstract displaces where we only meet as "specters."

This is how technics contributes to community formation. Even though examining technics in function of finite communities gets us further than examining them in relation to abstract humanity, some questions remain. Firstly, as Derrida mentions in *Echographies of Television*, community is not necessarily a positive term. Most of the time "community" is understood as we who identify ourselves as a community or who recognize each other reciprocally as members of the community. Such identifications and recognitions imply the identification of those who do not belong to the community and who are excluded from them, sometimes harmlessly but all too often with tragic consequences. Against such exclusionary situations Derrida looks for ways of welcoming the stranger and the newcomer. Hence, as we know that technics makes communities, can it also make hospitalities?[7] Secondly, and this is actually the same thing from the point of view of technics, as we saw, technics tends to make communities of its users or of those who benefit from it indirectly. How can technics be made welcoming, enabling, and liberating?

In what follows, I will look for answers to these questions in Bernard Stiegler's and Giorgio Agamben's different points of view on technical communities. Unlike the authors examined previously, these authors write in the context of twenty-first-century technology, which is marked especially by digitalization and biotechnology. Both emphasize the nefarious effects of these technologies on community formation in today's globalized situation, but neither is against technology as such as both also seek its emancipatory potential. For sure, their political readings of the contemporary situation do not amount to the same thing. What interests me more, however, is not technological politics but the philosophical deep structure of "technological humanity" that Stiegler and Agamben bring forth in complementary ways.

Bernard Stiegler

Drawing from all authors discussed previously, Bernard Stiegler presents the most comprehensive theory of technological humanity to date. It consists of showing why the very technicity that determines human existence prevents providing any definite figure of humanity. He therefore develops an existential analytic of technicity instead.

Unlike most authors that we have studied so far, Stiegler does not refuse the anthropological question of the human being but on the contrary poses it relentlessly. At the same time he rejects the possibility of ever giving a definitive answer to the question because he assimilates humanity integrally with technical artifices. In other words, the human being is not for him a clever creature who extends its intentions over things by producing technical equipment to mediate them. On the contrary, the human makes the technics that makes the human; the invention of technics ("what?") is also the invention of the human being ("who?"):

> "Who" or "what" does the inventing? "Who" or "what" is invented? The ambiguity of the subject, and in the same move the ambiguity of the object of the verb "invent," translates nothing else but the very sense of the verb.
>
> The relation binding the "who" and the "what" is invention. Apparently, the "who" and the "what" are named respectively: the human, and the technical. Nevertheless, the ambiguity of the genitive imposes at least the following question: what if the "who" were the technical? and the "what" the human? Or yet again must one not proceed down a path beyond or below every difference between a *who* and a *what*?[8]

Instead of following the old anthropocentric reflex, Stiegler not only invites us to think of technics as the origin of the human being, but more precisely he exhorts thinking along the lines of the double invention of technics by humanity and humanity by technics, such that both humanity and technics ultimately have their origin in a singular movement of *invention*.

If humanity *is* technicity, one cannot attribute inhumanity to the use of technics as such: technophobia makes as little sense as technophilia. However, there certainly are inhuman technological systems that degrade human and nonhuman existence, and Stiegler makes great effort to designate and to diagnose the dehumanizing and destructive aspects of contemporary technology. He thinks that certain technological configurations are powerful enough to determine entire epochs. While our epoch is still determined by the industrialization that exploits human and natural energy, the distinctive feature of the contemporary epoch is what Stiegler calls hyperindustrialization, which exploits the human spirit itself.

Together these produce the present-day globalization that is characterized by systemic stupidity and beastliness, by increasing economical exploitation, and by fatal ecological disaster. As these are the most urgent political problems today, the critical interpretation of the technological systems that produce these ills is according to Stiegler the most important political task at hand. This is how, contrary to the traditional consideration of politics in function of the history of human liberation, Stiegler considers politics in terms of the epochality determined by technology.

In this subchapter I will first sum up Stiegler's fundamental conception of technological humanity as presented in particular in *Technics and Time*. Then I will show what he says of the technical object or thing. Following that, we will see how the thing elucidates the structure of the human community, which Stiegler depicts as essentially produced by the epochally prevalent technics. As Stiegler thinks of the human being-with as not only mediated but really *produced* by the technical object, he foregrounds the collective, unconscious, and often alienated aspects the technological community—but as we shall see, he also indicates some leeways for collective and individual individualization. Finally I will summarize Stiegler's reading of our present epoch, which is both more up to date and more somber than anything we have read previously. In fact, Stiegler thinks that we are running headlong into a brick wall. This is probably the reason why Stiegler's later thinking has been received somewhat cautiously. After all, the public rarely likes to hear bad tidings.

Whose Invention?

We have said as first approximation that Bernard Stiegler develops a comprehensive theory of technological humanity. However, we have seen over and over again how imprecise and misleading the terms *technology* and *humanity* can be, evoking for example the Promethean figures of the Cartesian "master and possessor of nature" or of the monster technologically produced by doctor Frankenstein. Instead referring his work to the absent-minded Epimetheus, Stiegler does not think of the human relation to technics in terms of a willful conscious production but in terms of a reciprocal unconscious constitution in which the very senses of "technics" and "human" are also at stake. This is why, instead of labeling Stiegler's theory as a techno-anthropology, it is better to use his own interrogative terms "who?" and "what?" "What?" is technics examined neither as an epistemic object nor as a practical instrument but as an "elementary supplement."[9] "Who?" is disclosed by an "existential analytic, which should accordingly be interpreted *in terms of the question of prostheticity*." As we have seen, the prosthesis introjects an inanimate supplement into the animate body (or mind) in such a manner that they become inseparable although still distinct, the body animating the thing and the inanimate thing keeping the body alive. The "who?" and the "what?" open the question of this interlacing on a philosophical level.

Stiegler's first and maybe the most important work, *Technics and Time, volumes 1–3* (and volume 1 in particular), investigates the intertwining of "who?" and "what?" through a critical encounter with Derrida's *Of Grammatology* and Heidegger's *Being and Time*. As we saw, in *Of Grammatology*, Derrida reinterprets language, not in terms of the logos that language is supposed to serve but in terms of the differential structure of the trace that shows how language functions. Examined as writing, language comes forth in its technical function, which is also pertinent to philosophy of technics. In *Technics and Time 1*, Stiegler acknowledges his debt to his mentor's work but he also shows how he takes his distance from it. In the important section 3.1 "The Différance of the Human," he points out perceptively that Derrida relates his notions of *différance* and *gramme* to the general history of life. Instead of enforcing the anthropocentric ideal of the "speaking animal," they trouble the limit between the human and the animal by showing how the *gramme*, deployed as the program, organizes all, both human and nonhuman life.[10] As we saw in our study of Derrida, life itself is grammatical and programmatical; consequently, life itself, and not just human life, is intentional. By asking how the "*gramme* as such" emerges, Stiegler asks on the one hand how this life as such becomes thinkable. But on the other hand, and more importantly for him, he asks how the *gramme* gets inscribed on separate, material supports, organizing inanimate matter as technics; this is how he reinscribes the birth of the human (through the invention of technics) back into the history of life. Understood in this general sense as inanimate matter incorporated and introjected into the course of life, technics is everything from flint tools to reading and writing machines—including languages, through which life exteriorizes itself in sound and light vibrations and onto solid supports. Grammatical life invents grammatical technics. But at the same time, only grammatical technics makes grammatical life emerge as its own absolute past.

Following the work of paleoanthropologist André Leroi-Gourhan, whose importance for *Of Grammatology* Stiegler emphasizes, Stiegler stresses the qualitative leaps that take place when life extends itself over matter, which it organizes and programs so as to complement its own programs. In this conceptual evolution, life is not only determined by what is called a *genetic* program.[11] Furthermore, when life bends reflectively back on itself, it acquires an *epigenetic* dimension, that is, life becomes an experience of living, an experience that it can also hold and reflect upon, that is to say, remember (more or less consciously). But most importantly, according to Leroi-Gourhan, a distinctively human type of life emerges when life invents a way of deposing some of its memories on supplementary material supports. The way in which wood can be cut is inscribed in the form of a flint chopper, the way in which moving things are lighter and quicker is inscribed in the wheel, or the way in which I see the world today is deposed in a diary that I can reread later. In material supports, experiences are conserved and sedimented, passed on to later moments, to other people, and to later generations, as when other human animals use the flint chopper and a descendant of the Incas reads a text written in ancient Greece today. Epiphylogenetic memory is this supraindividual memory that makes the specific strength of the human species. Stiegler calls the memory deposed in material objects an *epiphylogenetic* memory:

> In this case, the individual develops out of three memories: genetic memory; memory of the central nervous system (epigenetic); and techno-logical memory (language and technics are here amalgamated in the process of exteriorization). . . . Epiphylogenesis, a recapitulating, dynamic, and morphogenetic *(phylogenetic)* accumulation of individual experience *(epi)*, designates the appearance of a new relation between the organism and its environment, which is also a new state of matter. If the individual is organic organized matter, then its relation to its environment (to matter in general, organic or inorganic), when it is a question of a *who*, is mediated by the organized but inorganic matter of the *organon*, the tool with its instructive role (its role *qua* instrument), the *what*. It is in this sense that the *what* invents the *who* just as much as it is invented by it.[12]

The invention of the human is this invention of the epiphylogenetic, technological supplement to genetic and epigenetic life.

This is how Stiegler continues Derrida's work by emphasizing the anthropological inspiration of *Of Grammatology* and by underlining the role of technics in the process of hominization. Now, in Stiegler, the emphasis on anthropology and technology results in the close attention being given to the concrete history of technologies, and this is also where Stiegler ends up turning against Derrida. In interview with Stiegler, Derrida admits that he analyzes the technologies of all epochs as instantiations of abstract structures of writing so that even "tele-technology" is not an epochal form of technology but above all a structure that can be found in all technologies regardless of their epoch.[13] Stiegler criticizes Derrida for forgetting the concrete history of technologies and distinguishes for his part a number of technological epochs, as we shall see later.[14] Is Derrida's philosophy too abstract and transcendental or is Stiegler's thinking too anthropological and historical? From my point of view, it would be unduly restrictive to choose only one of these approaches, since one functions as a salutary corrective to the other. Derrida prevents Stiegler from becoming simple techno-cultural history and Stiegler brings forth the ethical and political implications of Derrida's work, such that he shows what kind of a theory of human existence Derrida would have made if he had considered it possible, and Derrida reminds us of the risks of erecting a theory of the human. However, neither can be reduced to these oppositional positions.

If Stiegler's reading of Derrida shows the specificity of his interpretation of technics as a supplement to life, his reading of Heidegger develops certain of Derrida's suggestions much further and expands these into a theory of technics as time. Stiegler's deconstructive reading of Heidegger's *Being and Time* results, as he put it, in an existential analytic of prostheticity in which Dasein ("who?") is thought of as originarily supplemented and prosthetized by technics ("what?"). As we have seen earlier, Heidegger's thinking of technics can be seen to originate in *Being and Time*'s interpretation of "equipment" (*Zeug*). Although the equipment lays out the entire world-relation of Dasein, it does so in the mode of the Dasein's inauthentic existence where Dasein is still lost in "being-with" (*Mitsein*). In order to gain its authenticity—and thereby its philosophical role as the site of the question of being—Dasein must tear itself away from inauthentic existence in the familiar equipped and inhabited world and face its fundamental alienness and solitude in the experience of being-toward-death. Only this experience gives it access to what is according to Heidegger the sense of its existence: temporality. Dasein is temporalization and its temporalization is radically distinct from the time of the world of equipment, including the equipment for measuring time such as the clock. Stiegler learns from Heidegger how to think of being-in-the-world starting from the equipment and to think of existence as temporalization, but he criticizes Heidegger for leaving these two aspects of Dasein separate and for occulting the role of equipment in temporalization itself. Although the major consequence of *Being and Time* had been "the highlighting of the ontological singularity of beings that are 'ready-to-hand,' tools, that is, the *what* (and the sign is itself a tool),"[15] Heidegger remains within the traditional opposition between logos and techne and rejects techne as the domain or instrumentality and calculation. This is how Heidegger dissolves the sense of factical things whose importance he was the first to notice, such that "the prosthetic facticity of the already-there will have had no constitutive character, will never have taken part in the originarity of the phenomenon of time, indeed, and on the contrary, will have only figured as a destitution of the origin."[16]

Against this, Stiegler redefines Dasein decisively as a prosthetic being:

> Dasein is outside itself, in ec-stasis, temporal: its past lies outside it, yet it is nothing but this past, in the form of *not yet*. By being actually its past, it can do nothing but put itself outside itself, "ek-sist." But *how* does Dasein eksist in this way? Prosthetically, through pro-posing and pro-jecting itself outside itself, in front of itself. And this means that *it can only test its improbability pro-grammatically.*
>
> 1. Dasein, essentially factical, is pros-thetic. It is nothing either outside *what* is outside of it or *what it* is outside itself, since it is only through the prosthetic that it experiences, without ever proving so, its mortality, only through the prosthetic that it anticipates.
>
> 2. Dasein's access to its past, and its anticipation as such, is pros-thetic. In accordance with this condition, it accedes or does not accede to this past as it has been, or not been, durably fixed, and to which, at the same time, Dasein is to be found, or not found, durably fixed.[17]

Dasein's prosthesis includes equipment and language, as well as what Heidegger in later works analyzes as technics and art. All of these constitute the world in which Dasein finds itself thrown and which it cares for in its ordinary life. At the same time, these also constitute Dasein's historicity, which is made and sedimented not only in individual experiences but also in collective and transgenerational experiences. While Heidegger insists in differentiating between this collective time of the world and the authentic temporality of Dasein, Stiegler thinks that "the temporality of the human, which marks it off among the other living beings, presupposes exteriorization and prostheticity: there is time only because memory is 'artificial,' becoming constituted as already-there *since* [from the point of: *depuis*] its 'having been placed outside of the species.'"[18]

Stiegler's pursues his reading of Heidegger by tackling the latter's readings of Husserl and Kant in the volumes 2 and 3 of *Technics and Time*. While these reinterpretations show in detail how the technical supplement constitutes sense-making and consciousness, Stiegler does not return to Heidegger's original question concerning being. It could be said that Stiegler's interest in historicity and epochality develops one aspect of Heidegger's inheritance in a very useful way, but that on the other hand, he does not really say whether or not this changes Heidegger's original question concerning being. In Heidegger's work, the tension between the universal openness of the question of being and its historical, epochal "sendings" is indelible. Stiegler does not solve this philosophical tension. He seems to foreclose the question of being, but he also shows that Heidegger's thinking of historicity, rejected by some as toxic, is an inescapable dimension of any archaeology of the present.

WHAT?

Let us now focus on the "what?" of the technical object itself. We can see Stiegler's concept of the technical object by tracking how he highlights, discusses, and continues Derrida's, Simondon's, and Heidegger's pioneering ideas in particular. In addition to the concept of technics, Stiegler is interested in concrete historical technical systems. His use of history is closer to Heidegger's than to Simondon's because his aim is not to rewrite an entire history of technology but only to identify the rare fundamental epochs of technics brought to light by certain symptomatic technical systems.[19]

First and foremost, Stiegler thinks of the technical object as a memory. Like writing as studied by Derrida in his reading of Plato's *Phaedrus*, the technical object appears built to serve as an externalized memory, and it functions by incorporating this memory as a supplement to *noesis* (thinking). The technical object is a prosthesis and like any prosthesis has the ambivalence of a pharmakon. Stiegler extends the structure of memory to all technical objects. Like Leroi-Gourhan, he interprets the tool—from the simplest prehistoric flint tool to contemporary automatized factories—as the memory of a gesture. It records a certain know-how (i.e., how to make fire by drilling a stick on a piece of wood). As practical knowledge of the environing nature is inscribed in its very structure, the tool user can profit from this without being conscious and aware of it. Stiegler also interprets all kinds of pictures as technical objects, from cave paintings to cinema and not just the tools and machines that produce these pictures (paintbrush, camera). The picture is a technical support that holds the memory of a mental image and of an intuition. In addition to being a product of technical activity, it is itself a machine that produces human perception by orienting vision and sensibility and by thus preparing the possibility of any glance. For similar reasons, he counts all kinds of recordings of speech (in writing) and of sound and light frequencies (in discs, films, and digital files) among technical objects: they, too, make it possible to conserve, use, and share past experiences. Finally, today's digital network is a gigantic shared/shareable memory, but its algorithmic structure also tracks, memorizes, and produces individual behaviors, social relations, and collective and transindividual individuation.[20]

Second, like Heidegger, Stiegler relates the technical object to *calculation*. This term is often used but rarely defined by philosophers. In its original mathematical sense, a calculation is an operation or a set of operations on a set of numbers or subsequently some other given values. Even though calculation reorganizes the original set, it is limited to conserving the original value instead of inventing totally new contents. A calculation is a form of cognizing, but the philosophers studied in this book generally use the term in order to refer to the kind of cognizing that is not conscious of itself and that therefore cannot really be called thinking. As it is an operation that conserves the value of the original data, the calculus is also a kind of a memory, but for our philosophers it is a kind of a memory that cannot be said to remember in the proper sense of the word.[21] It is a memory that is not a memory *for itself*, nor is it conscious of itself as memory: it is just a repetitive automatism. It is a memory translated into a code that can provoke reminiscences in the human user, but whose own structure and functioning is only that of a repetition. The human user retrieves significations, the machine calculates values that are insignificant to it. We previously saw how Derrida analyzed texts and living beings in terms of texts, programs, and codes. Stiegler follows this lead, but he pays particular attention to digital technologies and stresses the element of calculation, which not only accounts for the *reproduction* of a code but also the *extension* of the initial codes into new areas and the recent applications in the *generation*, one could almost say *invention*, of new codes.

Artisan tools are not usually described in terms of calculation, except as a metaphor (i.e., a good hammer results from an "unconscious calculation" of a well-balanced form). Starting from the Renaissance, however, the development of machines went hand in hand with mathematics: the machine results from technical calculations and the calculator eventually became *ne plus ultra* of machine technology. The core problem of inventing a calculator lies no longer in the adjustment of material parts but in the programming of the calculative operation itself. The essence of the technical object is no longer in its matter but in the logical operation. From the nineteenth century onward, calculating machines gradually developed into computers, which calculate far more quickly and infallibly than human beings ever could. A computer is much more than a simple calculator, however. It hosts a cybernetic machine that can effectuate long series of commands, choose the appropriate commands in function of the feedback received from the exterior world, and today can even to some extent rewrite its own programs in function of such feedback. The cybernetic machine is an algorithmic being that mimes life. Like a living being, the algorithmic being appears as an autopoietic organism that not just reproduces the same commands ad infinitum but modulates the codes in function of external information.[22] (Can a code invent a new code? The terms *machine learning* and *artificial intelligence* refer to similar objects but give two different answers to this question: if one thinks that the production of new code by a code can only ever be an application of the original code in new areas, albeit in a much more complex form, one speaks of machine learning, and if one thinks that in such a situation a code really creates new code, one speaks of artificial intelligence.)

Despite the sophistication of digital algorithms, the digital machine remains a technical object. It lacks what is usually called consciousness, and that is precisely what characterizes human beings (and no doubt many other living beings). Calculation and computation are *nonconscious* modes of cognizing. Nonconscious does not mean of lesser importance, but being markedly different in kind. As explained especially by N. Katherine Hayles in *The Unthought*, the vast majority of all cognizing happens nonconsciously anyway, whether it takes place instinctually, mechanically, or computationally. Unthought cognizing characterizes both machines and living beings, including humans, whose thinking is largely based on unthought processes, for example in the unconsciousness, or again, externalized into machines.[23] Despite science fiction speculations about conscious machines—simulacra of the human—such a thing does not exist and probably cannot exist either because calculating is a fundamentally different mode of cognizing from conscious thinking. A calculation cannot lead to the type of evidence that characterizes the Cartesian *cogito ergo sum* because the latter is a fundamental and simultaneously inexplicable and indubitable intuition. As a calculation cannot prove an intuition—a demonstrated intuition would not be an intuition at all—these two types of cognizing are totally incommensurable, qualitatively speaking. Stiegler emphasizes the same idea as developed by Husserl, for whom the mathematization of science goes hand in hand with the *blinding* of intuition and by Heidegger, for whom the mathematization of thinking leads to forgetfulness of being.[24] Thought experiments like Alan Turing's imitation game or John Searle's "Chinese room experiment" can compare human thought and machine calculation because they bracket what happens in the "black boxes" of the human mind and of the calculating machine. But as soon as one opens the boxes and looks at the precise type of operations that take place in them, their nonidentity becomes evident.

However, although computational machines and systems do not know that they are thinking, they support, prosthetize, and parasitize human thinking. They contribute to our acting and thinking by providing data (memory stock), cognitive operations (algorithms), and social contacts (communication), and their design actually contributes to everything that Stiegler calls *noesis* (thinking) and psychical and social individuation. Their influence on our thinking need not be conscious—and mostly it is not—to be intense. Stiegler stresses that the kind of computational reality that constitutes the contemporary algorithmic governmentality in particular is the elementary, *khoratic* dimension in which we find ourselves today. As he says, it is like the water that constitutes the fish's element but which for this very reason the fish cannot see.[25] The englobing, invisible, elementary nature of technological computation explains why its effect on human existence is so intense, as we shall see below.

The third main characteristic of technical objects emphasized by Stiegler is the idea of the *associated milieu* of a technical object, which he adopts from Simondon. Stiegler explains the associated milieu as a technical milieu of a highly specific kind: a natural milieu (called techno-geographic milieu) is referred to as "associated" when the technical object of which it is the environment structurally and functionally 'associates' with the energies and elements of which this natural milieu is composed, in such a way that nature becomes functional for the technical system. This is the case for the Guimbal turbine, which, in tidal power plants, assigns to seawater, that is, to a natural element, a threefold technical function: providing the energy, cooling the turbine and, through water pressure, sealing the bearings.[26]

Simondon's famous example of the Guimbal turbine shows how the technical individual is fundamentally associated with an environment from which it draws energy and functional support and without which it could not function at all. In this narrow technical use of the term, the associated milieu appears mainly as the resource exploited by the technical system. In a more fundamental sense, however, the associated milieu is a hybrid of a technological milieu and a natural, "geographic" milieu that the technical object relates together. The technical object (or system) needs to adapt to its environment in order to *use* it, but in so doing it also *creates* a techno-geographic environment that previously existed only virtually but that is made effective by the invention of the technical object.[27] In other words, technology changes its natural environment. The notion of associated milieu extends still further. As it is the technological milieu that provides the elements of the constitution of the technical individual, it is technological culture that includes both purely technical elements such as the machine pieces—themselves invented for very different purposes but reassembled to make this machine—and technological principles discovered in techno-scientific projects that are here applied. The technological individual draws from the entire technological situation and reflects this in its structure. Finally, the associated milieu of a technical object includes the human society that the object uses and creates. This is the society that knows how to use it (i.e., how to use a Guimbal turbine) and that is organized for its production. It disposes of human beings in function of the available technology, for example as artisans, engineers, factory workers, consumers, and so on. In his work, Simondon is very concerned with the way in which each technical system alienates and emancipates the people who are involved in it.

Hence the technological individual not only uses its milieu as a direct material resource for its functioning or as an abstract intellectual resource for its conception but also presents its environment in a certain way and transforms it. If Heidegger characterized Dasein as a clearing (*Lichtung*), the technical object is like a lamp that illuminates the *Da*, highlighting certain features and leaving other features in the shadows. Illuminated from the point of view of a technical individual, the environment appears, for example, in terms of available raw materials, transport, and consumption. Like the Heideggerian work of art that opens up a historical world, the Simondonian technical object opens up the techno-geographic milieu of its functioning. Furthermore, the object changes its environment: a cell phone requires mines from which the raw materials are extracted, huge factories where the phones are assembled, worldwide distribution networks, intensive marketing campaigns, and so on. A technical object or system is thus not only the pharmakon of the human being who uses it, but it is also the pharmakon of the natural and social milieu in which it is inserted, drawing from the milieu and changing it, sometimes intensely. This is something that Simondon, according to Stiegler, did not notice: "Simondon himself never posed the question of pharmacology,"[28] and therefore, despite appearances, "Simondon never theorized the process of technical individuation" and so did not problematize the "arrangement between the psychic, the technical and the social." Hence "a reinterpretation of Simondon's thought with respect to contemporary realities is essential. [The present-day techno-geographic milieu] forms what Simondon called an associated milieu, but in this case of a new type, and one he did not envisage."[29]

In order to figure out how intensely technology changes the environment, we can begin by thinking of the enormous changes that the invention of agriculture brought about in the natural milieu and in the social organization already present in the Neolithic period. However, Stiegler is most interested in the changes brought about by two recent technological paradigms, the industrial revolution based on the "thermodynamic machine"[30] and the present digital revolution. The thermodynamic machine—which is the generic term for designating the technological principle behind the industrial revolution—aims at replacing muscular energy with industrially produced (steam, electricity, nuclear) energy that is used to further industrial production (factories, modern cities, modern life in general). According to Stiegler, "The thermodynamic machine, which posits in *physis* the new, specific problem of dissipation of energy, is also an industrial technical object that fundamentally disrupts *social* organisations."[31] The concrete consequences of the principle of the thermodynamic machine are too well known to need any extensive presentation. On the one hand, it has led to the present industrial civilization where everything (from land use to social organization to science) is aligned with industrial production and to the capitalist social order that best favors it; on the other hand, it has led to the phenomenon called anthropocene, which means a period in which the effect of human activity on the planetary nature is so great that it can be counted as a geological force.[32] Stiegler dates the digital revolution from the 1993 invention of the World Wide Web and defines it as an unprecedented situation caused by omnipresent digitalization.[33] Over the course of its evolution, digitalization has come to exploit not so much natural resources but psychic resources, and its aim is to produce not material things but types of behavior and social association.

This is how the notion of the associated milieu leads to consideration of the ecological impact of technology, which Simondon only vaguely hints at but that Stiegler examines in all senses of ecology distinguished by Félix Guattari in *The Three Ecologies*: the ecology of mind, of society, and of the natural environment.[34] In all his latest books, Stiegler's central aim is to see if the two disruptive tendencies characteristic of our era— the anthropocene and generalized stupidity occasioned by computational capitalism—can be made to divert toward new, empowering modes of individuations.[35] We will come back to Stiegler's epochal analysis below. But let us first see how technics contributes to individuations.

Who?

We have now seen what Stiegler means by the "what?": it is the *technical object* (or system) explained in terms of memory, calculus, and milieu. We now want to consider what he means by the "who?," which is his answer to our nagging question concerning technological *humanity*. Stiegler follows the deconstruction of "humanity" initiated by his predecessors and shows how technics contributes to historical existence and furthermore to the individuation of groups.

The technical "what?" contributes to the invention of the "who?" The "what?" does not identify the "who?" but partakes in its individuation—and indeed reveals it as a question of individuation, not of identification or definition. This is why even though Stiegler's "who?" includes anthropological elements from Leroi-Gourhan and Rousseau in particular, it is not identified with any given human essence but on the contrary undoes all essentialism insofar as "human nature consists in its technicity, in its denaturation."[36] Analogically, Stiegler's "who?" is informed by the philosophical question of subjectivity, but this is always rooted in a concrete technological epoch in which it is enunciated. This prevents subjectivity from being fully identified with anything like the pure *I*, the ego, or authentic Dasein and instead connects it to Foucault's approach to subjectivation. However, although it is not the subject, subjectivity is nevertheless one of the possible modes of being of the "who." In Stiegler's work, the "who?" is neither the anthropos nor the subject, but it is reflected on in terms of their intertwining, where the subject constantly transcends the anthropological-historical situation of its emergence and the anthropos constantly challenges the subject. In the end, however, the "who?" aims to deconstruct both the anthropos and the subject insofar as these give themselves as a priori essences.

Stiegler's "who?" is therefore closest to Heidegger's Dasein, which is also defined in section 9 of *Being and Time* as a "who?" in contrast to the designation of the being of the present-at-hand in terms of a "what?" Like Heidegger's Dasein, Stiegler's "who?" is the relation between the existent and its world, the world being first and foremost the everyday world that the equipment (*Zeug*) opens up as a relational totality (Heidegger) or associated milieu (Simondon). In *Being and Time*, the "who?" of this everyday world is inauthentic being-with, which can only be identified as the impersonal crowd of "the they" (*das Man*), that is, as anybody and nobody. From Stiegler's point of view, this downplays the factual role of equipment that individuates being-with because, for example, it allows differentiating between different human situations (Zinjanthropic gatherer-hunters, Neolithic cultivators, and people living in the contemporary digitalized environment do not amount to the same). Following Derrida, Stiegler thinks the technical "what?" as a prosthesis. Stiegler's "existential analytic of prostheticity"[37] shows how the prosthesis contributes to the individuation of the "who?"

Now, who exactly is the "who?" individuated by the prosthetic "what?"

Firstly, in order to emphasize that it has no natural or native identity at all, Stiegler defines the "who?" as the *originary default*. Being originary, the default is not the biological-anthropological fact that the human being is empirically too weak, dependent, and fragile to survive without the supplement of technics. Instead, being-in-default is the fundamental existential way of being of the "who?" just as being-at-fault (*Schuldigsein*) is the way of being of Dasein.[38] Unlike being-at-fault, the being-in-default accounts for prosthetic supplementation, which is therefore also an originary possibility of the "who?" "Religion, speech, politics, invention—each is but an effect of the default of origin."[39] Being the default, the "who?" is need and desire, the need and the desire for a satisfaction that no positive property will ever give but that will only ever be provisionally patched over by the prosthetic supplement. A prosthesis can never fulfil and satisfy the originary lack of the "who?" but can only provide a supplement that remains fundamentally alien to it. By retelling the default of origin first through the Platonic myth of Epimetheus (who, at the moment of the creation of all species, forgot to give proper characteristics to the human) and Prometheus (who supplemented to this accidental lack by giving them the prostheses of *techne* and political *sophia*), Stiegler illustrates the lack of essence defining the human and adds that even this essential non-essentiality is not a necessary feature of man but an accident that fell upon our kind and determined it accidentally forever. The human has no nature at all but is an originary default that calls for prosthetic supplementation.

In the end, the default is a fundamental existential characteristic of the "who?" because the sense of its existence is temporality. *Being and Time* has shown the ontological significance of the fact that because existence is temporal, it is necessarily incomplete as long as it lasts. This becomes the singular certainty-of-Dasein in being-toward-death, which grounds the discovery of all three temporal ecstasies. Like Heidegger, Stiegler thinks that the sense of existence is time, but unlike Heidegger, Stiegler situates the origin of temporality in technics. Existence is needy in relation to what might come (symbolized by the ills and hopes hidden in Pandora's jar). Existence is also care of existence. Existence appears in default before the future: careful, because conscious of being unsheltered before the future, or careless, because mindless of the future. Existence appears as default precisely in relation to what might come, and this is why it calls for supplementation. Technics is an effort to prepare for the future, and this preparation becomes desirable because existence is filled with "expectation, conjoncture, presumption and foresight."[40] This is why technics is for Stiegler what gives time and also what gives the world-time shunned by Heidegger.

Secondly, the "who?" is therefore individuated precisely by the prosthesis that is already there (*déjà là*). The technical prosthesis is already there as the epiphylogenetic memory (or tertiary retention, as Stiegler says in *Technics and Time 3*) that is inherited from people who have lived before and is shared by people in the same situation. The technical prosthesis implies its associated milieu. From the point of view of the "who?" the associated milieu is assimilated to the world in which the "who?" finds itself "thrown." It is already there like Dasein's *Da* is always already there, but Stiegler's explication of the world differs from Heidegger's insofar as Stiegler lays it out much more explicitly and decisively as the horizon of a technical object or system.

With the notion of the already-there, Stiegler is very close to Heidegger. To start with, already-there is a reinterpretation of the everyday being-in-the-world described in *Being and Time*. Like Dasein, the "who?" is always already thrown in the world it encounters in its everyday concerns. Like the Dasein, the "who?" encounters this world first and foremost as equipment that lays out the world as a work-world. But as we saw, while Heidegger thinks the equipmental work-world only defines Dasein in its impersonal inauthenticity, Stiegler is interested in how it can nonetheless contribute to the individuation of the they that the "who?" is for the most part. As the "they," people are individuated by the available totality of equipment. Absorbed in everyday life, they are not aware and conscious of the technical world in which they live, but it envelops them like the invisible element they are nonetheless adapted to.

Furthermore, the "already-there" is not simply the past but what Heidegger calls "having-been" (*da-gewesen*, *Being and Time* § 73), that is, the past that is given as a heritage to be inherited or abandoned (§ 74):

> The greatest point of proximity between existential analytic and Epimethean thanatology—one that also harbors the greatest divergence—is the theme of the "already-there": "Dasein has either chosen [its] possibilities itself, or got itself into them, or grown up in them already." Even those that Dasein chooses always originate from the world that is already there: every "understanding of the being of those beings . . . become[s] accessible within the world." To accede to the *who* (Dasein) is to approach this *who* in its "average everydayness."[41]

As section 74 of *Being and Time* states, Dasein exists natively because it is necessarily born into a historical world: this world is the having-been that Dasein can assume or not assume as inheritance. Stiegler agrees with the characterization of already-there as an inheritance, but he distances himself from Heidegger's notion of historiality in two respects. First, as we saw, he emphasizes that the past is only ever accessible to Dasein as a factical prosthesis.[42] The factical prosthesis is not only a great work of art or philosophy, as in Heidegger's works after *Being and Time*, but above all ordinary technical equipment. The prosthesis names a past that has not been experienced by Dasein nor recorded by it but by people before it and that it inherits imperceptibly. Second, historiality is not directed toward the history of being but toward everyday life: "Everydayness is the inauthentic modality of the historiality of Dasein that is neither in time nor in history, but is temporal and historial. Its being 'in' time is, however, possible, indeed unceasing: one need only think of the clock and the calendar. That said, its temporality makes intratemporality possible in the first place, and not the reverse—while being co-originarily intratemporal."[43]

Heidegger's Dasein tears itself away from the inauthentic world of technics in the solemn moments of anxiety and being-toward-death that give it access to its own authenticity and ultimately to the question of the meaning of being. In Heidegger's later works, the poetic and philosophical language tear themselves out of empty everyday babble in order to make a gesture toward the question of the meaning of being. Contrary to Heidegger, Stiegler thinks that precisely the inauthentic "what?" opens the way to the individuation of the "who?" It must not be pushed aside but its ways of signifying must instead be examined in detail.

Of course, as Heidegger says, being-in-the-world and having-been cannot individuate authentic Dasein, although they can determine or "destine" the being-with (*Mitsein*) as "people" (*Volk*, § 74). To put it in Stiegler's terms, the "what?" is the common epiphylogenetic memory that does not individuate an individual but a human group. The group gathered together by a shared epiphylogenetic memory is often situated in terms of its ethnicity and its historicity. Until our times, a technical community has generally coincided with an ethnic community (i.e., until the contemporary world started to develop toward a global mega-ethnie[44]). Stiegler's question is how ethnicity is related to technicity. Of course, a group's technics does not reflect its supposed common biological inheritance (as an anthill is thought to reflect the ants' genetic inheritance). It does not reflect an equally fantasmatic cultural inheritance either, a "spirit of a people" that would express itself in a cultural, artistic, and technical world (like the romantic philosophy of history had assumed and as Heidegger ended up reaffirming even though he claimed to "destroy" the romantic tradition).

Contrary to this, like Leroi-Gourhan, Stiegler thinks that although the technical fact can coincide with an ethnic situation, technics is fundamentally a relation between the human animal and the environing nature, which has its universal features, and this is why, although the technical fact has generally been local, the technical tendency is universal.[45] Because of its universality, technics has always provided ways of escaping from the narrow limits of ethnic communities and already in prehistoric times good technical inventions passed easily from one ethnic community to another while languages, for example, and the cultural formations transmitted by them (i.e., religions) were transmitted much more slowly.

The already-there determines the group as its past. The group's historicity is determined by the already-there that can be adopted thanks to the technical (epiphylogenetic) memory that gives "access to a past that was never lived, neither by someone whose past it was nor by any biological ancestor."[46] The adoption of technics enables the adoption of an artificial past "through which a common future can be projected."[47] However, the real sense of the already-there is not in the past but in the future. A human group is held together by the desire for a common future rather than simply by a common past.

> For Leroi-Gourhan, the unification process is one of adoption through which it is possible to construct, solidify, consolidate, perpetuate, and *extend* a *We*, to amass other I's and other We's. The general rule is to define this constitutive social-ethnic-group as sharing a *common past*, and this ethnic way of thinking is also how the ethnic, and the territorialized community in general, thinks (about) itself. Yet such a definition, giving credit to a myth of pure origin and coming from a past that is transmitted *locally*, is structurally and literally *phantasmagorical*: groups are founded through their common connection to a *future*. Ethnicity (and beyond that, all human social grouping) is above all the sharing and projection, through the group itself, of a *desire* for a common future. No human group is possible without desire; the link to the future controls ethnicity's "unifier-to-come."[48]

So far, we have seen how the adoption of a common already-there contributes to the individuation of the group, but the analysis has remained on the "inauthentic" level of the "they," such that the group is barely aware of itself as a group. How can it come to identify itself as a group and how do individuals relate to themselves and to each other within the group?

Third and finally, the "who?" can be individuated when it designates itself as the "I" or the "we," and this requires a new, inventive relation to the prosthesis. Let us first see what Stiegler means by individuation.

"My claim is that *I* and *We* are individuation processes in the Simondonian sense: the individual, whether psychological or social, and although the *We* is not indivisible as is the *I*, is an incomplete process of a metastable equilibrium: it is neither in stable equilibrium, which would be its completion, nor disequilibrium, which would be its decomposition—either leading to its disappearance."[49] The individuation processes in the Simondonian sense describes individuals in the making instead of individuations made by already constituted individuals. Individuation presupposes a preindividual reality whose virtualities lead to individuation and it produces both an individual and its milieu. Individuation itself is a becoming in which a problem remains in what Simondon calls a metastable equilibrium, that is to say, a provisional, changing, and unstable equilibrium that holds a system's energies in the face of the ever possible increase of entropy.[50]

For Stiegler, not only the affects and perceptions described by Simondon but above all the tertiary retentions (or epiphylogenetic memory) that constituted the group's unconscious already-there are the preindividual reality whose tensions and oversaturations can lead to individuation:

> This "charge of pre-individual reality" is a potential of adoption. The individuation process results from an irreducible inadequation at the heart of the individual, as always incomplete but also as the play of "pre-individual forces" in the individual: interiorized, interpretable tertiary retentions that are equally at play in the social individuation in which the psychic individual participates in the individuation process. Interpreted in this way, the pre-individual (different from Simondon's interpretation) is the "already-there," the potential for an inadequation instantiated by the *psychological* individual. But this also creates the *social* individuation of the group, in such a fashion that it is also the bearer of the same force of pre-individual reality as the potential differential of inadequation.[51]

For Stiegler, the already-there is a tertiary retention, a prosthetic memory, which is by nature external and at least partly nonconscious. It may operate (calculate) well, but it is also traversed by unnoticed tensions, inconsequences, and incoherences. These can become problematic if they disturb psychic, social, or natural equilibriums. Individuation happens when this kind of a problem is taken charge of. Individuations are metastable processes in which the inconsequences, *dephasages*, and in general entropic tendencies are dealt with (more or less successfully, of course). Like Simondon, Stiegler thinks that individuations are both psychological processes of the *I* and social processes of the *we*. Individuation happens essentially in relation to the preindividual milieu consisting of the tensions between these processes: "The inadequation animating both the *I* and the *We* is first and foremost an inadequation of the *I for the We* and of the *We for the I*; their ideal projective convergence is effectively an originary divergence in their individuating dynamic. *I* and *We*, in forming the two faces of the same individuation process, do not coincide."[52] The *we* is, so to speak, a part of the preindividual milieu of the *I*, whose tensions it incorporates and tries to bring to equilibrium, and similarly the *I*'s are part of the preindividual milieu out of which the *we* constitutes itself.

The individual and the collective who identify themselves as *I* and *we* provide a novel interpretation of the community. Although the previous description of a group could evoke a communitarian substance, this is not Stiegler's final word on the community. He certainly does not think of the *we* as the collective substance like the people described by Hegel in his philosophy of history or by Heidegger in his thinking of historicity in his middle period. The *I*'s are not given, either, such that the community could be their rational communication, as Habermas describes.[53] Instead, Stiegler follows Simondon's way of thinking both psychological and collective individuation in terms of *transindividuality*. Transindividuality means the reciprocal constitution of individuals. For Stiegler, they do so against the preindividual milieu, which can be psychological and social only because it is first of all a technical milieu consisting of artifacts and writings. In contemporary society, as we shall see in more detail in the next subchapter, the technical milieu has become so overpowering that it tends to choke both individual and collective individuation—but even in this society, the empowering of society depends on the free use of available technics.

Individuation means asking who I am or who we are. It does not mean giving a fixed and definitive answer to these questions. On the contrary, it does not matter if the resulting individualities are ephemeral and fragile because individuation precisely does not aim at establishing a definitive identity but at freely using the virtualities of the preindividual (technical, social) milieu in living, doing, and knowing. Individuation happens as life-knowledge, work-knowledge, and theoretical knowledge (*savoir-vivre, savoir-faire et savoirs théoriques*), and these are always in becoming.

Individuation draws from the virtualities of the already-there. For Stiegler, they are above all technical virtualities in the most general sense of the term. As such, they are always "pharmaceutical," they enable and care for life but they also incorporate toxic and invasive automatisms. An essential example of the former is education: there is no individuation without education, education being essentially an apprenticeship to the available technics. An important example of the latter is the overwhelming and dazing effect of contemporary commercial techniques.

Individuation needs education and habituation and these rely on tertiary retentions. The aim of individuation, however, is to convert tertiary retentions into secondary ones, that is, to change technical prostheses into personal experiences—and to make an experience of the prosthesis. An experience is necessarily aware of itself as an experience and this is why it is the starting point for *noesis* (thinking). The epigenetic experience is the invention of a way of combining the "who?" and the "what?" The experience in which the "who?" endures the "what?" then turns into invention, in which the "who?" produces the "what?" Invention is not a *creatio ex nihilo*, for it always draws from the virtualities of a preindividual milieu that are already there, but it is nonetheless a new way of using the already-there. Stiegler also describes this as the invention of a *style*: the "articulation of the *who* and the *what* implements an indetermination that is simultaneously the enigma of style—an enigma resulting from its indescribability."[54] Style is the extreme point of idiomaticity and singularity that allows the identification of an ethnic personality or a moral personality. It starts from the already-there, but it delocalizes it and deterritorializes it, transfers and translates it, until it finally reaches an untranslatable idiomaticity. For example, a musician necessarily starts from available instruments, scales, and sonorities, but the musician can create something unheard-of from this. As Stiegler says, this is why Mozart's style is at the same time the German musical style of his time and a unique movement within it and out of it. The style is born as a new relation to the available technics and it can lead to the invention of new technics altogether. The gesture of seizing the technics may be made by an *I*, but it always happens in relation to an at least potential *we*. This is how the *I* and the *we* are individuated together.

Today, however, the possibility of inventing and creating new styles is threatened by a new technical culture. Let us now see what Stiegler means by this.

Today

A large part of the numerous books and other interventions that Stiegler has produced after *Technics and Time, volumes 1–3*, is dedicated to a diagnostics of our world and to a kind of a Foucauldian metaphysics of the present. In what follows, I will not delve into his prolific and generally very somber readings of what our epoch consists of, but I will show how these analyses use, make explicit, and refine the philosophical concepts developed previously.

Stiegler's diagnosis of an epoch always starts from its technologies. According to him, the technologies that determine the contemporary hyperindustrial epoch are the *technologies of the spirit*: television, media, culture industries, program industries, and telecommunications. These were already influential when Adorno and Horkheimer first wrote of cultural industries in 1947, but they have gained an unprecedented power and unimagined new applications today thanks to new computational technologies. These technologies are cognitive technologies whose resource and end product is the human spirit itself (or mind, soul, or consciousness, if you prefer). According to Stiegler, the industrial, automatized production or spirit now composes the "what" and the "who?" in a new way and this determines the present phase of what we investigate under the name of technological humanity.

What is the "what?" of the new cognitive technologies, especially in their contemporary computational form? Firstly, these technologies can be called cognitive precisely because they have the role of memory. Media and culture technologies produce recordings of image, sound, and text, and the new digital supports in particular enable huge and, in principle, worldwide archiving and distributing of these recordings. They are *hypomnemata* that function like the writing Derrida analyzes and Foucault's description of the writing of the self. They contribute in many ways to psychological and social individuation. In this role they are functionally not that different from past mnemonic techniques such as the book. The worldwide hypomnematic archive may be more standardized than the libraries of old, but it is also bigger and quicker than any archive humanity has ever known. However, Stiegler thinks that their effect on consciousness has changed because this memory consists also in cinematic and sound recordings. Because these are isomorphous with the soul, they therefore affect it strongly by formatting ways of perception already on an unconscious level, as Stiegler shows in *Technics and Time 3* especially. He also thinks that this formatting has become far more intense as the production of memory does not need to reflect an author's own individuation but has become industrial.[55]

Secondly, the new computational technologies differ from older mnemonic supports because the digital archive not only conserves data but also organizes, recommends, and generates it. Moreover, new technologies address not only consciousness but human behavior more generally (rhythms of sleep and work, forms of community and communication, ways of moving and traveling, etc.). All of this is possible because computational technologies bring the principle of calculus to the fore. This principle belongs to all technical supports but it is the explicit core of these technologies. They do not just repeat the data, they realize entire algorithmic operations that result in new data (i.e., existing data organized in new ways). Realizing multiple noetic operations, they can manage whole areas of society and this is why they give the impression of "thinking." Computational agents seem to have far more noetic power than their human users because they calculate so much faster and more flawlessly than humans, thereby generating so much new data that a human memory is overwhelmed if it tries to follow and understand its generation. But at the same time, as we have seen, these machines still do not have the intuitive power characteristic of the human mind. Individuation can only happen if a human mind "sees" what the machine mind "calculates." When the former is overwhelmed, the latter produces for nothing.

Thirdly, the technologies of the spirit (internet, television, the digital management of life) provide the associated milieu in which practically everybody lives today, mostly directly but always at least indirectly (today, access to computational technologies is far from limited to wealthy populations, and the living conditions of all are affected by computerized globalization). The oldest and best known version of this milieu is the consciousness formed by the media especially in its cinematic and radiophonic versions. According to Stiegler, they tend to format consciousness more efficiently than text because cinematic image and sound are isomorphous with the soul itself. To use the expression coined by Derrida in his dialogue with Stiegler in *Echographies of Television*, teletechnologies create an "artifactual" presence.[56] Radio and television especially excel in the creation of a sense of topicality when they decide what is important news (and what can be left out). But they at the same time stage and frame this news by the way they present it in a certain setting and with certain rhythm, by submitting the initial choice of this very dispositif to commercial imperatives. Today, this milieu has been powerfully enlarged and enhanced into a digital presence, where it may be more difficult to track down all the interests behind an artifactual creation (Whose interests lie behind such and such a platform?), but it is obvious that the very architecture reflects the needs of corporations like the GAFAM (Google, Apple, Facebook, Amazon, Microsoft) or more generally what Stiegler calls 24/7 capitalism. The media presence that the public is aware of is but just one part of the newly organized digital reality. In *Automatic Society*, Stiegler refers frequently to the Foucault-inspired works of Antoinette Rouvroy and Guido Berns on "algorithmic governmentality" in order to show how contemporary society is profoundly and increasingly managed by automatic computational systems.[57] When public governance (e.g., taxation, education, justice, war) but also different services (e.g., commerce, banking, transports) comes to rely, to a massive extent, on automatic systems, and even if these systems are efficiently and cheaply organized, it implies a shift from public governmentality to private management, where it is no longer a matter of deciding together what should be done but simply how to unquestioningly realize decisions already made.

What happens to the "who?" that is supplemented by this kind of "what?" in the contemporary world? It is easy to see that its "already-there" is the associated milieu provided by the new noetic technologies. People, in a very general and global sense, find themselves thrown into a reality in which their consciousness is fed into by omnipresent cinematographic, phonographic, and digital media and their entire lives are embedded into an increasingly pervasive algorithmic governance. The computational "already-there" is very intense and it is increasingly difficult to escape from. At the same time, it is an artifactual stage, just a displace made of pure transitivity and not a *Da* that one could inhabit.[58] For Stiegler, the computational already-there is alienating because it is created by global capitalism at the expense of all local belonging.

The fundamental problem of the computational reality operated by "24/7 capitalism" is how it blocks processes of individuation. This reality is filled with industrially produced messages that seek only to increase consumption. This functions best when the messages solicit only those elementary drives that aim at immediate consumption and destruction at the expense of addressing desires that would lead to inventive and creative individuation. The messages that solicit the attention of elementary drives are produced so intensely that they fill the entire space of attention, not just increasingly our waking time but also the time that should be consecrated to sleeping, dreaming, and resting. This is how the noetic industry ends by exhausting attention and killing the capacity to desire. In such an apathetic state, individuation becomes impossible. This is tragic for the individual who no longer sees life as a desirable opportunity, but it is also dangerous on a social level, where it leads to what Stiegler calls "systemic stupidity and beastliness."

This diagnosis is harsh. It takes the cultural criticism Adorno and Horkheimer first formulated in an even more dismal direction. It is important to note, however, that contrary to these predecessors, Stiegler does not label the new technologies as being destructive as such. Like all technologies, they are *pharmaka*, open to both destructive and caring uses. If Stiegler focuses on identifying their toxic potential, he also calls for the development of their emancipatory potential. For Stiegler, technologies are always politically important because they are the milieu in which properly political individuation can take place. Contemporary computational technologies are a major political concern because they intervene directly in the psychic and social individuation, which is the very domain of politics. They can be used for education, a vital part of democracy since the Enlightenment. But for the same reason they can also be used for something that could be called anti-education: active and programmatic stupidification and beastification. It is therefore indispensable, Stiegler thinks, that we actively study their role in different processes of identification, counter their ravaging effect, and foster what Stiegler calls *ecology of spirit* that leads to a political economy of spirit. In other words, what Stiegler wants to fight is not digitality as such but its rapacious use in 24/7 capitalism, and what he wants to promote is its capacity to enhance desire and individuation. I want to conclude my reading of Stiegler by quickly summarizing the two basic challenges that his techno-politics identifies. Both indicate how the relation between the "what?" and the "who?" should evolve today. First, for many first readers of Heidegger's critique of the epoch of technology, Stiegler does not reject modern technology in general or media and computational technologies in particular, but rather he is for developing them in more inventive and creative directions. To put it in Marxist terms, people should revolt against noetic exploitation and seize the means of noetic production. On the level of technics itself, the root codes especially should be open and free and, further, computational know-how should be taught openly and widely. But independently of purely technical know-how—which is after all not what everybody wants to spend their time on—new technologies are already used well when they become means for the invention of new styles of using them and thereby of individuation. This is by nature a collaborative process in which people contribute to one another's individuation and thereby care for the common world. We can see that Stiegler's utopia has a Marxist undertone. The end of work is the leisure to work differently: to produce, to build, to create, and to invent.

The second challenge Stiegler identifies is the ecological destruction that is caused not only by computational capitalism but by the thermodynamic technology that made industrialization possible in the first place, as each Intergovernmental Panel on Climate Change (IPCC) report makes ever more obvious. Of course, this traditional heavy industrialization is strongly enhanced by computational capitalism, which continues to increase fossil-fuel–based pollution without much soul-searching and which extends its carelessly polluting ways of functioning into new areas, in particular into biotechnology. Stiegler analyzes the vast problem field of anthropic pollution and destruction under the general term *entropy*, which the present industrial-economic system increases and seeks to combat with new negentropic practices.

It seems to me, however, that Stiegler ultimately holds to an anthropocentric view of ecology in which non-inhuman existents should take care of the world over which they have extended so much of their technical control. This is of course true. But in order to understand how this could be done, it seems to me that we should step definitively beyond the problematics of technological humanity and extend the analysis of *bio-technical existence* further than the relation between human beings and their technical supplements. We should also pay attention to the bio-technical relations between humans and other living beings as well as between technical and living nonhuman beings. Bio-technicity is the common condition of human and nonhuman, living and nonliving, living and technical beings insofar as they supplement one another and therefore touch one another ever so curiously, skillfully, and technically.

From Foucault to Agamben: Apparatus and Use

Giorgio Agamben pushes Foucault's ideas of anthropo-technics to their most contemporary conclusions and interprets them on a resolutely ontological level. At stake here is not any specific historical constellation but the very notion of subjectivity defined in terms of apparatus and use. On the one hand, he reinterprets biopower in terms of apparatuses. These are not only social power dispositifs but any kinds of machines, starting from concrete appliances such as televisions, cell phones, and computers. As we shall see, according to Agamben, all apparatuses can determine communities by orienting subjectivation. However, like Stiegler, Agamben thinks that contemporary apparatuses especially tend to induce a desubjectivation that can insidiously turn forms of life into a barren "bare life."

On the other hand, we shall also see that Agamben reinterprets Foucauldian technologies of the self in terms of *use*. Traditionally, use denotes the subject's relation to instruments. In his reinterpretation of the term *use*, Agamben deconstructs the classical sense of instrumentality by examining the subject's own instrumentality and mediality. He extends the question of the use of instruments beyond the domain of dealings with technical objects into the domain of dealings with other human beings and gods. This is how he extends the Foucauldian idea of self-techniques into the field of hetero-techniques, in which human beings have an instrumental relation to one another. In use, the human being not only forms itself but it forms others and is formed by others. The notion of use also attracts attention to the skill needed in the manipulation of instruments—as we remember, skill was the first, Greek sense of *techne*. Skillful use is not simple utilization of exploitable objects because it can also be a learned interplay with the instrument or with the other person. If so, it does not necessarily aim at exploitation but at enjoyment: it does not consume the other but allows the flourishing of both. This is why use can enable the explication of Agamben's utopia of the "coming community" when developed into its full potential as a form-of-life.

Before explaining these charged terms and the structures they articulate, let us see how they follow from some of Heidegger's and Foucault's ideas. These motivate Agamben's political ontology, whose key terms are *apparatus* and *use*.

THE PROBLEM FIELD LEFT BY HEIDEGGER AND FOUCAULT

The most straightforward sense of technological humanity is humanity as a technological product. The temptation of producing humanity technologically has animated all modern totalitarian projects, and this has also been the motivation behind the most important critiques of modern technology, as Agamben emphasizes.

A given technics does more than produce a technological community by calling for cooperation between its users. As Stiegler has shown, whether people are conscious of it or not, a technology forms the community of people who use it habitually. Agamben learns this especially from Heidegger and Foucault. In his interpretations of the *Ge-stell* of modern technology, Heidegger shows how it captures people as its "human resources": modern industrial technology (rather than simple ideology) insidiously formats, for example, the Leninist or Soviet human type, the Nazi human type[59] and also the liberal American human type. The modern community is therefore not grounded on biological types or spiritual figures, as nineteenth- and twentieth-century nationalist ideologies had expected, but on a technological type that determines the epoch itself. For Heidegger this was a catastrophe, as for him a historical (national) *Geist* is closer to being than a technical (and essentially supranational) *Ge-stell*. We, however, have no reason to think that a technical type of human would be any worse than a national one—many at present would even think the opposite. Yet Heidegger is of course right in showing that technologies contribute to what a Dasein sees as its possibilities. For example, it is easy to see how the car makes the driver of a petrol-run car or how the computer makes the user of a computer connected to the Internet. Such a formatting obviously consists of specific skills needed to use the machine (knowing how to drive a car and how to use a computer) and more insidiously of ways of doing and thinking brought about by the machine (in a car culture, people believe that they need to travel fast and far; in a computer culture, people's very ways of thinking change so that they no longer think in a linear and focused fashion but in leaps and by multitasking. In such a culture, a school child is no longer taught as the subject of its thoughts but as a hub through which information transits). The type of human that these machines produce consists of conscious desires that wouldn't exist without the possibilities engendered by the machine, like moving fast and communicating over large distances. But there is even more at stake. The often overlooked infrastructures needed to keep the machines running actually build a whole new world, in this example a world consisting of an energetic system based predominantly on fossil and nuclear fuels and of a commercial system that assembles and sells the machines worldwide instead of locally. As Gilbert Simondon has shown, all technical objects produce their *associated milieu*,[60] and as Stiegler has said in a continuation of this, today this milieu has expanded into the worldwide system of assembling and using modern industrial goods in which technics also has a huge pharmakon effect on its milieu.

Foucault studied the effects of different technics on people in more detail than Heidegger did; furthermore, all techniques have social power in his readings and are integrated into power techniques. Disciplinary techniques and biopower especially turn individuals into populations, which are economical units (resources, as Heidegger would say) and objects of domination. Foucault explains this process particularly well in his Collège de France lectures from 1975 to 1976 entitled *Society Must be Defended*, where he elaborates on his distinction between sovereign power and biopower. For him, the distinction is first of all historical. Sovereign power dates from the Middle Ages and develops around the problem of the monarch and the monarchy,[61] while disciplinary power dates from seventeenth and eighteenth centuries and is a nonsovereign power invented by bourgeois society in order to extract time and labor from bodies,[62] and the "biopolitics of the human race" that has emerged from the second half of the eighteenth century onward is a nondisciplinary power applied by new power technologies "not to man-as-body but to the living man . . . to man-as-species."[63] More importantly, the distinction between sovereign power, disciplinary power, and biopower is a conceptual distinction between different forms of power: Sovereign power is the right of life and death—in practice the right to kill—while modern biopower is on the contrary the "right to make live and to let die."[64] This modern biopower controls and enhances things such as the birth and death rates by means of hygiene and medicine in particular but also fundamentally through racist measures. Modern biopower starts as a disciplinary power that dresses bodies to be more productive, but it flourishes properly as a biopower that aims at making life itself healthier and more robust by controlling "the relations between the human race, or human beings insofar as they are a species, insofar as they are living beings, and their environment, the milieu in which they live."[65] In Stiegler's terms, we could say that the pharmakon effect of modern biopower is on the positive side a healthier population and on the negative side a racist society. It is very important not simplify very complex realities, for example by equating such claims with the claim that the public health system is a totalitarian system (as Agamben's silly remarks on COVID-19 suggest[66]). Yet we should still be aware of the dangerous effects of the use of any well-intended techniques.

Sovereign power, disciplinary power, and biopower are power technologies, but interestingly in the present context they function differently. Sovereignty, says Foucault, "Is a theory that establishes the political relationship between subject and subject"[67] by assignating their capacities and powers in terms of legitimacy or law. Disciplinary and biopower, on the contrary, are not established between subjects at all, for they are effects of an asubjective technological power. As such they are forms of domination that contribute to subjectivation—concrete relations of subjugation that "manufacture subjects."[68] Foucault's objective is to "identify the technical instruments that guarantee that they function" by examining "the heterogeneity of techniques, and the subjugation-effects that make technologies of domination the real fabric of both power relations and the great apparatuses of power."[69] As Agamben puts it, disciplinary and biopower function as *apparatuses*, a term that draws both from Hegel's *positivity* and Heidegger's *Ge-stell* but which Foucault thinks of as a *dispositif*, a term that he uses to refer "to a set of practices and mechanisms (both linguistic and nonlinguistic, juridical, technical, and military) that aim to face an urgent need and to obtain an effect that is more or less immediate."[70] While sovereign power may *use* different techniques for governing its subjects, biopower really *is* nothing other than an assembly of heterogenous dispositifs that manufacture subjects. Sovereign power is truth because it is law, whereas biopower is mostly unconscious. Science is among its techniques, not as truth, however, but as another element of the technical dispositif. Because sovereign power belongs to subjects, classical juridical theory conceives of it as something that can be possessed like a commodity and therefore also transferred or alienated. Bio- and disciplinary power, however, are not commodities that a subject could possess or acquire but only relationships of force that contribute to subjectivation.[71] Agamben will open up the possibility of thinking power relations in terms of *use*, rather than possession, such that power uses techniques and is used by them, although it neither possesses nor really controls them. Yet Agamben himself tends to reserve the term *use* for the human use of things and humans.

In *Society Must be Defended*, Foucault describes the advent of the modern biopower that regards people as a population to be improved by means of hygiene and medicine especially. Biopower not only regulates the details of everyday life but it changes something in the human condition itself because it touches on its liminal conditions, death, and birth. As Foucault says, biopower "has no control over death, but it can control mortality."[72] Individual death becomes unimportant, invisible, or even shameful, whereas on the level of the population mortality becomes a cause for concern. On the other hand, Foucault claims that since the nineteenth century, scientific, medical, and educational institutions have been obsessed by sex—not as a source of pleasure but as the means for reproducing the population that is itself expected to be healthy and productive. Unproductive pleasure is thwarted by all means as an obstacle to productive reproduction. Seen especially through these limit conditions, the modern state is obsessed with the reproduction of ideal population, which in the worst case it ends up understanding as a superior race for whose sake the suppression of the inferior race becomes necessary. This is why, according to Foucault, the modern state is fundamentally racist.[73] According to his profound analysis, races are not a given, they are produced by splitting the population into good and bad elements such that in the most extreme cases the inferior element is killed off so that the superior element can live a fuller life. Racism animates all modern states, including socialism, but was of course intensified in the colonialist necropolitics Achille Mbembe describes[74] and reaches its paroxysm in Nazism, which combined murderous biopolitics—thanatopolitics—with a suicidal sovereign power, as Agamben in particular has shown.[75] We see that biopower formats people in a much more intimate way than the Enlightenment (which was happy to *educate* people first in the use of productive techniques and ultimately in the use of their reason and freedom as citizens) because biopower determines who, nay, what kind of bodies can live and what kind cannot and selects what bodies are deemed worthy of being educated to full citizenship. The object of biopower is not the individual but the species life. This is why it is so profoundy alienating and treats people only in terms of what Heidegger would call the inauthenticity of *the they*.

In what follows, we will retain only one element of Foucault's famous, thought-provoking analysis, namely that biopower is neither the will of a subject nor a law, but it is an *asubjective technique* or *mechanism*. To be precise, it is not one singular mechanism but the entire dispositif of an apparatus, an agglomeration of many simultaneous techniques that marks a given situation. It is a technique without an identifiable subject that wields it; it is a technique without an identifiable object; it consists purely in its functioning, which then results in subjectifications, in the gradual and always provisional production of a population that appears to be economically viable. At the same time, biopower is profoundly racist. In order to produce a well-formed life, it identifies deformed life and seeks to oppress and even suppress it so that the chosen population can better flourish. These ideas Foucault launched were later developed especially by Giorgio Agamben, Roberto Esposito, and Achille Mbembe. In what follows, I will concentrate on Giorgio Agamben because his consideration of use brings a wholly new dimension to our question of anthropo-technics.

Agamben's Political Ontology

Giorgio Agamben, whose early work was marked by a more or less implicit debate with deconstruction (the idea of Voice developed in *Language and Death* is a veiled counterargument to Derrida's idea of writing[76]), has developed Foucault's themes especially in his vast *Homo Sacer* series (1997–2015). Continuing Foucault's ideas, he also elaborates on aspects of their potential to provide direction that differs from and even contradicts Foucault. To put it historically, he claims that sovereignty and biopolitics do not represent two consecutive periods in history but are the gradual coming to light of the hidden foundations of existence that were always there. The real crux of political ontology is their intersection. Foucault left this obscure, according to Agamben. Agamben's project starts from this obscurity: "The present inquiry concerns precisely this hidden point of intersection between the juridico-institutional and the biopolitical models of power" and locates it as follows: "*It can even be said that the production of a biopolitical body is the original activity of sovereign power.*"[77] Although Foucault says that the two forms of power can sometimes be mixed,[78] Agamben says something quite different when he claims that sovereign power precisely produces the bare life that can be the object of biopower. Contrary to Foucault, Agamben does not identify the sovereign with the law but relegates it to a paradoxical position in which it is at the same time outside and inside the juridical order.[79] The sovereign would not be sovereign if it was simply submitted to law, but nor is it outside of the law as an exterior political power that guarantees it (Schmitt) or as the supreme norm of a juridical order (Kelsen). It is an exception that is included "by means of the suspension of the juridical order's validity—by letting the juridical order, that is, withdraw from the exception and abandon it."[80] The sovereign maintains the validity of law through its capacity to suspend it. Symmetrically it can assign bare life to being both interior and exterior to law,[81] interiorized as exteriorized, defined by law as being beyond law's protection and precisely for this reason exposed to direct lawless biopower. Bare life is produced by the ban that does not exactly outlaw life but abandons it: "He who has been banned is not, in fact, simply set outside the law and made indifferent to it but rather *abandoned* by it, that is, exposed and threatened on the threshold in which life and law, outside and inside, become indistinguishable."[82] Bare life is human life that has lost the protection of its community and become vulnerable, exposed to violence. While the Foucauldian sovereign is the power to kill, the Agambenian sovereign designates the life that biopower can let die.

Bare life is not the same thing as natural life. It is the product of an "anthropological machine" that defines the human by excluding a part of itself that it judges to be nonhuman. In *The Open*, which is his treatise of the notion of nonhuman life (which the philosophical tradition since Aristotle has interpreted above all as animality), Agamben summarizes this point: "Precisely because the human is already presupposed every time, the machine actually produces a kind of state of exception, a zone of indeterminacy in which the outside is nothing but the exclusion of an inside and the inside is in turn only the inclusion of an outside. . . . What would thus be obtained, however, is neither an animal life nor a human life, but only a life that is separated and excluded from itself—only a bare life."[83] Bare life is thus the abstraction of a rational life deprived of its rationality, of political life divested of its political power. It is the helpless animality of the political animal deprived of politics: *inhuman life*. According to Agamben's famous claim "the Greeks had no single term to express what we mean by the word 'life.' They used two terms . . . *zoe*, which expresses the simple fact of living common to all living beings (animals, men, or gods), and *bios*, which indicated the form of way of living proper to an individual or a group."[84] Following this division, life in a community would therefore be *bios* and bare life excluded from community would not be the natural zoe but the zoe that bios excludes from itself.[85] For biopower to find life, sovereign power must firstly produce it by inhumanizing some part of human life such that it can be seized by governmental techniques. Bare life is therefore never given but always produced.

Agamben's theory of bare life has aroused intense debates, most of which were provoked by the historical examples Agamben gives of this structure. The first paradigm for the production of bare life is the *homo sacer* defined in ancient Roman law as the one who has been punished by banning it such that it can be killed without this act being homicide and without celebrating a sacrifice.[86] Agamben identifies the same structure in many situations throughout history: the *wargus* of ancient Germanic law,[87] the *Muselmann* at a concentration camp,[88] the Nazi euthanasia programs,[89] the refugee, the modern hospital patients in "overcoma,"[90] and the camp at Guantanamo.[91] To claim—as Agamben does—that the Nazi concentration camp is the biopolitical paradigm for the entirety of modernity[92] seems like a historical exaggeration, nay, a real mistake. According to Mika

Ojakangas, this is a misunderstanding of Foucauldian biopolitics, which does not capture bare life but *cares* for life such that it might even be said to produce what Agamben calls a form-of-life.[93] Provocatively, but in my opinion justly, Ojakangas says that Agamben's claim is an oversimplification of Western modernity whose fundamental biopolitical paradigm is not the concentration camp but the present-day welfare society (which is not a fascist invention but the sign of the excellence of a society). Its paradigmatic figure could well be the middle-class Swedish social democrat, who is obviously a product of a huge biopolitical machine but who certainly cannot be killed without committing homicide.[94] This being said, as Mathew Abbott observes, the reduction of Agamben's thinking to a political critique of modernity is an uncharitable reading,[95] for the true import of his theory is not a reading of political history or contemporary reality (although his interpretations of the states of exception proclaimed in the aftermath of September 11, 2001, were certainly sobering) but a political ontology of everyday existence as life in exception. Figures like *homo sacer* and the Muselmann are not universal ideas but only paradigms that help make certain historical phenomena intelligible. On the contrary, "Bare life is something very different. It is precisely not a paradigm: rather, it is a metaphysical problem that prompts the construction of Agamben's paradigms, which are the exemplary ontic figures that bring it to light. . . . Agamben is more primarily concerned with the historically contingent quasi-transcendental conditions of the biopolitical as such."[96]

In Agamben's ontology of political life, bare life is only one pathological aspect of human life, while its other, more fulfilling aspect is *form-of-life*. "Political power as we know it always founds itself—in the last instance—on the separation of a sphere of naked life from the context of forms of life."[97] If sovereignty produces the bare life that biopolitical technologies can capture, it produces it by blocking existing forms of life that also contain genuine possibilities of a form-of-life, which is the other originary modality of human existence. In *Means without End*, Agamben defines form-of-life as follows:

> By the term *form-of-life*, on the other hand, I mean a life that can never be separated from its form, a life in which it is never possible to isolate something such as naked life. A life that cannot be separated from its form is a life for which what is at stake in its way of living is living itself. What does this formulation mean? It defines a life—human life—in which

Humanity and Inhumanity of Technical Communities   237

the single ways, acts and processes of living are never simply
facts but always and above all possibilities of life, always and
above all power.[98]

This definition is clear but problematic. How could form-of-life be a life
in which it is never possible to isolate naked life if, for bare life to take
place, it suffices that a sovereign glance divests a person of its form?
Agamben holds to this dense definition—he repeats it and develops it
further in *The Use of Bodies*, where he refers the notion of form-of-life
to the *anthropogenetic* event and, further still, to the fundamental ontological structure of human existence.[99] It seems to me, however, that the
terms *form-of-life* and *bare life* should really be interpreted as two possible
modalities of existence.

It is instructive to read the anthropogenic event of the emergence
of the possibility of a form-of-life against the possibility of bare life as
Agamben's critical repetition (or even "deconstruction") of Heidegger's
idea of Dasein. Similar to how Heidegger's Dasein is not a substance
but a possibility of being, Agamben's form-of-life and bare life are not
human life as given, substantial facts but as spheres of possibilities of life
that articulate life as power and potentiality. Like Heidegger's existential
analytic where Dasein is divided into two modalities, *Uneigentlichkeit* and
Eigentlichkeit (inauthenticity and authenticity, or literally nonproper and
proper modes of existence), in Agamben's political ontology, life is divided
into bare life and form-of-life. These are not two distinct beings but two
modalities of life in which living itself is at stake. Living is taking and
giving form to itself: when it is separated from its form, it becomes bare
life, and when it is freed for form-giving, it is form-of-life, but both of
these are always possible as modalities of form-of-life. In opposition to
Heidegger, Agamben thinks of human existence as life. His definition of
life is interesting because life is for him both potentiality and form, such
that potentiality must be thought as the potentiality of a form, and form,
as ever renewed potentiality. Form is not for Agamben anything like a
stable *morphe* that can be imposed onto a *hyle*. Form is itself a potentiality. Not potentiality as a pure absence of determination (that would
be *hyle*) but as a form's inherent potentiality to be realized or not (i.e.,
the doctor's capacity for healing or not, the pianist's capacity for playing
or not playing).[100] Such a potentiality is not emptied in its realization
but becomes a new potentiality. On the other hand, the form not only
conditions its realization but results from it: it is a trace of experience

that life has made. This is properly realized as form-of-life, which is not a predetermined form but only ever a process of its own formation. It is not a form that a human being could decide to assume but the capacity for realizing its potentialities or not. The form in form-of-life is the form of the experiment that life makes with itself, without model, without aim but in such a way that at stake is its unknowable, unforeseeable possibility of happiness. For Agamben, human life is essentially a life of thinking redefined as "an experience, and *experimentum* that has as its object the potential character of life and human intelligence."[101]

As in the case of bare life, Agamben provides historical examples of form-of-life. Although he finds the term *eidos zoe* in Plotinus and *forma vitae* in Cicero, Seneca, and Quintilian, he presents medieval Franciscan monasticism in particular as a form-of-life that tried to conceive of life not as the application of a rule but as coincidence with it.[102] Although Franciscanism can indeed provide a *paradeigma* of life that withdraws from public political life and its sovereign law in the search of a more originary form of community, and that rejects property in favor of a simpler commonality of things, it is difficult to see how a religious community could exemplify an emancipatory form-of-life pointing at a contemporary form of coming community. But this empirical difficulty is once again secondary. Like the bare life, the form-of-life cannot be reduced to its paradigms but must be thought above all as the other modality of the biopolitical life as such.

Agamben's account of the human being—anthropogenesis—thus comprises two modalities: bare life and form-of-life. They constitute a political ontology in which bare life emerges where a form-of-life is banned from itself by sovereign power and form-of-life emerges when life escapes from sovereign power and liberates new potentialities in extant forms of life, thereby creating forms-of-life that open up the possibility of a coming community of whatever singularities.[103] These are not two historically or institutionally distinct political systems but two modalities that are possible in all real forms of life. Both push Foucault's idea of governmentality further. This idea was, as we saw, divided between the governmentality of disciplinary and biopolitical machines on the one hand and the technologies of the self on the other hand. On the one hand, Agamben situates the governmental machines in biopower, which was first made possible by the sovereign act of creating the exception in which the biopolitical machine can function. On the other hand, because form-of-life is a creation of self through experience and reflection, it must be read as a development of Foucault's idea of the technology of self. Both account for ethics.

APPARATUS AND USE

After this lengthy reminder of the main categories operative in Agamben's *Homo Sacer* series, it is high time to explicate their pertinence for our question of technics and especially of anthropotechnics. How does Agamben develop Foucault's original ideas of biopolitical and ethical technologies? I agree with Timothy Campbell when he says that Agamben's notion of bare life must be thought of together with technology, but I do not think that this means simply expanding Agamben's idea of the state of exception to also include situations in which the empirical "terrifying world of technology" "takes the center stage."[104] In Agamben as in Stiegler, technics must be examined as an element of an existential analytic. Agamben develops it further by bringing forth the ontology that underpins both what Foucault calls the governmental technologies that are constitutive of all situations of power and what he calls technologies of the self, which ultimately produce an ethical subject. Agamben reinterprets both these types of technologies in the context of his political ontology. But more importantly, he brings forth their common condition that he claims was left underdeveloped or even ignored by Foucault, namely *use*.

On the one hand, then, Agamben develops Foucault's idea of governmental technologies that he reinterprets in terms of *apparatuses*. In Agamben's work, the apparatuses are visible far beyond situations of social power in all kinds of apparatuses including techniques, works of spirit, and technological equipment. As Agamben says in "What is an Apparatus?": "I shall call an apparatus literally anything that has in some way the capacity to capture, orient, determine, intercept, model, control or secure the gestures, behaviors, opinions or discourses of living beings . . . also pen, writing, literature, philosophy, agriculture, cigarettes, navigation, computers, cellular telephones and—why not—language itself."[105] Like Bernard Stiegler, Agamben thinks that "apparatuses are not a mere accident in which humans are caught by chance, but rather are rooted in the very process of 'humanization.'"[106] Apparatuses belong to the anthropogenic event because they operate subjectification: instead of being simple tools of power technologies, they precede them and make their use possible. "Every apparatus implies a process of subjectification without which it cannot function as an apparatus of governance, but is rather reduced to a mere exercise of violence. . . . Apparatus, then, is first of all a machine that produces subjectifications, and only as such is it also a machine of governance."[107] Agamben's most extensive reading of governmentality is his study of the early Christian conception of the divine *oikonomia* in *The Kingdom and the Glory*, but more interesting for us is his reading of the governmentality of our contemporary world. Like Stiegler, Agamben thinks that today, governmental technics has less influence on people's subjectification than the material apparatuses that are characteristic of our phase of capitalism, especially cellular phones, television, and computers. According to both thinkers, these mainly generate stupidity and a mindless compliance with rapacious capitalism. The new governmental machine increasingly relies on further applications of these technologies in appliances such as biometric apparatuses, anthropomorphic technologies, and video surveillance.[108] Analyzing technical objects precisely as apparatuses and not simply as objects entails interpreting them as factors in subjectification and not simply as objects available for more or less clever use. The danger of contemporary apparatuses is that they interrupt subjectification and induce desubjectification, such that they generate what Agamben also calls a larval or spectral subject.[109] (This is not what Derrida calls *spectrality* because Derrida's specter acts without presence, whereas Agamben's spectral subject is a present subject divested of its capacity to act.) "What defines the apparatuses that we have to deal with in the current phase of capitalism is that they no longer act as much through the production of a subject, as through the processes of what can be called desubjectification."[110] Agamben's way of resisting this situation is different from Stiegler's: whereas Stiegler calls for a new industrial politics, Agamben calls for the *profanation* of divinized technologies in a way that owes much to contemporary art and literature, which is used to divert technics from its ordinary use.[111] Agamben's irritation with stupefying apparatuses is not a political program that aims at the creation of new institutions, it is a way of fleeing institutional politics toward an inoperative coming community of singulars. Like Jean-Luc Nancy's inoperative community, it is a description of being-with as a fundamental dimension of human existence, not as an economical or political organization.

For Agamben, technological communities are thus generated by apparatuses that can induce subjectification toward genuine forms-of-life (as when a flute transforms a music student into a real musician) but that can also produce desubjectification (as when people get hypnotized by idiotic reality TV shows). Now, how can the disabling tendencies of technological communities be resisted? Foucault sought answers to oppressive governmentality and desubjectifying apparatuses by studying the technologies of the self that developed into an aesthetics of existence and an ethics. But how should such technologies of the self be understood? In the *Intermezzo I* of the last volume of *Homo Sacer* series, *The Use of Bodies*, Agamben objects to Pierre Hadot's idea that to describe technologies of the self as an aesthetics of existence does not mean that life should be understood as a work of art that we make. According to Agamben, Foucault explicitly refuses this interpretation and thinks of the aesthetics of existence as a constant process of subjectivation that does not aim at realizing any given norms but that happens throughout the gradual process of the care of self. The aesthetics of existence is nothing other than the ethical life itself.[112] However, although Foucault's idea of the care of self is groundbreaking, Agamben thinks that Foucault fails to see the real meaning of the *care* that it is based on. Foucault still thinks of subjectivation to a certain form-of-life in terms of a power relation,[113] such that in the best case it amounts to the constitution of a free subject capable of parrhesia even against the sovereign political power. Contrary to this, Agamben thinks that what care of self really constitutes is "a possibility of a relation with the self and of a form of life that never assumes the figure of a free subject—which is to say, if power relations necessarily refer to a subject, of a zone of ethics entirely subtracted from strategic relationships, of an Ungovernable that is situated beyond states of domination and power relations."[114] Agamben reaches beyond subjectivity, which is always a function of a power that it uses or that uses it, toward a life that has its own powers that no sovereignty is capable of comprehending or capturing. He calls this the Ungovernable. It is an asubjective being that resembles Lyotard's description of inhumanity quoted in the beginning of this section: it is the intimate and secret form of inhumanity that is also the only possible source of resistance to systemic inhumanity.

Form-of-life is existence as detached from relations of power and governmentality, hence, from the desubjectifying effect of apparatuses. Although detached from power, the form-of-life is not detached from the world in which power functions, but it is a different relation to the same world. This means that form-of-life is also *in* a form-of-life or in a *bios* and uses *its* potentialities. In a sense, the form-of-life does not redouble the *bios* as such, but it echoes the *bios* as expelled from itself and banished into the condition of bare life. In fact, it is neither a form-of-life nor a bare life alone, but it emerges in the place of their relation.

Under sovereign power, this relation is the ban, and form-of-life emerges when the very relation of the ban is destituted and made inoperative.[115] "Form-of-life is this ban that no longer has the form of a bond or an exclusion-inclusion of bare life but that of an intimacy without relation."[116] How this destitution happens is not so clear: life under sovereign power is always potentially that of naked life. In worst cases, this is exemplified by life in a concentration camp, while in ordinary cases it can come forth as a simple political subjectivity reduced to formal citizenship defined by freedom and equality but deprived of singularizing features.[117] In the same world as these forms of bare life, but relating to it differently, Agamben sketches a totally different type of political subjectivation that is compatible with the coming community. This not only allows but requires the singularization of all forms-of-life, the sharing of commons instead of property and inoperativity instead of productivity. How and even whether this could happen in reality is unclear—maybe it is never anything other than another look at the same world. But why Agamben describes the liberation of a form-of-life from the sovereign ban is clear. He wants to show how a biopolitical de/subjectification can turn into a singular life that can no longer be reduced to the grasp of the sovereign. The joint that enables this turning between de/subjectification and singularization is the concept of *use*. This is Agamben's original contribution to our question of anthropotechnics. Use functions in many senses here. On the one hand, when people use apparatuses, they are actually used by them, thus formatted to their mechanisms. But on the other hand, a form-of-life is born when a self learns how to use itself in ways that are not reducible to apparatuses. In a form-of-life, "*the self is nothing other than use-of-oneself.*"[118]

The term *use* opens up yet another perspective in the question of technics that forms the horizon of this book. Use is a relation to instruments, the know-how of relating to means that is at the core of all technical activity. Through the notion of use, Agamben thinks on the one hand how humans use apparatuses and are used by them and on the other hand how they use other people and finally their own self. This allows him to describe the human self beneath the level of the free and rational consciousness studied by traditional philosophy and to articulate its normally hidden core in terms of the multiple forms of use. This is why his theory of use is probably the most elaborate anthropotechnical theory to date. It is not a theory of technologically produced or productive humans, but it is an existential ontological account of how bare life and form-of-life relate to one another in the use of bare life by form-of-life.

Agamben is by no means the first to pay attention to the term *use* while formulating a theory of selfness in terms of care of self, but he is the first to put it at the center of the analysis. As he emphasizes, the Latin *usus* translates the Greek *chresis*. Heidegger translated it into German with *Brauch*, which he uses in two important contexts: first in *Being and Time*, where Dasein's use of tools is its first world-opening activity and second in "The Anaximander Fragment," where *Brauch* is the way in which being is given as Destining: it is *moira* that uses the mortals and the immortals. The general structure of *moira*'s use of mortals is repeated in Foucault's dispositif and Agamben's apparatus, except that *moira* is the truth of being while the dispositif and apparatus are only technological functionality. But instead of reducing the question of instrumentality to the simple use of tools and more generally of the world, as in Heidegger's analysis of *Zeug* in *Being and Time*,[119] Agamben interprets it in all the senses given to the term in Foucault's reading of Plato's *Alcibiades*. As we have seen, in order to live well, Alcibiades must learn to take care of himself. According to Agamben, Foucault expresses this well, but he does not really problematize his own term *care*—which is, according to Agamben, based on the relation of use, *chresis*. Hence Alcibiades must learn how to use instruments (of gymnastics, arts, and war), but more importantly still, he must learn to relate to other people and to himself. All three of these types of relation to the world are characterized with the same word, *chresthai*. In order to care for himself, Alcibiades must therefore use things, use other people, and use his own body and mind in the good way.[120] Ethics means care of self, and care of self presupposes the use of things, world, and oneself. Let us now examine these three dimensions' of use individually.

First, it is easiest to see what Agamben means by use when we think of the use of instruments, although he does not elaborate on this example extensively. In order to get a better grasp of what Agamben means by use, we should choose another example than the hammer, which was the principal example in Heidegger's theory of the tool. I suggest the example of a flute, which was already used by Aristotle, because it is compatible with Agamben's frequent references to the arts.[121] A flute shows what Agamben has in mind because it is an instrument that does not produce anything separate from itself: it is actually what Aristotle calls a *praxis*, an activity the aim of which is in the activity itself and that produces the agent itself. Its use aims only at itself, at playing the flute. On the contrary, the hammer is a means of a *poiein* whose aim is outside of the agent as well as of the hammer. As Heidegger pointed out, its sense is the reference to that which it produces but which it is not, such that it is a transitional inapparent thing that only becomes apparent as such when it breaks down and its use is interrupted. The flute, on the contrary, becomes visible precisely in its use. The flute comes to the fore when it is used, when it resounds, because the music cannot be separated from the instrument that plays it. The aim of using a musical instrument is not to *produce* separate objects but to play music: *use* itself. The use of a flute is a relation between a human being and an instrument that cannot be explained as the expression of a human subject's musical ideas through a tool understood as a neutral means of expression. On the contrary, when the flutist learns to play well, it learns the flute's proprieties and resounds the instrument's potentiality (its specific sound and timber). If it is a very good flutist, it might even find new ways of producing the sounds that belong to the flute's hitherto unknown potentialities. At the same time, the flutist is affected by the flute's particular structure (it needs to learn new ways of breathing, blowing, moving fingers) and by its sonorous potential (not only its ears but its musical imagination is affected by the flute's soundscape). This is how the flutist exemplifies the sense Agamben gives to use, *chresis*: "We can therefore attempt to define the meaning of *chresthai*: *it expresses the relation that one has with oneself, the affection that one receives insofar as one is in relation with a determinate being*."[122] Using something—in my example using a musical instrument—affects the user, here the player, who only becomes a musician by playing and who can become a greater musician if it lets itself be affected more deeply by the instrument because only then can the musician can also discover the hidden potentialities of the instrument, which are also the potentialities of the player. At some point, the musician and the instrument become one musical form-of-life, in which the instrument is as though grafted onto the musician and the musician is adapted to the instrument in multiple ways.

One could say that the previous description of the use of an instrument was still limited to the ideal form of of use. Any real situation of use contains further aspects that Agamben does not really pay attention to but that Stiegler could have described as the pharmakon effect of the technical object. Because the instrument is separate from its user, it contains its own possibilities and its own becomings that its user does not need to master, let alone perceive. It is not as easy to provide good examples of such becomings as in the case of a flute, but we could think, for example, of a sound that is shrill instead of sweet or of using a flute to play the part of a saxophone. On the other hand, the flute surely makes its player, but this is not necessarily a successful nor a pleasant process.

Growing into an instrument changes a body in both blissful and painful ways, as Peter Szendy shows in his *Phantom Limbs: On Musical Bodies*.[123] In an excellent article, Naomi Waltham-Smith has shown in a more general manner that although Agamben certainly draws on Derrida's analyses of use and use value, he is curiously silent about them and especially about the elements of *usure* and *usury* that Derrida has stressed. For Agamben, "*Use is the potential to be used without being used up and the use of a potential without using it up—the potential for use without usure and the use of potential without its usure.*"[124] In short, Agamben's theory of use succeeds in including instrumentality in the care of self, but he does not develop its negative consequences, which Derrida and Stiegler place under the heading of prostheticity.

The second, more ambiguous dimension of Agamben's theory of use is the use of other people. The particularity of Agamben's theory of use is that it is not limited to the use of material instruments but on the contrary draws its most fundamental sense from the strange relations of use between humans. In the same passage, Agamben continues: "*Somatos chresthai*, 'to use the body,' will then mean *the affection that one receives insofar as one is in relation with one or more bodies*. Ethical—and political—is the subject who is constituted in this use, the subject who testifies of the affection that he receives insofar as he is in relation with a body."[125] In other words, use is essentially the use of other people, first and foremost of their bodies. And the right use of bodies is the condition sine qua non of the form-of-life capable of an inoperative community. But what can the use of other bodies/people mean and how can using other people open up the utopic space of the inoperative community?

In his description of the use of other people, Agamben is manifestly inspired by Foucault's readings of ancient Greek texts in *The Use of Pleasure*, where the free man's ethical care of self was explained as his capacity to use other people's bodies (women, boys, slaves, etc.) as instruments of his sexual pleasures. In the texts Foucault studies, the good use of pleasures did not depend on the other body's response or even consent but only on the pleasure it procured the master. The use of other bodies contributed to the ethos of the free man, who was expected to learn a moderate use of pleasures that avoids both excessive debauchery and sterile prudishness. In *The Use of Bodies*, Agamben complements these readings through an interpretation of the definition of the slave in Aristotle's *Politics*. He emphasizes that contrary to current perception, the slave did not appear to Aristotle primarily as a *property* and the aim of having slaves was not production, *poiesis*, but inoperative *praxis*. Agamben is fascinated by Aristotle's definition of the nature of the slave as "the use of body, *he tou somatos chresis*."[126] For Aristotle, the slave is an animate instrument that the *oikonomia* of the house needs much like it needs inanimate utensils. Some instruments are used to produce works, but others, such the flute and the slave, are only used for *praxis*. The master uses the slave's body as it uses its own body, from which the slave's body is not really separated. The use of the body is practical, not productive. The slave is like the eyes that the master uses to see and the ears it uses to hear. Fundamentally it is comparable to the organs of a body, to equipment, to animate instruments, and to *automata*.[127] Agamben draws a remarkable thesis from Aristotle's few remarks on the slave: "One can ask, however, whether mediating one's own relation with nature through the relation with another human being is not from the very beginning what is properly human and whether slavery does not contain a memory of this original anthropogenetic operation."[128]

As Arthur Bradley has noted, calling slavery the original anthropogenic operation, and furthermore the operation that lets us perceive the use of bodies that is characteristic of form-of-life, is a scandalous idea.[129] Like Foucault's *The Use of Pleasure*, this account is told from the master's point of view, the male master of economical relations, and it pays no attention whatsoever to the people—slaves, women, children—who are reduced to being *instruments* of the master's pleasures and practices. Today it is impossible not to react against such an instrumentalizing view of human existence. This is why Gert-Jan van der Heiden has reason to say that instead of just contrasting use to possession, Agamben should have contrasted it more clearly to *abuse*.

> "Use" now names the particular relation of the good life to the (bare) life that it requires. . . . Clearly this use of bare life is not the use that Agamben concludes his homo sacer series with. . . . It is better to speak here of an abuse of life, and it seems to make sense to describe the wager of *The Use of Bodies* exactly in these terms: how to distinguish use from abuse of life. . . . Agamben does not claim that we can live without using each other's lives or bodies—we continuously do so and need to do so—but he rather raises the question of how to think such a use without any exclusion of lives from the good life or, what amounts to the same, without reducing human lives to mere means.[130]

However, although Agamben does not elaborate on the possibility of abuse, it must be noted that he does not exactly defend slavery, for in the same passage he continues: "The perversion begins only when the reciprocal relation of use is appropriated and reified in juridical terms through the constitution of slavery as a social institution."[131] In the end, slavery belongs to the figures of bare life standing at the threshold between *zoe* and *bios*; it is a condition that Western culture has generally repressed until it finally reemerged in pathological form in the figure of the worker.[132] The slave is a liminal figure that belongs to both bios, whose praxis it does, and banned bare life. It is not that the slave would be anything like a model of life but that its shunned figure is one of the rare places where the use of bodies has historically become visible. The slave stands for pure instrumentality, which is an originary modality of human existence. This is why the question of the slave and the question of the instrument must be thought of together as two dimensions of the fundamental human condition of use.

> Slavery (as a juridical institution) and the machine represent in a certain sense the capture and parodic realization within social institutions of this "use of the body," of which we have sought to delineate the essential characteristics. Every attempt to think use must necessarily engage with them, because perhaps only an archeology of slavery and, *at the same time*, of technology will be able to free the archaic nucleus that has remained imprisoned in them.
>   It is necessary, at this point, to restore to the slave the decisive meaning that belongs to him in the process of anthropogenesis. The slave is, on the one hand, a human animal (or an animal-human) and, on the other hand and to the same extent, a living instrument (or an instrument-human). That is to say, the slave constitutes in the history of anthropogenesis a double threshold, in which animal life crosses over to the human just as the living (the human) crosses over into the inorganic (into the instrument), and vice versa.[133]

The positive counterpart to the scandalous example of the slave is love.[134] Although in *The Use of Bodies* Agamben treats love only through a reference to Foucault (Alcibiades, but also de Sade), it seems that a different reading of love could be imagined here. Love may be the "use" of the beloved—whose mind and body the lover wishes to get to know, like the flutist wants to learn to know its instrument. Furthermore, love is also the affect of the lover who is profoundly touched by the beloved. But one is thus affected only if one knows that one cannot possess the beloved and is hardly capable of knowing it well enough. After all, Agamben's paradigm of love is the love sung by medieval troubadours, the most touching example being Joffrey Rudel's love of the dame he had not possessed, let alone seen or touched. The self who encounters such a love is not the cold-hearted master who uses a person that he possesses as if it were just an inanimate instrument of the master's pleasures. It is on the contrary a heart open and exposed to the beloved who appears fundamentally inaccessible even when minds and bodies can actually touch one another (as Derrida says in *On Touching—Jean-Luc Nancy*: one only touches the untouchable). The most positive example of the use of another person is precisely this kind of a love relation.

The third dimension of Agamben's theory of use is finally the use of self: ethics. Like all other forms of use, the use of self is a relation to something that one cannot possess but that is inappropriable and that (therefore) one cannot master but can only learn to use. This can be explained in two ways. First, as if interpreting Hölderlin's famous verse "the most difficult is the free use of the proper," that Heidegger comments on powerfully in *Andenken*, what one would like to appropriate or at least use is one's history, one's homeplace, and above all one's language. Agamben adds to this the dimension of the body, which Heidegger always leaves aside. Indeed, Agamben's three major examples of what one uses are the body, language, and landscape.[135] All of them are profoundly "ours," for we would not be who we are, were it not by the body in which we live, by the languages that we speak, and by the landscape in which we inhabit or that we traverse. Still, none of these has been produced by us, but all are inherited from a history into which we are "thrown" (as Heidegger says). Using our body, language, and landscape, we use inappropriables that will keep something of their secrecy even when we are deeply habituated to them.

By body, language, and landscape we are also tied to others, for these are commons that would not mean anything if they were not shared. Paradoxically even the body is a common insofar as it is an opaque element of myself that I cannot master entirely but that is something through which other people encounter me in the world. Language is a common, for we can only use a language that we share with others: it is nobody's possession but the first common that makes any possession possible in the first place. Similarly we inhabit a landscape together with others, and it would not really be a landcape (but only a slice of nature) without their (potential) presence in it. Our form-of-life is our unique way of using our body, language, and landscape; our singular style of moving in them; our singular gesture of using these inappropriable commons.

Another, more abstract way of explaining the free use of the self is to say that in a form-of-life one relates to the form that one already has: it is the form sedimented by our past experiences with our body, language, and landscape. One relates to one's form as to still another inappropriable that one cannot possess or control (how could one, since it is the trace of *past* experiences)—but that one can use as a source of new activities and inventions. Our past is the principal instrument of our ethical activity. We use it, but we also miss it constantly, and we tear ourselves away from it in order to encounter other things.

> Perhaps the only way to understand this free use of *self*, a way that does not, however, treat existence as property, is to think of it as *habitus*, as *ethos*. Being engendered from one's own manner of being is, in effect, the very definition of habit (this is why the Greeks spoke of a second nature): *The manner is ethical that does not befall us and does not found us but engenders us.* . . . The impropriety, which we expose as our proper being, which we *use*, engenders us. It is our second, happier nature.[136]

In free use of ourselves we do not use ourselves as property but as the situation in which we are thrown. This is how we can find the ethos of singular life that is also the ethos compatible with the coming community.

We have now examined the principal dimensions of Agamben's anthropotechnics. His fundamental aim is not to say how some empirical appliances affect human minds and bodies (although he occasionally writes of this too, sometimes with reason and sometimes less reasonably, as in his inconsiderate remarks on the COVID-19 restrictions). Instead, he develops a political ontology compatible with the contemporary world. Fundamentally, it has the scope of an existential ontology that corresponds to his general modal ontology. It interests us here because it is integrally articulated through a certain technicity. First, collectives are subjectified and also desubjectified by apparatuses, which are not only the social machines Foucault describes nor the industrial technological systems Stiegler designates but really all kinds of machines and appliances, including playful toys and instruments, that orient people's lives. Second, singular subjectivation takes place through complex relations of use, where use denotes a certain instrumentality and mediality. What the Agambenian subject is in the end is this pure mediality and instrumentality. In extreme cases, it is only pure instrumentality at the service of apparatuses. In such cases it is desubjectified, finally reduced to bare life. But it can also liberate itself from given apparatuses, inactivate them, and grow into singular gestures, singular manners of using instruments people, oneself, one's body, language, and landscape. In such free use of world and self, a form-of-life emerges. As Agamben says in "Notes on Gesture," ethics is really the sphere of gesture that learns to use apparatuses differently and thereby open new situations.

To conclude, we can note that notwithstanding Agamben's important remarks on animality on the one hand and landscape on the other, he interprets these apparently natural dimensions of reality as dimensions of human existence. In his works on bare life and form-of-life, Agamben examines the foundations of Foucauldian biopolitics and he also attempts to show what is pushed away from it and what tries to escape from it. However, when he speaks of life, he still means human life and although he works on the concept of animality, he does not really extend his reflection beyond different, albeit marginalized figures of what we must still call humanity. He does not look beyond the human into the continuity between human and other forms of life any more than Stiegler does. This is why he, like Stiegler, ultimately regards technics as a primordial anthropotechnical force and not as a force that connects human and nonhuman forms of life. Both show how a careful inquiry into technological humanity deconstructs the very image of technological humanity it thinks it can raise, but they do not step beyond this deconstruction to study the generalized "bio-technicity" that connects living and nonliving, human and nonhuman beings.

Stiegler or Agamben?

There is a clear family resemblance between Bernard Stiegler and Giorgio Agamben. No doubt this reflects their common philosophical origins in Heidegger, Foucault, and I daresay Derrida especially, who had a talent for showing how apparently secondary questions (secondary to the philosophical tradition, like supplementary artificial structures such as writing, or imaginary communities that are born in encounters with specters or animals) hold decisive philosophical problems. Stiegler's and Agamben's family resemblance also reflects the epoch because both of them feel urged to confront it incessantly in order to follow a Foucauldian call to work on a metaphysics of the present, which is why their work has a strong political dimension.

On the other hand, Stiegler and Agamben are also as different as brothers can be. They look at the same reality but see it differently and respond to it in two very different styles. They do not collaborate nor even comment on one another, such that it falls to their readers to detect the philosophical reasons for the disparity and the allergy between their respective styles. They share the same question but approach it from opposite directions and with different philosophical tools. To the reader, their two approaches betray more of a complementarity than fundamental discord.

What holds Stiegler and Agamben close to one another is an urge to rethink the human soul instead of just deconstructing it like Foucault and Derrida did. They do not speak in favor of any new humanism because they are very aware of the reasons for this deconstruction and especially of its political reasons, namely the catastrophic consequences of Enlightenment humanism that surely fostered democracy and science but that also lead to, or at least did not prevent from deteriorating into, the major catastrophes of the twentieth century: murderous totalitarianism, despicable colonialism, and the insane destruction of nature. But like Lyotard, they prefer to resist this inhumanity instead of rejecting humanity altogether by stepping toward something else that would come after or beyond it. Stepping beyond humanity, as in transhumanism and in some versions of posthumanism, shows the temptation to forget and even to repress the painful history of humanism as if this too could be overcome. But a repressed history always risks returning, not because of some magical repetition compulsion of past demons but because this history reflects a fundamental, indelible possibility rooted in the human soul itself that cannot be just stepped over. This is why, like Lyotard, both Stiegler and Agamben concentrate on identifying inhumanity and finding ways of resisting it. Like Foucault says, one must constantly exert one's critical capacities in order to identify toxic developments and to find ever new ways of thwarting them. Life is here and now, among the humans that we still are, not in a utopia that may subsequently come. For Stiegler, the inhumanity of contemporary life is rooted in the thermodynamic and computational technologies left in the hands of what he calls 24/7 capitalism. For Agamben, the inhumanity of contemporary life is rooted in the degradation of democracy into biopolitical administration that thrives in a political state of exception, the historical models of which are found in the first half of the twentieth century but that keep reemerging in new political, biopolitical, and biotechnological constellations.

What makes Stiegler and Agamben so interesting to us is not this political analysis, which is by no means unique, but their philosophical work that shows what makes such politics unavoidable. The true philosophical issue between Stiegler and Agamben might be best named the *soul*. Both remove the soul from the old spiritualist and religious traditions that have refined it into a fine, diaphanous vapor that hardly touches the earth whose vocation it is to leave in any case. They go back to the more robust Platonic-Aristotelian concept of the soul, but they also rethink it all over again in the light of Heidegger's important reinterpretation of existence as being-in-the-world. Being the form of this existence, the soul is not separate from the world, it is *in* the world, it *is* being-in-the-world, one could even say that it is just the world affecting, sensing, building, thinking itself. The soul is the place of the world's auto-affection.

But this fundamental phenomenological intuition is not all that Stiegler and Agamben share. Unlike Heidegger, or at least much more explicitly than him, both Stiegler and Agamben think of this world relation and being-in-the-world itself in terms of technics. In order to do so, the word *technics* must of course be understood in as open a sense as the term *soul*: it is *techne* in its relation to *psyche*, not only a technical object encountered in the world but also the subject's technical skill and artistic capacity that has an intentional relation to this world in all forms that we have previously seen. The new concept of the soul is the soul as a technical capacity, as an originary technicity. Technics is the way in which the soul gets a hold of the world—and the way in which the world captures the soul and leaves its mark on it. The technical soul is much more earthly than the old spiritualist soul; it is engaged in hard, dirty, and ambivalent actions. The soul incorporates technical objects and principles, and these imprint and implant the world's forms into the soul. However, technical things cannot invade the soul entirely but remain embedded in it as its intimate alienness: they are at most the alien automaton that works in the soul, twisting the soul from itself ever so slightly. On the other hand, the soul grabs the world, uses and exploits it, and transforms it either into a tender homeworld or into a barren displace, if not a desert. The soul

Humanity and Inhumanity of Technical Communities   253

is the site of the ambivalent pharmakon effect of technics. On the one hand, the soul's technical formatting accounts for human suffering and evil. Changing under external technical pressure is a potentially violent and hurtful experience. But on the other hand, the very same potentiality for experiencing that can endure violence can also exert its own potentiality when the soul deploys its techniques of living and creating—when it *is* inventiveness and creation. By deploying the originary technicity of the soul, both Stiegler and Agamben have thus given very rich and, one would even like to say, definitive answers to our original question of technological humanity.

As capacities of the soul, technics and art are connected and practically synonymous. Similar to how the Greek word *techne* has the double sense of art and technology, the Latin word *ars* too has acquired the double signification of art and technique that qualify the subject's capacity for *artifice*. Both *techne* and *ars* both name the activity of producing and the result of production. Hence it is possible to say that for Stiegler and Agamben the soul is an artificial thing. As soon as it leaves the almost fantasmatic prime infancy, it is never again an innocent, natural capacity but always a formed, educated, formatted, and constructed being. It can always be submitted to external forces and suffer their oppression, and under this pressure it can lose its native capacity for experience and desire, becoming just the site of blunt, blind, and deaf drives to consumption and destruction. But as they also show, for the very same reasons the soul is also an artistic capacity for freedom and curiosity: it is the perceptive, affective capacity for encountering the world and other people and therefore for inventing and creating the world anew. The soul is neither an alienated artifice nor creative art, it is the Janus-faced capacity for both, they are the two originary modalities of its existence, just as authenticity and inauthenticity are the originary modes of existence of Dasein.

Despite these profound similarities, Stiegler and Agamben have two different approaches to this soul, which is of interest to us here. Stiegler thinks of the human soul as the relation between the "who?" and the "what?" He insists that it is neither of the two terms as such but their relation such that one should constantly follow the way in which the "who?" invents the "what?" and the "what?" invents the "who?" However, as he names the "who?" and the "what?" much more firmly than their relation, this constantly draws attention back to these terms: the "who?" is *I* or *we*, our fragile community, and the "what?" is the technical object and ultimately the technological system in which we live. Their relation is the invention of the one by the other, which invention happens as technics (in all senses presented previously). Now Agamben's great invention is the very name of the relation between the "who?" and the "what?" *use*. With this term he draws attention to the relation itself, which he understands above all as something that the human being does (consciously or unconsciously). Use is precisely what Stiegler too speaks of: the relation between the "who?" and the "what?" It is the multiple event of this relation in which the "what?" uses the "who?" (enslaving its body and mind, manipulating it, but also educating it and giving it the instruments for a finer use of the world) and in which, in return, the "who?" uses the "what?" (use of instruments and techniques, use of other bodies and minds, the invention of art, science, love, and politics). Agamben's term *use* is important because it allows speaking of the relation between the "who?" and the "what?" in terms of having, not of being, or having as constitutive of being. You *use* the techniques that you *are* not, but this is how you become who you can be. Furthermore, as using is not possessing (englobing, appropriating, devouring) but precisely using things whose core remains inappropriable, use helps to maintain the distance between the "who?" and the "what?" that Stiegler also emphasizes: the "who?" cannot ever appropriate the "what?" entirely but holds it as the inappropriable, and vice versa.

One way of seeing how Agamben's term *use* and Stiegler's term "what?" coincide is relating both to Heidegger's reading of Hölderlin's poem "Remembrance." In the eponymous article "Remembrance," Heidegger studies the term *use* in Hölderlin's dense expression "the free use of what is proper to one is the most difficult."[137] For Hölderlin, this is a law of poetic expression.[138] But for Heidegger, Hölderlin's research into his own poetic language through his conflictual readings of Greek poetic works also explicates the law of historicity.[139] It is a richer version of the kind of historicity first presented in *Being and Time* because here historicity consists explicitly in the remembrance of one's own past, which is the past of another, a past that has never been one's own present experience and whose appropriation is therefore waiting as a coming task. This is also how Stiegler defines the "already-there": not just the past poetic work but the entire technical world is one's "own" but in such a way that it has never been presently experienced by one, and in this sense the "own" is also fundamentally evasive and inappropriable. This is how the relation of "use" is also the relation that opens up temporalization as historicity.

In sum, for Stiegler the "own" that is used in the technical use of the world is the "already-there" and for Agamben it is the multiplicity of works that we find in the world. Our own inheritance is whatever work comes to us from the past where its moment of invention is lost. Under such conditions, how should one use freely what is proper to one? How can one inherit without enslaving oneself to past people's desires—inheriting, instead, the capacity for questioning, inventing, creating? And also, inversely, how can one use one's world not only freely but also with respect for the others' freedom? That is, how can one use other bodies without enslaving them, educate other minds without manipulating them, use other people in the conflictual relation to politics and other living beings in our relation to nature? Use names a technical relation to things and people and it carries the indelible ambiguity of technics. It does more than touch the world; it grabs, holds, and transforms it and this is why it cannot but lead to ethical and political challenges.

To put it otherwise, Stiegler's and Agamben's conceptions of the soul elaborate on the Greek inheritance that leads to a down-to-earth ethics and politics, rather than the Christian inheritance that leads primarily to the contemplation of the supreme being.

Both Stiegler and Agamben are deeply political thinkers. They do not, of course, write programs for political parties, but considering that critique is a part of the philosopher's task, they are passionately engaged in what Foucault called a metaphysics of the present, diagnosing this life and interpreting this world. Stiegler has developed inheritance of Heidegger's diagnosis of the technological epoch on the one hand and of Adorno's and Horkheimer's diagnosis of the epoch of cultural industries on the other hand. We live in an epoch of technology that is framed by hypercapitalism and their combination has ignited entropic tendencies that are absolutely ruinous to the human spirit and the natural environment. Agamben has developed Benjamin's and Arendt's inheritance by showing why biopolitical oppression and a fundamentally totalitarian state of exeption have not disappeared with the historical forms of totalitarianism, but they reemerge constantly as the flip side of democracy. While these problems are of course primordial political concerns today and while Stiegler's and Agamben's analyses are mostly right,[140] the fundamental philosophical question opened by Stiegler and Agamben lies elsewhere. How have these tendencies come to be such an indelible part of the modern supposedly democratic and scientifically enlightened time?

The answer provided by both Stiegler and Agamben is rooted in the very structure of the soul. The modern soul, we could say, is no longer a simple reflection of the purity of the ontotheological logos. It is a reflection of technical reason, which is always ambiguous, both reasonable and unreasonable, both imaginative and apathetic, both enlightening and bestial. Both Stiegler and Agamben consider that the worst that can happen to a soul is the loss of experience, which renders the soul apathetic, idiotic, brutal, and cruel. These harsh terms should not be heard as insults but as pathologies that can befall anybody. The concentration camp on the one hand and the anthropocene on the other hand are brutal realities that are not imputable to others who are brutal and idiotic, but to people like us, to a brutal element in anybody's soul. Loss of experience is an originary possibility of the human soul too, a possibility that is used by the potentially totalitarian forces relentlessly criticized by Stiegler (contemporary technology submitted to hyperindustrial capitalism) and by Agamben (biopolitical forces). This is why the originary political act is resistance to such forces, as Foucault showed, and this resistance is rooted in the very capacity for experience.

Stiegler and Agamben share a fundamental presupposition according to which the possibility of resistance is based on the capacity for experience, and the possibility of experience depends on the capacity to separate individual individuation from collective individuation. Experience makes it possible to cultivate freedom and inventiveness. Freedom is not an empty capacity but a freedom to live, to act, and to create. It needs leisure but the active leisure to work happily, lovingly, freely, and creatively. Fundamentally, freedom is experienced in the free exercise of what is proper to us, as becomes visible in art, philosophy, science, invention, love, and politics. In his descriptions of this happier, more inventive life, Stiegler looks for ways of using technology more inventively. Among his examples are music and film that test new technologies, collaborative networking projects that test new ways of socialization, and projects in which technological invention is pursued on open source technologies outside of big companies. Agamben, on the contrary, looks somewhat more dreamily back toward the ancient works created by the troubadours, poets, and writers of distant times. His community is an inoperative community of singular individuals who also cultivate their solitude in order to work. The moments of creation he describes are always steps back from the noisy crowd and its blinking gadgets, steps toward an intimacy that faces just one's soul and love. Contrary to this, Stiegler seeks the crowd and its latest machines, not necessarily because he likes them but because each one of them poses a new question to the thinker. This is how the two thinkers cultivate very different kinds of pathos that lead to very different styles of writing. While Agamben's style of thinking and writing draws upon irony, is sophisticated, and develops in refined literary ways, Stiegler's style of philosophizing is engaged in economical and political enterprises where the tone is more that of explaining, convincing, and debating.

These two authors also face two further questions that also divide them. The first is the question of animality, regarding which both thinkers also oppose Derrida, albeit in different ways. Stiegler overlooks the question of animality because he provides such a strong defense of human specificity. He does not, of course, defend cruelty toward animals, but he does not feel the ambiguity of the human-animal-relation that intrigues Derrida either. Agamben, on the other hand, has written a lot about animals (*The Open* is entirely consecrated to the question) and in doing so he ended up in stubborn debates with Derrida. After all, Agamben speaks of animality mainly as a metaphor for what is banned from human society whereas Derrida studies the fundamental alienness of the animal, which also stands for the alienness that breaks all humanist dreams. The animal is alien to the human in a way that is similar to, but not identical with, the way in which technics is alien to the human. Human use cares and exploits, liberates and enslaves. Both the animal and the technical object flee from this use at some fundamental ontological level. This is why Derrida is ultimately the only of the thinkers examined in this book who develops a conception of bio-technics that connects human and nonhuman, living and nonliving kinds of existence through an originary technicity that is shared by them all.

The second question faced by Stiegler and Agamben is the question of being, which both read in terms of its historicity. One could say that because they are so interested in techne, being itself appears to them in terms of what Heidegger called the *Ge-stell*: it is epochal and the epoch is fundamentally that of technics. Heidegger never thought the *Ge-stell* apart from the question of being that there is (*es gibt Sein, es gibt Zeit*) or that "enowns" (*Ereignis*). Contrary to this, both Stiegler and Agamben are more interested in the future dimension of the historical time, whether it is the possibility of a future created by negentropic care of the world (Stiegler) or the possibility of a messianic advent (Agamben). Both have gone much further than Heidegger in the description of the epoch determined by *Ge-stell*, but they have not for that matter responded to the question of being.

Chapter 6

From Technological Humanity to Bio-technics

Over the course of this book I have followed the theme of *technological humanity* in the works of a series of philosophers who have discovered the importance of technics to the becoming of the human and who have at the same time shown how technics hollows out humanity—or more precisely, how the concept of technics allows showing the hollowness of the term *humanity*. Technological humanity is therefore not an ideal figure that this philosophical discussion aims to erect, but it is on the contrary an ambient and distorted image of the human that philosophy reveals in order to undo, dismantle, and deconstruct it. The aim of this critical work is not to restore an alternative figure of humanity (or posthumanity) either but to clear out the ground for another way of thinking of existence. Unlike the philosophers studied in this book but drawing from their works nonetheless, I call this other way of thinking of existence *bio-technics*. This term is meant to refer to a philosophical question rather than to the contemporary technological situation. Bio-technics is quasi-homonymical with the narrow technological sense of biotechnology, but it distances itself from it by using the hyphen, by suppressing the "-logy," and by changing the ending into technics. The homonym is a disadvantage rather than an advantage, but it of course also reflects the inevitable intertwining of philosophy and its epochal context.

The question of technics raises precisely the question of humanity (instead of its theoretical doubles "consciousness" or "subjectivity"). The question of the human being coincides inevitably with the question of the subject of philosophy, which is at the same time the object of interrogation and the subject who interrogates, and furthermore, the exemplary being in which the entire situation of thinking is reflected. When the situation of thinking is only that of reason, the subject of philosophy still carries the ideal form of the *cogito*—of reason reflecting on the certainty of reason. But when the situation of thinking is technics, the subject of philosophy loses this ideal form and interrogates its condition, the human situation in a technological world. It asks what is the human being insofar as it is at the same time the technician, the product of technics, and the technical production of the technician itself, and moreover, what is the entire technical situation that conditions and shapes the existence of the "human"—who puts this situation in question.

Throughout this book we have seen that the human being cannot be a product of technics in the same sense as an ordinary technical object because it is never just passive matter (*hyle*): it can be formatted in the first place only because it is fundamentally an open potentiality, a lively plasticity, and a capacity for transformation. This is why the anthropotechnical idea of producing a new enhanced humanity simply by applying the instruments given by bio- and information technologies to living material is based on an illusion that can dissipated by a human sciences' perspective on the human that can be formed and educated—also in the use of technics. Technical production of the human aims at the human being as a potentiality and therefore primarily makes a producer instead of just a product. This is why the different senses of the word *technics* examined at the outset of this book always determine a technical *activity*. The first sense of technics, the *techne* of antiquity, is a skillful person's know-how, the first illustration of which is a craftsperson but the most elaborate form of which is a well-educated person skillful in the art of living well in the polis. The second sense of technics opens the modern era. Modernity thinks of technics in terms of the machine, which it understands as an assemblage of disparate parts making up a provisional whole capable of carrying out relatively autonomous activity. The machine is built in the image of a living organism and ultimately of the human mind, but it also becomes the specular image in which the human being and human society can contemplate their functioning. The organism and the machine develop in a mimetic relation with one another: the more one tries to resemble the other, the more obvious their dissimilarity becomes, pushing the parties to a constant readjustment. If technical skill and the machine originally appeared as ingenious instruments for positive aims, the twentieth- and twenty-first-century thinkers we have analyzed more closely in this book have underlined the ambiguity of technics and of the modern technology especially, which is also a major source of pollution and alienation.

Finally, the contemporary conception of technics that we called *bio-technics* designates a technics that goes further than building an artificial organism: it does not mime the living *organism* but the living *matter* itself (of body or of thought) with which it aims to fuse. Today the human being thinks of itself as a bio-technical being. It is difficult to grasp this kind of existence philosophically because bio-technicity does not coincide with the self-conscious subject of philosophy but on the contrary with the obscure domain of bodily and mental materiality that is the situation of the subject. Bio-technical materiality is not the mindless matter of the res extensa nor the innocent state of nature before the invention of culture and its techniques. Instead, it is the technical relation between these two forms of nature—that actually projects these two poles as objective and human nature. Its materiality consists of the nonconscious functioning that rationality and consciousness rely on. This functioning does not reflect rationality, not to speak of consciousness. It is the technical functioning that supports them like the code supports user interfaces in computers. Here "matter" is almost immaterial, it *matters* insofar as it *functions*: calculations in a computer, metabolism in a cell, cerebral activity in a brain, and so forth. Those who have pushed the analogies between computer, brain, and biological processes far have sometimes ended by postulating an ontology of the code that would be instantiated in biological and informational entities alike[1]: being is code, not in the sense of a final formula of everything but in the sense of infinite computational operativity. The bio-technics that I am speaking of is not such a common explicative ground, however. Bio-technics speaks of life in a wider sense than just humanity and of technics in a wider sense than instrumental or machine technology. It speaks generally of life that prolongs itself in technics and of technics that ambitions to be alive. Yet these two do not amount to the same. The hyphen of bio-technics symbolizes the impossibility of reducing the two aspects of existence to a common ground, for example to an ontology of code or to an ontology of autopoietic being. If our time needs to think existence in terms of technics, technicity does not have the ideal clarity of an a priori condition of reality but only the historical form of a quasi-transcendental configuration. The quasi-transcendental configuration of our time deploys as a changing dimension of comparisons between life and technics: each draws its sense from its similarity with the other, but as soon as a similarity is established, the difference it contains appears too. The hyphen thus stands for both the absence of an ontological ground and for the quasi-transcendental activity of comparison between life and technics.

A quasi-transcendental configuration is experienced as a situation. Bio-technical materiality is not a thing but the situation created by the originary technicity that traverses both human and nonhuman life and shows their profound continuity. Bio-technics characterizes the "in" of being-in-the-world by indicating that being is not just *thrown* into a given world but is the very *doing* of this world—a doing that is by definition artful, skillful, and technical. It is not an intentional technical activity undertaken by human consciousness. It is the originary technicity that translates the simple fact that even nonconscious living and technical beings do not leave their situations as they are but affect them because they are always related to their situation via some kind of interpretative and transformative activity. Bio-technics has an ontological sense because it characterizes the mode of being of what is, but this is an ontology of relation, not of a unitary principle or substance. To some extent bio-technics resembles Merleau-Ponty's "flesh," but it is not the natural touching between sensing and sensed. It resembles even more Nancy's "singular plural being" where singulars touch one another technically. Bio-technics is the artificial, skillful, technical touching between living and nonliving things that can be called originary technicity.

Technics imitates life and life reinvents itself in the mirror of technics; technics is the capacity of life itself and life is originary technicity; existence is bio-technical. Bio-technical existence is the endless doubling between *bios* and *techne* insofar as they are indistinct. But at the same time their distinctiveness is revealed when *bios* becomes apparent as the negativity that withdraws from *techne* and when *techne* becomes apparent as the alienation of *bios*. The subject of philosophy sees bio-technicity as the *relation* that endlessly produces the two poles of life and technics that do not coincide with the subject of philosophy. But the bio-technical relation between life and technics is the situation in which the subject of philosophy nonetheless finds itself today.

The question of the passage from technological humanity to biotechnical existence arises from out of the present historical situation and not solely from philosophical considerations. Of course, as we have seen throughout this book, the technical production of the human is nothing new, but recent technological developments—especially progress in biotechnology and information technology driven by the globalized capitalist economy—invite a reconsideration of its meaning. Martin Heidegger and more recently Bernard Stiegler think that technological change can be so fundamental that it inaugurates a whole new epoch. In a parallel manner, today's trans- and posthumanists suggest that new technologies do not simply pursue the old humanist dream of improving the human condition but they point toward a transition beyond humanity where the old ideal of humanity finds itself upgraded or degraded, troubling the figure of the anthropos and inviting a reconsideration of the subject of philosophy. In this book I have suggested that if we want to speak of epoch-making changes, these cannot be described in terms of changing figures of humanity but they require the undoing of the entire idea of humanity and the formulation of a different thinking of existence, one that is not exclusively "human" but that accounts for the continuities between all kinds of living beings. Some develop this line of thought in terms of posthumanity. I have inscribed these continuities in the term *bio-technics*. Ultimately—but this would be the subject of another book—existence should not be thought of only against inhumanity (Jean-François Lyotard says) but also against hostility to life and against hostility to the world (what Jean-Luc Nancy calls *l'immonde*).

Contemporary technologies certainly impel a reexamination of the sense of "human." They are strongly anthropotechnical and this makes them self-reflective. In the most general sense of the word, anthropotechnics is a technical transformation of the human mind and body. As Peter Sloterdijk in particular has shown, some kinds of anthropotechnics have always existed, for humans have always taken care of themselves and their offspring within the contexts of family, clan, and political community. All of these reproduce themselves through medical, educational, and other techniques that aim at cultivating human beings. The formation of a body primarily comprises the acquisition of physical skills, but this has always also included the elaboration of the body and even of its very flesh with the help of diverse technical supplements: clothes and ornaments have protected and hindered it, nourishment and drugs have strengthened or weakened it, technical and aesthetical prostheses in general have supported and hurt it. The formation of a mind includes all noetic learning processes—languages, behaviors, customs, knowledges—that can be either imposed by the society or freely chosen by the individual. As Derrida and Foucault have suggested and as Stiegler has emphasized, noetic processes are also mediated by technical prostheses or dispositifs.

Modern biotechnology (in the narrow technical sense and not in the general philosophical sense that we have introduced) continues these processes but it also imports a new element into them insofar as it consists of the possibility of acting directly on living matter in laboratory conditions. Its most striking achievements are based on the manipulation of stem cells and especially gametes, which represent a nonformed nonindividuated bodily potential. In such biotechnological interventions, the individual is neither the subject nor even the object of its own formation but the result of complex intermingled medical-scientific-social-industrial processes that make up the "matter" that has come into the place of "nature." Analogically, the information technologies, which are increasingly grafted upon memory and cognitive activities, comprise an intensive noetic activity that is much less that of the individual than that of the virtual field that frames individual and collective noetic activity. While classical anthropotechnics aimed at the individual, contemporary anthropotechnics aims at the physical or the noetic matter that makes up individuals and that is as distant from the individual as res extensa is from res cogitans. Using Simondon's terms, it can be said that new technologies increasingly act already on the level of preindividual virtualities instead of acting on constituted individuals. This is by no means unprecedented because a technological situation can always be interpreted as a part of the preindividual virtual field of human individuations. But with the intensive development of what we have called bio-technics—a technics that acts on life and acts like life—it can also be claimed that a qualitative change has taken place.

In this book I have traced the ways in which contemporary philosophy has prepared—and reacted to—the new technological situation.

The traditional view of technological humanity—an old program reaching to Enlightenment, Renaissance, and Antiquity—consisted in using all available techniques and technologies in order to realize a New Man who is healthier, stronger, more intelligent, and morally superior. This formation process can consist in freely chosen exercises, the likes of which are described in Sloterdijk's *You Must Change Your Life* or it can consist in medical, pharmaceutical, and prosthetic operations practiced by specialists on human bodies, either at their demand or at the demand of others (parents, society). When the aim of medical interventions on the human body is not just to heal but also to overcome a human body or mind's ordinary capacities, this means taking a step toward human enhancement.

The realization of the promise of technological humanity that calls for human betterment and enhancement is at the core of so-called transhumanism. We have seen why human enhancement is neither morally unproblematic nor philosophically self-evident. It is morally problematic because before promoting the enhancement of human health and intelligence, one should ask whether this amounts to deeming less healthy and less intelligent life as less valuable, which is morally untenable, and because of the question of who defines what counts as healthy and intelligent, which is a politically sensitive topic (e.g., Who would profit from the reduction of sleeping time, the person capable of working more or their employer?). As the defenders of human enhancement justify their projects by appealing to the individual's right to choose what it does with itself, it is important to note that technical possibilities characterize entire socioeconomic situations and never just a single individual. The most radical projects do not even concern individuals but their offspring, the "designer babies" who are not given a choice regarding their own design. We should also notice that medical human enhancement treats the human body and mind as simple matter—*hyle*—and overlooks its own internal needs and capacities. If it goes so far as to suppress them, much may be lost. Apart from these moral considerations, we can note that the transhumanist project also overlooks its own philosophical presuppositions. The first of these concerns is temporality, which the transhumanist will to mastery paradoxically tends to close off instead of opening it up to new and unexpected becomings. By projecting today's ideal of humanity into the future, it extends the rule of these same ideals instead of interrogating them. Yet the very idea of transforming humanity would seem to first of all require an undoing of the existent form of humanity and deconstructing its present ideal in order to liberate the very possibility of transformation. The transformation of the subject presupposes a subject who is not fixed regarding its idea of itself. Working on the human tacitly presupposes a bio-psychical plasticity by virtue of which the human can be educated, formed, and transformed in the first place—a plasticity that is possible because the very being of the human is potentiality and capacity. This bio-psychical plasticity goes together with an originary technicity owing to which the human is capable of working on itself with the help of various practices and instruments that can be described as technics. The being of the human lies also in this potentiality to use its potentialities, the capacity to use its capacities. The human is therefore always divided into what can act and what can be acted on, and it is the relation between these two aspects of its own character of possibility. If the transhumanist will to mastery relates to the human as if to a neutral *hyle* on which a *morphe* can be imposed, we can object to it as follows: this hylemorphic attitude presupposes an originary plasticity and an originary technicity that can never be the products of a technical production because both are potentialities-to-be that such a hylemorphism precisely extinguishes.

This is why, perhaps somewhat paradoxically, the reflection on "technological humanity" is also its deconstruction, and one that brings forth the dimensions of existence that escape anthropotechnics. The authors whom I examined in this book have all contributed to this deconstruction. The hollowing out of technological humanity also prepares the possibility of thinking existence in terms of bio-technics.

I first examined a number of German thinkers who, starting from the 1920s, started to speculate about the human as determined by technics and not (only) by reason. On the one hand, the philosophical anthropologists Scheler, Plessner, and Gehlen thought of the human being as the originarily artificial being who realizes itself by producing an artificial technical world. The essence of the human is this technicity capable of taking all forms without having any of them as its proper essence. On the other hand, Heidegger rejected the notions of the human and of the anthropos totally and instead developed a new thinking of existence in terms of Dasein. In the existential analytic developed in *Being and Time*, Dasein finds its world first and foremost through its use of tools and itself in working community. In later texts and especially in "The Question Concerning Technology," he further showed why the entirety of the modern world horizon must be thought of as a technical *Ge-stell*. Even though Heidegger was critical of this horizon, he also sketched the first ontotechnology with it. Despite their profound disagreements—after all the Philosophical anthropologists think in terms of the human being whereas Heidegger rejects this notion and speaks of existence instead— these German thinkers share the idea of a technical world horizon and the idea of a lack of human essence. Nothingness is the core of (human) existence: the world in which it believes it is grounded is just an artificial world image, just a veil of the fundamental nothingness of the human situation. Nothingness is more fundamental than any attempt to impose an essence, a form, or a figure on the human being—so fundamental that if the uncertainty of human existence were to be replaced with such a full essence, existence would also lose its possibility character.

Secondly, from the French thinkers from the 1960s onward, I chose to study Foucault and Derrida, who open up complementary possibilities for thinking of the way in which technics marks human existence. Despite their profound differences, Foucault and Derrida share the idea of the human being as the effect of language and not as its origin and this is why they have both been situated within the so-called linguistic turn in philosophy. Furthermore, because they do not think language as the expression of the logos but as a signifying or power technics, their philosophy has recently been reinterpreted in terms of a technological turn in philosophy. When the human being is examined as an effect of signifying or of power technics instead of logos, philosophy discards the beautiful human form erected by classical humanism and announces the end of the human. Considered from the perspective of the end of the human, philosophy deconstructs the question of the human being instead of posing it, but it also deconstructs Heidegger's thinking of existence by presenting finite existence as the trace of ambient and inherited discourses and current practices and technologies of power. However, the existent is not their direct product but just their effect and this is why it can also resist ambient discourses and reinvent itself anew. The core of the existent is not just the nothingness discovered by Heidegger and by Philosophical anthropologists or, to put it differently, it is a *fertile* nothingness, a fold between receiving and giving a form, or better still, a suspension in which imposed techniques lose their self-evidence such that they can be reinvented. Some commentators have tried to establish a connection between the antihumanism of the 1960s (that claims the end of the human) and contemporary transhumanism (that wants to overcome the human) but this link is weak. While the former deconstructs the figure of the human integrally, the latter confirms it in its overcoming. More important for me is the way in which Foucault (with his notion of biopolitics) and Derrida (with his work on the parallelisms between life and writing) prepare the thinking of bio-technics, which appears to me to epitomize the contemporary situation of existence. They do not directly affirm bio-technical existence and maybe they would be skeptical of it as all direct affirmations typifying existence, but they surely provide all elements for thinking in terms of bio-technicity. Especially Derrida underlines the parallelism between life and language, both of which are thought of today in the technical terms of code and program. However, instead of constructing anything like an ontology of code, he deconstructs the ontological question by imagining the materiality of the *khora* and the ideality of spectrality, which show how to face the question of being in the epoch of bio-technics.

Thirdly, I have studied two continental theoreticians from the turn of the twenty-first century. These are less interested in defining a humanism or an antihumanism than in resistance to inhumanity such as it was already named by Lyotard. Instead of examining signifying or power practices that can be interpreted as abstract techniques, Bernard Stiegler and Giorgio Agamben examine the markings of concrete technologies on existence in all kinds of material technologies starting from the toys and artworks that charm Agamben and ending with the huge thermodynamic and digital systems tirelessly analyzed by Stiegler. Close and very different at the same time, these two thinkers agree in thinking that technics and the human must be thought of together as a singular movement of reciprocal invention. Stiegler investigates technics (*what?*) as the prosthesis of the "human" or, more exactly, of the *who?*, whom he also calls the "non-inhuman living being." Being an originary default, the *who?* desires its technical supplement—whereas in being just a supplement, the *what?* can never fully satisfy the lack of the *who?* but engenders it ever anew. While showing that "the animal" is created by banning something from "humanity," Agamben has also shown how "the human" is each time defined by this exclusion. But the "human" is also made in its use of different power, signifying, and material techniques. In his late works on the notion of *use*, Agamben examines the dynamic between the human and its instrument—also when the instrument is another human being—as a reciprocal relation of use in which the positions of user and utensil change ceaselessly. Stiegler and Agamben differ insofar as Agamben develops a beautiful ontology of potency (whereas Stiegler does not really have an ontology, unless one gives an ontological sense to his late considerations on neg/anthtropy) and on the other hand Agamben underestimates what Stiegler calls the pharmakon effect of technics, which can always lead to the misuse of human and nonhuman life. What interests me specifically is the way in which both Stiegler and Agamben emphasize the artificial nature of any image of the human, including technological humanity, and reveal interdependencies and continuities between human, animal, and technical "life." Artificial images of the human seem unavoidable in their works, but they also help in casting a glance at the bio-technical existence underlying all such images.

As opposed to the dreams of a life that is integrally determined or produced by technics, all authors cited above stress the insurmountable gap between life and technics by virtue of which something of life always escapes anthropotechnics. The "life" that our authors have studied mainly as human existence is always marked by lack and negativity. It is a nothingness; it is not only a nothingness of being but also a nothingness of determination that turns into indetermination and originary plasticity; it is not only a default of determination but also a desire and a potency for ever new determinations—an originary technicity. Technics, on the contrary, is full and positive; it is integrally determined by its functions and programs; it ignores the tension between lack and desire because it realizes itself entirely while following the automatism that defines it. Its occasional failures are due to contingencies of matter (*automaton*), not to the existential capacity of facing chance (*tyche*). Even though its operations can be extremely complicated and sophisticated, it does not amount to the *same* as life—and precisely for this reason technics is useful to life. It accomplishes what life alone cannot do, as Aristotle said. Life needs the supplement of technics in order to reflect itself in what it is not—and technics can at most mimic life but this does not make it alive. This reciprocal mimesis and utilization of life and technics is the logic of what we have called bio-technics.

The authors studied in this book have deconstructed the Enlightenment conception of the human and the instrumental concept of technics, but to a great extent they still study the relation between these two poles, whose specter remains despite its deconstruction. The term *bio-technics* is an effort to draw attention to their relation, which tends to disappear like the hyphen that keeps *bio* and *technics* apart, although its very function is to ask what keeps the poles together despite their apartness. The *bio* of bio-technics denotes a life, not the life of the mind defined by rationality, nor the existence of Dasein defined by its relation to the world-as-world, but a singular, finite life as a deeper and wider capacity for touching what is not (its) life. Life, "bio," is therefore differential: it is the *split* into life and (not) (its) life and the *relation* between them. What a life relates to is condensed into the admittedly heavy expression *(not) (its) life*, which includes the different possibilities of *life*, *not*-life, *its* own life at another time, or *not its* life but some other life. Let me explain this a bit. *(Not) (its) life* can be *its* life not now: its own past or future life. What it was does not live anymore, what it will be does not live yet, its past and its future are only material to its present life, which is the only real sparkle of life's aliveness to itself. (*Not*) (*its*) *life* can also be *not its* life: anything in its environment and even the environment itself. Living is relating to an environment or to a world, where finite life finds what it takes to be the nonliving material of its life, although the surroundings can also have a life of their own. And finally, (*not*) (*its*) *life* can be the life of another living being: for a simple cell that might be another cell; for an animal organism that might be food, shelter, prey, predator, partner, progeny, parasite, host, just an indifferent passer-by, and sometimes even its own body. Living is touching and relating to all these forms of (not) (its) life. It is important to see that this touching and relating is nothing immediate and natural: the contact between life and (not) (its) life is not easy and direct. When life touches (not) (its) life, it senses and *uses* it, and this double relation is *technical*. *Life is originary technicity because it consists in the technics of using (not) (its) life.*

This is why the "technics" included in the term "bio-technics" is neither a subject's skill nor an object's function, although it makes these terms possible, too. Much more fundamentally, it is the art and the technics of living in the most general sense: life as art and technics. "Technics" is the way in which a life touches and alters what is (not) (its) life, and the way in which what is (not) (its) life touches and alters a life in return: the reciprocal relation between a life and what it is not. Speaking of this touching in terms of technics amounts to emphasizing the gap that every touching must bridge, and the difficulty or at least the non-obvious character of doing it. We say of life what Heidegger said of Dasein and death: like one cannot die for another, one cannot live another's life, and this is why living beings may be similar, but they are still in the very last instance unique and irreplaceable. A life and (not) (its) life are ontologically distinct, which is why, even if they were next to one another, their touching of one another is not simple continuity but a gap, whose crossing requires skill: touching is possibilizing an impossible passage. Touching does not have the clarity and the evidence of Cartesian truths; it reaches over obscurity and uncertainty like the art of crossing a difficult gap, like the artifice of bridging it. A life can use (another) life as a technical means, like a skillful subject uses a technical object. But much more fundamentally, technics is what bridges the gap between life and life, whether the latter has a life of its own or not.

Understood in this way, technics needs sensing and is the art of sensing. Sense, too, bridges a gap between a life and (not) (its) life. From Aristotle up to Merleau-Ponty, philosophers have thought that the life of a living being consists in sensing itself, its world, and other living beings. What they have generally ignored or underestimated is the complexity of the act of sensing. Like touching, of which it is a kind, sensing is neither direct nor simple. It requires both the capacity of *receiving* whatever alien stuff appears and gives itself as sense, and the capacity of *making sense* of it. Sensing as receiving a sensible appears passive. However, it is not only obtaining a unit of information and stocking it, but it is being *able* to be *sensitive to* the sense that gives itself. As Uexküll has shown, a living being is never sensitive to everything that surrounds it but only to what is significant to it. Sensing is possible because a life is *capable* of being exposed to what is not its life, to the point of making itself vulnerable up to the point of risking the exquisit annihilation of its auto-affective completeness. On the other hand, *making sense* is obviously active, but instead of being simply an expressive activity of producing significations, it invents new sense only to the extent that it interprets *given* sense and responds to it. There is no sensation without both of these two aspects: sensing and sense-making. They go together, activity and passivity intertwined, sensibility and intelligence meshed in a common gesture. The capacity of sensing and making sense is not just given, it is the primary technics of life that must be *done*: practiced, exercised and refined. This is the primary technics of life: the bio-technics of sensing.

Now, if "bio" is life touching what is (not) (its) life and if "technics" is the art and manner of this touching, what is their relation, that I try to mark with the inaudible hyphen? Or, to pose the question only in the terms of this investigation concerning human existence under technological condition, what is existence as bio-technics? It engages life in all three dimensions that make up life: relation to self, to the world and to other living beings. Let us summarize them one by one.

The easiest of these relations is life's relation to its world (or actually to the situation that "world"'s deconstruction brings forth). As Heidegger shows, existence is being-in-the-world. Interpreted in these terms, technics is the internal structure of the "in" of being-*in*-the-world, that Heidegger actually explained in technical terms as the *inhabiting* and the *building* of a world. In the framework of my book, existence is thought in more general terms as life, and world as its situation. The situation overflows the span of the term "world," because a "world" is in some sense a meaningful whole, whereas the bio-technical situation of a life is neither meaningful nor a whole but only an element out of which a meaningful world can be built. One could also use another Heideggerian term and say that bio-technics is the *Ge-stell* of the contemporary world. Expressed in this way, the *Ge-stell* is not articulated like a machine-like totality but as an indefinite tissue or matter. *Ge-stell* is not really a "world," it is the horizon out of which a world can take shape, an elemental techno-nature that existence can never grasp as a totality. Because this bio-technical situation is not one (nor many but the innumerable element) and because it is not particularly human, it might be easier to grap it in terms of Merleau-Ponty's elemental "flesh." Technics is the internal structure of what Merleau-Ponty named the *flesh* and what Nancy analyzed in terms of *technics of sense*. Interpreting flesh as a technics of touching means paying particular attention to the complication that resides in the contact that makes a flesh, and studying, instead of a contact still marked by the naturalness of perception, the artful, skillful touching between a life and what is not (its) life. The "in," which is really the zone of touch and contact, is here examined in terms of the *originary technicity of life*.

A life is always exposed to its surroundings, which exert pressures and impose changes upon it. Inversely, these surroundings are always exposed to a life, which exerts pressures on them and imposes its fantasy on them. Living is this double exposition. Examining *living as bio-technics* means extending, stretching, and studying in detail the gap between a life and its surroundings: examining the mediations that reside in the apparent immediacy of "touching" and "exposition." Any contact is already a solution to a problem of separation: in time, in place, in rhythm, in figure, in desire, in understanding. A life's technical relation to its surroundings certainly touches the surroundings, but touching actually takes place over a distance that makes direct adaptation difficult and calls for the invention of "technical" solutions to the problem presented by the surroundings. The surroundings also touch the life that tries to feel at home in it, but at the same time the surroundings exert a pressure over the life so that its form is changed, dressed, educated, formatted. All efforts at technical formatting necessarily remain partial.

The philosophers studied in this book only pay attention to human technics, although they are certainly not the only ones nor necessarily the most interesting ones (e.g., because, contrary to an old prejudice, many animals' constructions not only repeat the same model as if the design was determined genetically, but they actually invent new solutions in new circumstances caused, for example, by climatic changes and displacements). Although I do not believe in human exceptionality in this field, I have limited myself to humans as well, for the sake of familiarity and the ensuing simplicity. As Agamben shows, the technical relation is not that of production but that of *use*. Use says that whatever is used is not appropriated but only diverted from its original course: it always has a reverse side that remains untouched by use. This is how a life conserves the core of its singularity, which in the case of human life is also called freedom. This is how the environment conserves its elemental ground, which Heidegger designated with the old word *physis* that no *Ge-stell* can definitively suppress. As Stiegler shows, any technical solution is also a source of new problems—a pharmakon that heals and intoxicates, a solution that cultivates and pollutes. This is why technics is never a simple step over a gap: it is the slow traversing of an uncharted distance that causes both joy and pain. Because life is the technics of traversing ever changing distances, it is also destruction, discovery, and invention.

Having cast a look at bio-technics as life's relation to its world or actually its situation, what about life's relation to itself and to other living beings? Here, too, technics is life's way of bridging the gap to (not) (its) life: one's own past and future life, others' lives. The technics of crossing the gap can be an immaterial technique, a material inscription or a technical object: in each case it is a *supplement* to life. As Stiegler has shown, technics can never get a definitive hold of the "who?" ("the non-inhuman living being"): technics is a supplement that answers to a lack without being able to fulfill it. At the same time, the technical supplement is, according to Stiegler, always a pharmakon that invades the "who?" with its alienating automatisms, contaminating *noesis* with its inhuman logics, and poisoning life with its non-life. Its effects do not affect consciousness but act on a "nonconscious" biological, psychological, social, and we should also add, ecological level. The toxic effects of technics are partly due to its nonliving materiality and partly to the fact that each technology carries the trace of a collective to which it tends to assimilate the individual. As Agamben shows, the relation between the human and its tools is that of use. When the tool is used, it is not possessed and assimilated to the user; it is not known by the user either because technical know-how is not yet theoretical understanding. Agamben names the relation of use better than Stiegler, but Stiegler shows better than Agamben how, in a relation of use, the user certainly continues its power in the tool, but also inversely, the user is used by the tool.

Technics poisons the living being when it takes life only as *hyle*, matter that is reduced to simple a resource. According to Stiegler and Agamben, anthropotechnics treats human beings as resources when they take away their capacity for *experience* by invading the time, the space, and the imagination required by experience: this happens especially when it robs the human beings of their own experiences of anthropotechnics. Here experience is essentially an experience of technics.

On the contrary, technics emancipates the living being when it leaves space for a real experience of technics that both Stiegler and Agamben name but that they do not really develop. This can happen in an ordinary instance where an object or a situation is being *produced* and technics is only an instrument for the production of a world of aesthetic or ethical beauty. More intimately, technics emancipates the living being when it becomes *its technique*: when technics liberates life's potencies and changes them into capacities for doing, acting, and knowing. This is where a life develops a technics of relating to its own life. A simple example that we used previously is that of an instrumentalist, let us say a violinist this time. The violinist does not learn to play if it does not adapt its body and soul to its instrument. It must train its fingers, tune its ears, cultivate its musical taste. The violin awakens different possibilities of being in the player, these would not exist without the instrument; they are not only musical possibilities but also dimensions of the soul that music can sound out, make resonate, and create. Learning to play takes a very long time and includes also boredom and even pain. Pain shows the player is undergoing an experience—*experiri*. It is the sign of the resistance of the body and the soul to the change brought about by the instrument. The pain of resistance indicates that some "bio-technical" change is happening in the sensible system when it plays, although here the term *bio-technics* is used in quite a metaphorical sense. By and by the musician grows together with the instrument such that they become a kind of a musical *cyborg*. At the same time, the musician grows together with the music that it plays such that it become a kind of a *medium* for musical thoughts consigned to the score. Only if the musician agrees to this becoming-cyborg and becoming-medium can it become a true musician who not only plays well but can also create a personal style in which the possibilities of a violin, the possibilities of the score, and the possibilities of the musician's soul are reinvented. Here—in a moment where the violinist touches the composer's mind and the listener's mind—we have an example of a life touching other lives by means of technics.

But what makes the turn or the fold in which painful submission to the technique (instrument, score) becomes blissful mastery and finally original interpretation possible? The gap remains between the musician and its instrument even in the case of a "musical cyborg," where the musician and the instrument seem so complicit and closely welded. The gap is revealed first of all in the *pain* of learning and habituation, pain in which a technique's unnaturalness and strangeness becomes manifest. The gap is also indicated by the possibility of *invention* in which the musician escapes from the instrument's automatisms and in the best of cases even develops a personal style. The pain is an experience of the instrument that shows the ultimate incommensurability between life and the technics that imposes its form on life. The invention is an experience of the player that indicates the possibility of departing from the automatisms suggested by technique.

How is it possible to invent, that is, to divert the technical automatisms or to turn them elsewhere? This presupposes the capacity for suspending the activity dictated and rhythmed by the instrument. This is the moment when an "I can do" becomes an "I cannot do," more precisely, an "I can choose not to" that Agamben emphasized in his famous reading of Melville's "Bartelby" where the suspension of a potency brings it forth in its purity. But in order to invent a new way of doing, it is not enough to cease following old ways: one must also find new ways, new manners of doing, for example a new extended technique in which the violin is played with a peacock feather. Sometimes Stiegler describes this using terms inherited from Simondon, such as saying that invention takes place when a problem that resides in a field of virtualities is taken charge of and solved in a new manner—which is always provisional, transitory, and unstable. Invention is the invention of a new arrangement of a field of virtualities.

But it seems to me that invention is not always and not only the solution to a technical imbalance. It can also demand abandoning the instrument, forgetting its techniques, and paying attention to something that escapes its technicity. In our example of a violin, this could be a dry pitchless sound, the woodenness of the instrument, the airiness of the space, the noise without music, the violin welded into a visual work, and so on. Heidegger calls this the *phusis* that withdraws from *techne*. For him, *phusis* was a name of being. In our study of technics, what withdraws from technics is better called its elementary materiality, the "stuff" of the world before it was captured as a resource for such and such a technique.

This materiality is not the substantial matter discovered and elaborated by new technologies: it is the virtual matter in which the potencies of matter are still being invented. It is not a truth of being: it is the element of life in which possibilities of life are still being invented because it is itself a bio-technical reality of its own. Here the life can properly touch itself and other lives: it seeks the default in technics as an invitation to let life come forth untamed by technics, and therefore desiring ever new technics that allow it to come forth.

This is how technical humanity finally gives way to bio-technical existence. Bio-technicity does not come forth in the classical paradigms of technics, although it can be shown to be their unspoken condition. When technics is interpreted as *techne* and skill—and especially when the skill can be formalized and programmed into a machine—then the humanity that wields this technics appears formed to this skill, programmed like a machine, and finally educated along the same lines as the rationality animating the machine. Then technological humanity incarnates the technological rationality that is one modern interpretation of human rationality, as the Frankfurt School in particular emphasizes. But the technological humanity defined by this technological rationality is only a surface phenomenon. Behind it is a vast and obscure domain that came forth with "bio-technical" interpretations of technics. Technics interpreted as bio-technics does not reflect theoretical rationality but mimes, exhibits, and exploits "life itself" (actually a specific kind of natural potency that the technics in question brings to the fore). Of course, life is not irrational, but it manifests a different kind of rationality. While the modern idea of the machine stemmed from mathematical rationality, contemporary bio-technics stems from the rationality pertaining to biology, linguistics, and information theory that the theories of autopoiesis, autoimmunity, cybernetics, and genetic and informational code, among others, have tried to model while remaining for the most part highly aware of the approximative character of these models. When the human being can reflect upon its own technical skill, it can also define its own rational form. Contrary to this, bio-technical existence lurks beneath such definable skills. As such it is pure undirected activity, unproductive productivity that becomes useful production only if it is *used* (seized and directed) by a skill. Without the resource of this primary capacity, a skill is empty and unproductive. A skill must use (exploit and direct) bio-technical forces in order to be a properly practical technical skill, yet it must first find them, for it cannot create them out of itself.

Now, is bio-technics the contemporary form of *Ge-stell* that disposes of both human and nonhuman nature as material productive forces? Should the Dasein therefore seek to free itself from its framework and try to prepare a space for thinking and poetisizing (*Dichten und Denken*) capable of questioning the bio-technicity that proposes itself as a new name of being? Let us say that the bio-technical can always become such a rigid interpretation of being-in-the-world. But in the end, the impossibility of enclosing bio-technics in a rigid figure is inscribed in the fact that its *bios* is never just life but also the unfathomable elemental ground of life and in the fact that *techne* is never just a determined technics but the possibility of imagining, reinventing technics as art. This is why bio-technicity should finally appear as a groundless ground, albeit in the form of an insignificant bustling of biological, digital, electrical processes. Such is the elemental ground of technics: an elemental materiality that is certainly material but on a level where one does not know what matter is. It is not a substance, not a homogenous quality; it is nothing tangible, controllable, knowable. It is a dimension without fixed forms but with a potency for forming and deforming; it is a dimension that is not useful as such but *fertile*. It is not accessible as such but it becomes accessible when the technical activity is suspended and diverted such that other activities can emerge. This is why it is not only the source of the seriousness of technics, but it is also the source of beautiful, nonprofitable play, art, pain, and bliss.

Notes

Introduction

1. As it is well known, English, this otherwise wonderful language of contemporary learning, is marked by the inconvenience of having to refer to human beings in general using the gendered pronoun *he* or *she*. This is a problem because the human being is both and using only one of these pronouns is misleading. This problem affects not only those individuals who do not identify themselves with a view of humanity organized along the gender binary and who want to be otherwise designated. The problem of pronouns also sends seismic waves across the language because pronouns are everywhere. They are like the nerves of the language. For example, how should we refer to historical individuals who did not know of this problem? Normally one refers to Plato by *he*, and so do I, but should one exclude Plato from a general de-genderification)? How should we refer to modes of human existence that are not gender-specific, for example the child, the philosopher, the president, the winner, the soldier, the beloved? (If you hear these words in a gendered sense, your ears are contaminated by language, aren't they?) Recently it has become customary to circumvent the he/she problem by using *they* as a nongendered pronoun. But this leads to a huge amount of confusion because it blurs the distinction between the singular and the plural. I could cite a hundred additional problems with this—and in writing the first draft of this book I encountered them all—but here it suffices to say that I really need to distinguish between the concept of the human being and the plurality of human beings who exist empirically. For all these reasons, and drawing inspiration from my mother tongue in which the he/she distinction does not exist and the whole debate seems superfluous, I decided to use the singular pronoun *it* instead. This solution accords with my sense of grammar without hurting my moral sense. To those readers who find it shocking to see human beings referred to as if they were just things, I answer that as this pronoun is good enough for animals and robots, it should be good enough for human beings too.

2. The following passage on trans- and posthumanism summarizes a more detailed explication published in my "On Prosthetic Existence: What Differentiates

Deconstruction from Transhumanism and Posthumanism" in *Humanism and Its Discontents: The Rise of Transhumanism and Posthumanism*, ed. Paul Jorion (New York: Palgrave Macmillan, 2022). Transhumanism is based on the claim that technology is the motor of the future evolution of humanity; see Max More and Natasha Vita-More, eds., "Transhumanist Declaration," in *The Transhumanist Reader* (Oxford: Wiley-Blackwell, 2013), 54–55. Technology is also central to a more general experience of posthumanity that does not identify itself with transhumanism, such as N. Katherine Hayles's foundational book *How We Became Posthuman: Virtual Bodies in Cybernetics, Literature, and Informatics* (Chicago: University of Chicago Press, 1999). For how technology contributes to the emergence of posthumanity, see Ivan Callus and Stefan Herbrechter, "Extroduction. The Irresistibility of the Posthuman: Questioning 'New Cultural Theory' " in *Discipline and Practice: The (Ir)resistibility of Theory*, eds. Stefan Herbrechter and Ivan Callus (Lewisburg, PA: Bucknell University Press, 2004), 226–57. Posthumanism is a label that groups together very heterogenous positions. The possibility of a technological overcoming of classical humanism was first imagined in Donna Haraway's famous "Cyborg Manifesto: Science, Technology, and Socialist-Feminism in the Late Twentieth Century" in *Simians, Cyborgs, and Women: The Reinvention of Nature* (New York: Routledge, 1991), 149–82. Other forms of posthumanism study the possibility of an animal or ecological overcoming of humanism.

3. Robert Ranisch and Stefan Lorenz Sorgner, eds., *Post- and Transhumanism: An Introduction* (Bern: Peter Lang, 2014), 34, 301.

4. Jean-Michel Besnier, *Demain les posthumains: Le futur a-t-il encore besoin de nous?* (Paris: Pluriel, 2012), 165; Ranisch and Sorgner, *Post- and Transhumanism*, 40–43, 50.

5. Nick Bostrom, "Why I Want to Be Posthuman When I Grow Up" in *The Transhumanist Reader*, 28–53.

6. More and Vita-More, *The Transhumanist Reader*, 10.

7. Most of the previously mentioned "poststructuralists" or "postmodernists" are "antihumanists" not because they defend anything like inhumanity but because they criticize all theories that claim to know what the humanity is that is capable of justifying humanism and, furthermore, what else is invited into enlarged forms of humanity (this is how some forms of trans- or posthumanism understand their own task). Humanity and humanism are not rejected because of their ethos (respect the humanity in every human being) but because of the philosophical status of these concepts, which leaves them capable of betraying their own ethos (i.e., render some people less human than others).

8. Rosi Braidotti, *The Posthuman* (Cambridge: Polity Press, 2013); Stefan Herbrechter, *Posthumanism: A Critical Analysis* (London: Bloomsbury, 2013); Cary Wolfe, *What Is Posthumanism?* (Minneapolis: University of Minnesota Press, 2010); Frédéric Neyrat, *Homo labyrinthus: Humanisme, antihumanisme, posthumanisme* (Paris: Éditions Dehors, 2015).

9. Humanity+ website, quoted by Stephen Lilley, *Transhumanism and Society: The Social Debate over Human Enhancement*. SpringerBriefs in Philosophy (Dordrecht: Springer, 2013), 1. https://doi.org/10.1007/978-94-007-4981-8.

10. Andrew Pilsch, *Transhumanism: Evolutionary Futurism and the Human Technologies of Utopia* (Minneapolis: University of Minnesota Press, 2017), 1.

11. Ranisch and Sorgner, *Post- and Transhumanism*, 7–8; compare with More and Vita-More, *The Transhumanist Reader*, 4, 28–53.

12. Wolfe, *What Is Posthumanism?*, xv.

13. Jane Bennett, *Vibrant Matter: A Political Ecology of Things* (Durham, NC: Duke University Press, 2010).

14. Ranisch and Sorgner, *Post- and Transhumanism*, 9; Pilsch, *Transhumanism*, 114–37. Julian Huxley—Aldous's brother—was a biologist engaged in humanist, eugenics, and nature conservation movements. He was also a founding member of WWF and the first director of UNESCO. Pierre Teilhard de Chardin was a Jesuit priest and intellectual who combined the theory of evolution, the theory of the biosphere, and the Christian theodicy into the notion of a unique progression toward "noosphere." This would be a coalescence of consciousness that prepares for the advent of the cosmical Christ. This idea is isomorphous with today's idea of the becoming-conscious of the Internet as Singularity.

15. Ray Kurzweil, *The Singularity Is Near: When Humans Transcend Biology* (Penguin Books, 2005).

16. Cf. Tirosh-Samuelson in Ranisch and Sorgner, *Post- and Transhumanism*; Hayles, *How We Became Posthuman*.

17. Michael Hagner and Erich Hörl, eds., *Die Transformation des Humanen: Beiträge zur Kulturgeschichte der Kybernetik* (Frankfurt am Main: Suhrkamp, 2008), 10; Neyrat, *Homo labyrinthus*, 1.

18. More and Vita-More, *The Transhumanist Reader*, 4.

19. More and Vita-More, 10.

20. Martin Heidegger, *The Question Concerning Technology and Other Essays*, trans. William Levitt (New York: Harper and Row, 1977), 24–25.

21. Gilles Deleuze and Félix Guattari, *A Thousand Plateaus: Capitalism and Schizophrenia*, trans. Brian Massumi (Minneapolis: University of Minnesota Press, 1987), 313; Deleuze and Guattari, *Capitalisme et schizophrénie I: L'Anti-Œdipe* (Paris: Minuit, 1972).

Chapter 1

1. Erich Hörl, "The Technological Condition," *Parrhesia* 22 (2015): 1–15.

2. Gilbert Hottois claims to have invented the term *anthropotechnics*, which he uses to designate technical activities that aim to alter the human being instead of the environment. Gilbert Hottois, *Species technica*, suivi d'un, *Dialogue vingt ans*

plus tard (Paris: Vrin, 2002). Peter Sloterdijk studies anthropotechnics, that is, both imposed and freely adapted technics of individual and collective transformation in *You Must Change Your Life* (Malden, MA: Polity Press, 2013), which complements the Elmau Rede *Rules for a Human Zoo: A Response to the Letter on Humanism* (*Regeln für den Menschenpark*) (Frankfurt am Main: Suhrkamp Verlag, 1999).

3. Martin Heidegger, *Sein und Zeit* (Tübingen: Max Niemeyer, 1984) § 52, 256; translated as *Being and Time* by John Macquarrie and Edward Robinson (Oxford: Blackwell, 1985). The original page numbers are indicated in the margins of this edition.

4. Margaret Lock, *Twice Dead: Organ Transplantation and the Reinvention of Death* (Berkeley: University of California Press, 2002); Lesley A. Sharp, *Strange Harvest: Organ Transplants, Denatured Bodies, and the Transformed Self* (Berkeley: University of California Press, 2006); Susanna Lindberg, "The Obligatory Gift of Organ Transplants: The Case of the Finnish Law on the Medical Use of Human Organs, Tissues and Cells," *Alternatives: Global, Local, Political* 38, no. 3 (2013): 245–55; Kartina Choong, "Death Before Organ Donation: Exploring the Chasm between Lay and Professional Knowledge," International Conference: Phenomenology of Medicine and Bioethics, Södertörn University, Sweden, June 13–15, 2018; Fredrik Svenaeus, "To Die Well: The Phenomenology of Suffering and End of Life Ethics," *Medicine, Health Care, and Philosophy* 23 (2020): 335–42.

5. See the discussion of the peculiar status of the brain-dead in Giorgio Agamben, *Homo Sacer: Sovereign Power and Bare Life*, trans. Daniel Heller-Roazen (Stanford: Stanford University Press, 1998), chap. 6.

6. To quote the well-chosen title of Buchanan et al., *From Chance to Choice: Genetics and Justice* (Cambridge: Cambridge University Press, 2000).

7. As Jürgen Habermas mistakenly thought in his otherwise useful book *Die Zukunft der menschlichen Natur: Auf dem Weg zu einer liberalen Eugenik?* (Frankfurt am Main: Suhrkamp Verlag, 2001).

8. The first genetically modified human babies were "made" by Chinese scientist He Jiankui in 2018, provoking a worldwide outcry. Nonetheless, a third such baby is being gestated as at the time of this writing. Among numerous articles and news reports on the subject, see for example James Gallagher, "He Jiankui: Baby Gene Experiment 'Foolish and Dangerous,'" *BBC News*, June 3, 2019, https://www.bbc.com/news/health-48496652.

9. For concise explication and vigorous rejection of the eugenics movement, see André Pichot, *L'eugénisme, ou les généticiens saisis par la philantropie* (Paris: Hatier, 1995); and Pichot, *La société pure: De Darwin a Hitler* (Paris: Flammarion, 2000).

10. Eduardo Kac, *GFP Bunny*, 2000. GFP Bunny is (was) a living green, fluorescent rabbit that its conceptor and orderer, the artist Eduardo Kac, presented as a work of art. See Kac's website, https://www.ekac.org/transgenicindex.html.

11. The philosophical questions concerning chimaira are also discussed by the philosopher of science Gilbert Hottois in the novel *Species technica*, as well as in a discussion with Jean-Noël Missam titled "Dialogue philosophique autour de 'Species technica' vingt ans plus tard," both appearing in Gilbert Hottois, *Species technica* (Paris: Vrin, 2002).

12. Andrew Pilsch, *Transhumanism: Evolutionary Futurism and the Human Technologies of Utopia* (Minneapolis: University of Minnesota Press, 2017), 65.

13. Hans Jonas, *Technik, Medizin und Ethik* (Frankfurt am Main: Suhrkamp Verlag, 1985). A phenomenological description of the change of perspective in biotechnology is provided by Bernhard Waldenfels, *Bruchlinien der Erfahrung* (Frankfurt am Main: Suhrkamp Verlag, 2002), 422–55. It is historically understandable that German philosophers are particularly sensitive to the issue of eugenics. However, one should not forget that eugenistic measures such as forced sterilizations were imposed on defenseless individuals in many other countries too, such as the United States and Sweden, even up to the 1970s.

14. For a presentation of the discussion, see Michael A. Cerullo, "Uploading and Branching Identity," *Minds and Machines* 25 (2015): 17–36.

15. Stelarc is a performance artist known for works in which he extends his body through technical interventions, like in his iconic Third Arm (1980). Neil Harbison calls himself the world's first cyborg artist because in 2004 he had an antenna implanted in his skull that allows him to "hear" colors that he cannot see, being colorblind. The philosopher Jean-Luc Nancy has written a very important philosophical account of his receiving a heart implant, *L'intrus* (Paris: Galilée, 2000), English trans. Richard A. Rand, "The Intruder," in *Corpus* (New York: Fordham University Press, 2008), 161–70. Francisco Varela has described his experience of a liver transplantation in "Intimate Distances: Fragments for a Phenomenology of Organ Transplantation," *Journal of Consciousness Studies* 8, no. 5–7 (2001): 259–71.

16. See in particular Martin Heidegger, *The Question Concerning Technology and Other Essays*, trans. William Lovitt (New York: Garland Publishing, 1977); Max Horkheimer and Theodor Adorno, *Dialectic of Enlightenment: Philosophical Fragments*, trans. Edmund Jephcott (Stanford: Stanford University Press, 2002), Jacques Ellul, *The Technological System*, trans. Joachim Neugroschel (New York: Continuum, 1980); Félix Guattari, *The Three Ecologies*, trans. Ian Pindar and Paul Sutton (London: Athlone Press, 2000).

17. Yuk Hui, *Recursivity and Contingency* (London: Rowman and Littlefield, 2019), 185. The notion of algorithmic life was formulated by Éric Sadin in *La vie algorithmique: Critique de la raison numérique* (Paris: L'échappée, 2015). The notion of algorithmic governmentality, inspired by Foucault, comes from Antoinette Rouvroy and Thomas Berns, "Le nouveau pouvoir statistique, Ou quand le contrôle s'exerce sur un réel normé, docile et sans événement car constitué

de corps 'numériques,'" *Multitudes* 40 (2010): 88–103; and Rouvroy and Berns, "Détecter et prévenir: De la digitalisation des corps et de la docilité des normes," in *(Se) gouverner: entre souci de soi et action publique*, ed. Guy Lebeer and Jacques Moriau (Brussels: Peter Lang, 2010), 157–84. Bernard Stiegler develops Rouvroy and Berns's idea further in *Automatic Society Vol. 1: The Future of Work*, trans. Dan Ross (Cambridge: Polity Press, 2016).

18. Alan Turing, "Computing Machinery and Intelligence," *Mind* 59, no. 236 (1950): 433–60.

19. Michael L. Anderson, "Embodied Cognition: A Field Guide," *Artificial Intelligence* 149 (2003): 3, 91–130.

20. John Searle's famous criticism of Strong AI ignited huge discussions when it was first published in 1980: "Minds, Brains, and Programs," *Behavioral and Brain Sciences* 3 (1980): 417–57. Equally influential is Hubert L. Dreyfus's *What Computers Still Can't Do: A Critique of Artificial Reason* (Cambridge, MA: MIT Press, 1992). This is revised from the original 1972 edition.

21. For a good overview of this discussion, see Anderson, "Embodied Cognition," 1–40.

22. Andy Clark and David Chalmers, "The Extended Mind," *Analysis* 58, no. 1 (1998): 7–19, especially 8.

23. André Leroi-Gourhan, *L'homme et la matière* (Paris: Albin Michel, 1943); Leroi-Gourhan, *Milieu et technique* (Paris: Albin Michel, 1945); Leroi-Gourhan, *Le geste et la parole 1: Technique et langage* (Paris: Albin Michel, 1964); Leroi-Gourhan, *Le geste et la parole II: La mémoire et les rythmes* (Paris: Albin Michel, 1965).

24. Jacques Derrida, "Plato's Pharmacy," in *Dissemination*, trans. Barbara Johnson (Chicago: University of Chicago Press, 1981), 63–171. Derrida's early theorization of writing can also be found in *Of Grammatology*, trans. G. C. Spivak (Baltimore: Johns Hopkins University Press, 1976); *Margins of Philosophy*, trans. Alan Bass (Chicago: University of Chicago Press, 1982); *Writing and Difference*, trans. Alan Bass (Chicago: University of Chicago Press, 1978).

25. N. Katherine Hayles, *Unthought: The Power of the Cognitive Nonconscious* (Chicago: University of Chicago Press, 2017). See also Hayles, *How We Became Posthuman: Virtual Bodies in Cybernetics, Literature, and Informatics* (Chicago: University of Chicago Press, 1999).

26. Bernard Stiegler, *Technics and Time 1: The Fault of Epimetheus*, trans. Richard Beardsworth and George Collings (Stanford: Stanford University Press, 1998), 177.

27. Yuval Noah Harari, *Homo Deus: A Brief History of Tomorrow* (New York: Harper, 2017).

28. Martin Heidegger, "The Age of the World Picture," in *The Question Concerning Technology and Other Essays*, 153; "Die Zeit des Weltbildes" 1938, in *Holzwege* (Frankfurt am Main: Vittorio Klostermann, 1980), 109.

29. G. W. F. Hegel, *The Phenomenology of Spirit*, trans. Terry Pinkard (Cambridge: Cambridge University Press, 2018), 201.

30. Sigmund Freud, *Totem and Taboo* (London: Ark Paperback, 1983), 1–17.
31. Claude Lévi-Strauss, *The Elementary Structures of Kinship* (London: Eyre and Spottiswoode, 1970), 12–25.
32. Philippe Descola, *Beyond Nature and Culture*, trans. Janet Lloyd (Chicago: University of Chicago Press, 2013), 173.
33. Eduardo Viveiros de Castro, *Cannibal Metaphysics*, trans. Peter Skafish (Minneapolis: Univocal, 2014), 68. See also Viveiros de Castro, "Perspectivisme et multinaturalisme en Amérique indigène," *Journal des anthropologues* 138–139 (2014): 161–81; Descola, *Beyond Nature and Culture*, 129, 138–39.
34. Descola, 145, 150.
35. Descola, 212.
36. David Gé Bartoli and Sophie Gosselin, *Le toucher du monde: Techniques du naturer* (Paris: Dehors, 2019), 362–63. My translation.
37. In *Infancy and History*, Giorgio Agamben presents infancy as the condition for properly human language: "Because of his infancy, because he does not speak from the very start, man cannot enter into language as a system of signs without radically transforming it, without constituting it in discourse." Agamben, *Infancy and History: On the Destruction of Experience*, trans. Liz Heron (London: Verso, 2007), 63. "Contrary to ancient traditional beliefs, from this point of view man is not the 'animal possessing language,' but instead the animal deprived of language and obliged, therefore, to receive it from outside himself" (65). While infancy is interesting as a phase of human speech itself, real infants are of course quite different to nonspeakers. The necessity to learn an adult language does not mean that infancy was not already characterized by signification. Furthermore, the autistic children who stay in a nonverbal world do not live in a nonsignifying world either, as shown for example in the works of the special educator Ferdinand Deligny, described by Gé Bartoli and Gosselin, *Le toucher du monde*, 289–314.
38. Giorgio Agamben, *The Open: Man and Animal*, trans. Kevin Attell (Stanford: Stanford University Press, 2004), 21.
39. *Animot* is pronounced like the plural *animaux* but it indexes animality to the singular word *mot* that catches the plurality of animals. Jacques Derrida has written a number of important texts on the role of animality in the definition of humanity, the most famous of which is *The Animal That Therefore I Am*, trans. David Wills (New York: Fordham University Press, 2008). Derrida's discussion of animality has been further developed and notably criticized in Cary Wolfe, ed., *Zootologies: The Question of the Animal* (Minneapolis: University of Minnesota Press, 2003); and Leonard Lawlor, *This Is Not Sufficient: An Essay on Animality and Human Nature in Derrida* (New York: Columbia University Press, 2007). Lawlor stresses the animal condition for itself rather than just deconstructing the human condition.
40. Some species are our "companion species," as Donna Haraway puts it in *The Companion Species Manifesto: Dogs, People, and Significant Otherness* (Chicago: Prickly Paradigm Press, 2003).

41. Jef Akst reports recent studies that reveal that about 2 percent of the DNA of modern-day people with Eurasian ancestry is Neanderthal in origin. Akst, "Neandertal DNA in Modern Human Genome Is Not Silent," *The Scientist*, September 2019. https://www.the-scientist.com/features/neanderthal-dna-in-modern-human-genomes-is-not-silent-66299. "What is being rapidly established is that the human body, far from having one exceptional genome that marks it as superior to other organisms, is a complex admixture of bacterial, fungal, parasitical and viral components on a cellular level in which the strictly human cell (or rather the human as previously understood) is vastly outnumbered." Margrit Shildrick, "Microchimerism, Immunity and Temporality: Rethinking the Ecology of Life and Death," *Australian Feminist Studies* 34/99, no. 10–24 (2019): 12–13.

42. See in particular Elisabeth de Fontenay, *Le silence des bêtes: La philosophie à l'épreuve de l'animalité* (Paris: Fayard, 1998).

43. Haraway calls it the cyborg condition in her earlier "A Cyborg Manifesto: Science, Technology, and Socialist-Feminism in the Late Twentieth Century," in *Simians, Cyborgs, and Women: The Reinvention of Nature* (New York: Routledge, 1991), 149–81.

44. Up to now, philosophers have not studied to any great extent the moral implications of the use of artificial intelligence. But the need for such works is becoming increasingly clear, as is shown by the emergence of reports such as "EU Guidelines on Ethics in Artificial Intelligence: Context and Implementation," European Parliamentary Research Service PE 640.163, September 2019; or Cédric Villani, "For a Meaningful Artificial Intelligence: Towards a French and European Strategy," Parliamentary mission assigned by the French Government, 2018. Far from really seizing the problem at its root, these reports indicate the birth of a problem.

45. Jean-François Lyotard, *The Inhuman: Reflections on Time*, trans. Geoffrey Bennington and Rachel Bowlby (Cambridge: Polity Press, 1991), 2.

46. Immanuel Kant, *Kritik der reinen Vernunft 1–2* (Frankfurt am Main: Suhrkamp Verlag, 1995), B 832.

47. Kant, *Logic*, AA vol. IX, 25: "What can I know? What should I do? What may I hope for? What is a human being? The first question is answered in *metaphysics*, the second in *morals*, the third in *religion*, and the fourth in *anthropology*." Quoted in Michel Foucault, *Introduction to Kant's Anthropology*, ed. and trans. Roberto Nigro and Kate Briggs, Semiotext(e) Foreign Agents Series (Cambridge, MA: MIT Press, 2008), 75.

48. "Nothing straight can be constructed from such warped wood as that which man is made of" (or crooked wood, *krumme Holz*). Kant, *Idea for a Universal History with a Cosmopolitan Purpose*, in *Political Writings*, trans. H. B. Nisbet (Cambridge: Cambridge University Press), 46.

49. René Descartes, *Meditations on First Philosophy*, in *The Philosophical Works of Descartes*, trans. Elizabeth S. Haldane (Cambridge: Cambridge University

Press, 1911), 1–12. For a playful commentary see Andrzej Warminski, "Spectre Shapes: The Body of Descartes?" *Qui Parle* 6, no. 1 (1992), 93–112, especially 93.

50. When Kant speaks of "human knowledge," he could also for the most part instead say "finite knowledge" if he took seriously his own claim that there is no reason to deny the existence of extraterrestrials who have the same faculties. On this curiosity, see Robert B. Louden's note in *Kant's Human Being: Essays on his Theory of Human Nature* (Oxford: Oxford University Press, 2011), xix–xx; and the analysis of Peter Szendy, *Kant in the Land of Extraterrestrials: Cosmopolitical Philosophifictions* (New York: Fordham University Press, 2013).

51. Kant, *Kritik der reinen Vernunft*, B 25, A 11.

52. Kant, B 131–32.

53. Kant, B 428–32.

54. Jacob Rogozinski, "Je suis l'être même," in *Kanten: Esquisses kantiennes* (Paris: Kimé, 1996), 78. My translation.

55. Rogozinski, "Je suis l'être même," 83.

56. This is why Kantian morality is according to Hegel hypocritical. Hegel, *The Phenomenology of Spirit*, 332–57, 356–82. For an explication of this debate, see for example R. Z. Friedman, "Hypocrisy and the Highest Good: Hegel on Kant's Transition from Morality to Religion," *Journal of the History of Philosophy* 24, no. 4 (1986): 503–33; or Nicolás García Mills, "Realizing the Good: Hegel's Critique of Kantian Morality," *European Journal of Philosophy* 26, no. 1 (2017): 195–212.

57. Commentary of J.-M. Muglioni to Immanuel Kant, *Idée d'une histoire universelle au point de vue cosmopolitique* (Paris: Bordas, 1988), 48.

58. Christian Krijnen, "Kants Subjektstheorie und die Grundlegung einer Philosophischen Anthropologie," *Zeitschrift für philosophische Forschung*, Bd 62, H 2 (April–Jun 2008): 254–73, 268.

59. Christian Krijnen, "Kants Subjektstheorie und die Grundlegung einer Philosophischen Anthropologie," *Zeitschrift für philosophische Forschung*, Bd 62, H 2 (April–Jun 2008): 254–73, especially 259–60.

60. According to Foucault, this text goes beyond the abstraction characteristic of the Cartesian answer to the question "Who am I? (everybody and nobody)" by asking who we are in reality, here and now, at this precise moment of history. Michel Foucault, "The Subject and Power," in *Michel Foucault: Beyond Structuralism and Hermeneutics*, Hubert L. Dreyfus and Paul Rabinow (Chicago: University of Chicago Press, 1982), 208–26; and in Michel Foucault, "Le sujet et le pouvoir," in *Dits et écrits II (1976–1988)* (Paris: Gallimard, 2017), 1050–51. Foucault presents Kant as a direct predecessor of his own work in a detailed analysis found in one of his last texts, "Qu'est-ce que les lumières?" in *Dits et écrits II (1976–1988)* (Paris: Gallimard, 2017), 1381–97; "What Is Enlightenment?," trans. Catherine Porter, in *Ethics: Subjectivity and Truth: The Essential Works of Michel Foucault 1954–1984*, ed. Paul Rabinow (New York: New Press, 1994), 303–20.

61. Edmund Husserl, *Ideas Pertaining to a Pure Phenomenology and to Phenomenological Philosophy. First Book: General Introduction to a Pure Phenomenology*, trans. F. Kersten (The Hague: Nijhoff, 1982).

62. For more extensive references, see my *Techniques en Philosophie* (Paris: Hermann, 2020).

63. The theme of plasticity traverses all of Malabou's books. See also Juan Manuel Garrido Wainer, *La formation des formes* (Paris: Galilée, 2008); Boyan Manchev, *La métamorphose et l'instant: La désorganisation de la vie* (Paris: La Phocide, 2009).

64. As we shall see in chapter 5, both Bernard Stiegler and Giorgio Agamben emphasize on the one hand the lack or default of form and on the other hand the pure dynamis that makes it possible.

Chapter 2

1. According to the *Oxford English Dictionary*, technology means (if we rule out obsolete uses such as a "discourse or treatise on an art or arts" or "terminology of a particular art of subject") "the branch of knowledge dealing with mechanical arts and applied sciences; the study of this," "the application of such knowledge for practical purposes, esp. in industry, manufacturing, etc.; the sphere of activity concerned with this; the mechanical arts and applied sciences applied collectively"; "the product of such application; technological knowledge or know-how; a technological process, method or technique. Also: machinery, equipment, etc., developed from the practical application of scientific and technical knowledge, an example of this"; "a particular practical or industrial art; a branch of the mechanical arts or applied sciences; a technological discipline." *Technique*, probably a word borrowed from French, has a different sense, for it means "the formal or practical aspect of any art, occupation, or field; manner of execution or performance in regard with this. Also more generally: way of doing something." It also means "practical skill or ability in a formal or practical aspect of a particular field"; "a particular way of carrying out an experiment, procedure or task, esp. in a scientific discipline or a craft; a technical or scientific method. Also more generally: a skilful or efficient means of achieving a purpose; a strategy, a knack." Finally *technics* is a rare and in many uses obsolete term, but it has the advantage of combining the two senses. Either it is, mostly as the singular *technic*, a "technical method, a scientific procedure," a largely obsolete word superseded by technique, or it is, with singular or more rarely plural agreement, *technics*, "the mechanical or applied arts, esp. as a subject of study, the branch of knowledge dealing with such arts = technology." When these terms are separated, technology refers to applied sciences and technics to hand crafts. Nonetheless, most of the time, I will follow Mumford (*Art and Technics*) and the translators of Bernard

Stiegler's *Technics and Time* in using the term *technics* as a general philosophical term referring to the entire problematic.

By technics, I refer to the problematic that came to philosophy with the Greek word *techne*, which derives from the early word *tekton* (master builder, carpenter, smith), the root of which is *teks* or *tek*, which means among others weaving, fabricating, uniting, and joining. Wolfgang Schädewaldt defines the ancient sense of *tekhne* as the "art or skill of the carpenter and master builder and more generally the art of every kind of production," and by and by both craftmanship and trade and clever, crafty machination." Wolfgang Schädewaldt, "The Greek Concepts of 'Nature' and 'Technique,'" in *Philosophy of Technology. The Technological Condition. An Anthology*, eds. Robert C. Scharff and Val Dusek (Oxford: Wiley Blackwell, 2014), 25–32.

The adjective *technikon* "in addition to aptitude for the art of production also designates the general aggregate of what is in accord with and suited to art or skill." The sense of *techne* should also be delimited against *mechané* ("machine") and *organon* ("instrument, tool"), but let us not complicate things yet. The Latin *technica, ars* then yields the French *technique* (seventeenth century) and the German *Technik* (eighteenth century), which are the totality of means and modes of producing. These terms cover all of the main senses of the three English terms. Today, French, German, and many other languages also use the term *Technologie*, but it is only a calque of English that is mainly used to refer to industrial technologies, not to the phenomenon as a whole.

2. See Tom Angier and Anne Balansard, *Techné dans les dialogues de Platon: L'empreinte de la sophistique* (Sankt Augustin: Akademia Verlag, 2001).

3. Plato, *Prot* 319–20; *Men* 94 b.

4. *Eth. Nic.* IV, 4, 5; cf. Angier and Balansard, *Techné*, 35; 41 sqq.

5. Plato, *Ion*, in *Complete Works*, trans. J. M. Cooper and D. S. Hutchingson (Indianapolis: Hackett, 1997).

6. Plato, *Phaedrus*, trans. A. Nehamas and P. Woodruff (Indianapolis: Hackett, 1995); see Angier and Balansard, *Techné*, 16–17.

7. Heidegger emphasizes that Aristotle cannot define the *physei onta* without recourse to the *technui onta* in his important text "On the Essence and Concept of *Physis*, in Aristotle's *Physics*, B 1," in *Pathmarks*, trans. Thomas Sheehan and William McNeill (Cambridge: Cambridge University Press, 1998), 183–230.

8. Angier and Balansard, *Techné*, 4.

9. Martin Heidegger, *Introduction to Metaphysics*, trans. Gregory Fried and Richard Polt (New Haven, CT: Yale University Press, 2000), 122–28, 169–78.

10. For example Plato's *Alcibiades* mentions and distinguishes women's and men's work, the former being work such as weaving wool and the latter work such as fighting in war as a hoplite (*Alcibiades*, 127a).

11. Even though the aim of Foucault's readings of Plato (and of other classics) is not to write an ordinary contribution to the history of ancient philosophy

but to establish the genealogy of a contemporary question, meaning that he reads the dialogues themselves without paying much attention to Plato studies, a Plato scholar such as Anissa Castel-Bouchouchi finds his readings pertinent and innovative. Anissa Castel-Bouchouchi, "Foucault et le paradoxe du platonisme," in *Foucault et la philosophie antique*, eds. Frédéric Gros and Carlos Lévy (Paris: Kimé, 2003), 176–77.

12. Michel Foucault, "Subjectivité et vérité," in *Dits et écrits II, 1976–1988* (Paris: Gallimard, 2017), 1032; trans. Robert Hurley as "Subjectivity and Truth," in *Ethics: Subjectivity and Truth: The Essential Foucault 1* (New York: New Press, 1997), 87. Foucault studies the theme of techniques of self especially in volumes two and three of the *History of Sexuality*. He does not use the term but describes the idea of an ethical formation of the subject and the task of "making one's life into an œuvre" already in *The Use of Pleasure, Vol. 2 of the History of Sexuality*, trans. Robert Hurley (New York: Vintage Books, 1990), 139. The technics of the self are explained in detail in *The Care of the Self, Vol. 3 of the History of Sexuality*, trans. Robert Hurley (New York: Pantheon Books, 1986), 43 sqq. Technics of the self are also at the heart of Foucault's last lecture courses delivered in 1981–1984: *The Hermeneutics of the Subject*: *The Government of Self and Others* and *The Courage of Truth*.

13. Michel Foucault, *L'herméneutique du sujet: Cours au collège de France. 1981–1982* (Paris: Gallimard, 2001), 19; trans. Graham Burchell as *The Hermeneutics of the Subject: Lectures at the Collège de France 1918–1982* (New York: Palgrave Macmillan, 2005), 17.

14. Foucault, *L'herméneutique du sujet*, 57; Burchell, *The Hermeneutics of the Subject*, 58.

15. Foucault, 55; Burchell, 56.

16. Giorgio Agamben, *The Use of Bodies*, trans. Adam Kotsko (Stanford: Stanford University Press, 2016).

17. Foucault, "L'herméneutique du sujet," *Dits et écrits II*, 1174–78; Hurley et al., *Ethics*, 93–108.

18. See more precisely Foucault, "L'écriture de soi," in *Dits et écrits II*, 1234–49; Hurley et al., 207–16.

19. See Judith Revel, "Michel Foucault: repenser la technique," in *Tracés: Revue des Sciences Humaines* 16 (2009): 139–49. Revel's comparison between Heidegger's and Foucault's readings of *techne* is instructive.

20. The word *machine*, derived from the ancient Greek *mehane* meaning "remedy," "clever expedient," "cleverly contrived means," "means by which one gets." See the etymological explication by Wolfgang Schädewaldt, in "The Greek Concepts of 'Nature' and 'Technique,'" 28–29. Already in ancient Greece *mehane* also denotes the concrete machine, for example a war machine or a stage machine. *Mehane* gives the Latin *machina*, which gives the word *machine* to many modern languages (e.g., English, French, German). *Mehane* can also be heard in "mechanic," "mechanical," and "mechanistic." In the Middle Ages, "mechanical arts" desig-

nated ordered practices and skills (e.g., architecture, metallurgy, agriculture) as opposed to the liberal arts (i.e., arithmetic, grammar, and music that belong to the trivium and quadrivium curricula). Depreciated in antiquity and the Middle Ages, the mechanical arts became esteemed in the Renaissance (which celebrates artist-engineers such as Leonardo da Vinci) and they inspired an entire worldview as can be seen in the works of philosophers such as Newton, Hobbes, and Descartes. Among the numerous senses of "machine" listed by the *Oxford English Dictionary*, what is of particular significance for us are (1) "a structure regarded as functioning as an independent body," (2) "a material structure designed for a specific purpose," and (3) "an apparatus constructed to perform a task, or for some other purpose," especially "a complex device, consisting of a number of interrelated parts, each having a definite function, together applying, using, or generating mechanical or (later) electrical power to perform a certain kind of work." The machine is different from the Greek *techne* because the machine does not designate primarily the artisan's craft but the mechanical apparatus itself, to which its human operator gradually becomes integrated. Of course the machine is also an artifice produced by a crafty engineer—but the fabricator's work does not end in the product (e.g., a house that stands on its own) because the product is itself an instrument that will perform another kind of a work (e.g., a clock that measures time). From the point of view of human activity the machine is essentially a means, a tool, an instrument of an ongoing work. But considered as such it is a relatively autonomous assemblage that functions on its own, hence it can assist the human being in its work (a microscope), give the illusion of working on its own (an automat), and absorb its human operators into the machine (industrial complex).

21. Lewis Mumford, *Technics and Civilization* (New York: Harbinger Books, 1964), 12–18.

22. Mumford, 46–51, 132 sqq.

23. Alexandre Koyré, *From the Closed World to the Infinite Universe* (Baltimore: Johns Hopkins University Press, 1968), 90. However, in my opinion, Pythagoras's monocord is an early scientific instrument as well, but it is not connected with a theory of the experimental method.

24. Mumford, *Technics and Civilization*, 12.

25. Ernst Kapp in particular develops this idea. He is regarded as the inventor of the discipline of the philosophy of technology. He interprets tools in general as organ projections and instruments, for example telescopes and pianos, as projections of sense organs (eye, ear). Ernst Kapp, *Elements of a Philosophy of Technology, 1877* (Minneapolis: University of Minnesota Press, 2018). The anthropologist André Leroy-Gourhan, emphasizing the key role of technics in the very process of hominization, interprets technical evolution as the continuation of the evolution of the brain in external objects. André Leroi-Gourhan, *Gesture and Speech* (Cambridge, MA: MIT Press, 1993).

26. Mumford, *Technics and Civilization*, 10–11.

27. Francis Bacon, *The New Organon*, in *Cambridge Texts in the History of Philosophy*, eds. Lisa Jardine and Michael Silverstone (Cambridge: Cambridge University Press, 2000), LXII–LXV, LXXXIV, especially 69, 51. See also 11.

28. Bacon, LXXXII, 67. The technical metaphor is important for Bacon: "Neither the bare hand nor the unaided intellect has much power; the work is done by tools and assistance, and the intellect needs them as much as the hand." Bacon, 33.

29. Bacon, 18.

30. Bacon, CI, CII, 82. Ever since the fifteenth and sixteenth centuries, artisans and engineers have written rich technical treatises. See Paolo Rossi, *The Birth of Modern Science* (Oxford: Blackwell, 2001), 31. Jürgen Klein recounts how Bacon's idea of empirical science depended on his knowledge of the work of artisans and workshops, for example the master potter Bernard Palissy and the metallurgist Georg Agricola, in "Francis Bacon's Scientia Operativa: The Tradition of the Workshops, and the Secrets of Nature," in *Philosophes of Technology: Francis Bacon and his Contemporaries*, ed. Claus Zittel (Leiden: Brill, 2008).

31. Rossi, *The Birth of Modern Science*, 186–91.

32. Bacon, *The New Organon*, 21; cf. "The secrets of nature reveal themselves better through harassments applied by the arts than when they go their own way." Bacon, XCVIII, 81.

33. Bacon, XXXI, 151.

34. Bacon, CVIII, 85.

35. Sophie Weeks, "Mechanics in Bacon's Great Instauration," in *Philosophies of Technology*, 140–41, Weeks's citations.

36. Weeks, 153.

37. Contrary to Rossi, Weeks argues that unlike Boyle and Descartes, Bacon does not think of the universe as a "great piece of clockwork." Weeks, 156.

38. The machine could assume this role in early modernity because cultivated people gradually became familiar with increasingly ingenious apparatuses. Some of these were useful, like clocks and telescopes, but many of the most elaborate were showpieces, like stage machineries, anatomical theaters, and mechanical spectacles. Andrés Vaccari reports that Descartes probably had the chance to visit Royal Château of Saint-Germain-en-Laye park, which was equipped with self-moving figures, such as a statue of a nymph whose fingers moved such that she could play an organ, another of Mercury capable of blowing into a trumpet, artificial dragons belching water, and birds singing beautifully. Reported by Andrés Vaccari, "Legitimating the Machine: The Epistemological Foundation of Technological Metaphor in the Natural Philosophy of René Descartes," in *Philosophes of Technology*, 320–25. Later, in the eighteenth century, the most famous pieces of the art of automats were produced by Jacques de Vaucanson, renowned for example for the *Flute Player* (1737) and especially for the *Digesting Duck* (c. 1734) that could eat and defecate.

39. René Descartes, *Traité de l'homme* (1662) in *Œuvres et lettres* (Paris: Bibliothèque de la Pléiade, 1953), 807, 873; Stephen Gaukroger, ed. and trans., in *Descartes: The World and Other Writings* (Cambridge: Cambridge University Press, 1998), 119, 196.

40. Georges Canguilhem, *Knowledge of Life*, trans. Stefanos Geroulanos and Daniela Ginsburg (New York: Fordham University Press, 2008), 87. "[I]f the functioning of a machine is *explained* by relations of pure causality, the construction of a machine can be understood neither without purpose nor without man. A machine is made by man and for man, with a view towards certain ends to be obtained, in the form of effects to be produced" (86).

41. Julien Offray de la Mettrie, *Machine Man*, in *Machine Man and Other Writings*, ed. Ann Thomson (Cambridge: Cambridge University Press, 2012).

42. Denis Diderot, *D'Alembert's Dream*, in *Rameau's Nephew and D'Alembert's Dream*, ed. and trans. Leonard Tancock (London: Penguin Classics, 1976).

43. René Descartes, *Discours de la méthode*, in *Œuvres et lettres*, 164–65; Elizabeth F. Haldane, ed., David Weissman, trans., *Discourse on the Method and Meditations on First Philosophy* (New Haven, CT: Yale University Press, 1996), 34–35.

44. Quoted in Vaccari, "Legitimating the Machine," 300.

45. Vaccari, 326.

46. Vaccari, 332.

47. Vaccari, 327.

48. Vaccari, 334.

49. Jacques Ellul's idea of the technological society describes a technical system based on rationalization, automatization, and exclusive universalism becoming autonomous and organizing society as a whole, forcing human beings to adapt to it rather than serving them as was originally intended. Jacques Ellul, *The Technological Society*, trans. John Wilkinson (New York: Knopf, 1964). However, he speaks in terms of a "technical system," which is more than a machine or a megamachine. Ellul, *Le système technicien* (Paris: Cherche-midi, 2012), 195, 239. It is a system in the sense of a totality whose parts are connected, interdependent and have a shared regularity (163). It is organized as a closed world that is no longer controlled by any human being but rather controls humans (237). Martin Heidegger's idea of *Ge-stell* is the extreme philosophical formulation of the epoch structured not by technical apparatuses themselves but by the underlying logic of totalitarian reason and by the interpretation of natural and human beings as simple resources. Gilbert Simondon also regards the technical structure of the world as more fundamental than its political consequences, yet he is the most optimistic as he calls for a better use of technics. Gilbert Simondon, *On the Mode of Existence of Technical Objects* (Minneapolis: Univocal Publishing, 2016).

50. Already Mumford calls for such a utopia "basic communism," which he sharply distinguishes from both Marxism and Soviet Russia. Mumford, *Technics and Civilization*, 403.

51. Heidegger, "Brief über den 'Humanismus,'" in *Wegmarken* (Frankfurt am Main: Vittorio Klostermann, 1978), 337; see translation "Letter on Humanism," in *Basic Writings*, 243–45; Heidegger, "Überwindung der Metaphysik," in *Vorträge und Aufsätze* (Stuttgart: Neske, 1994); trans. Joan Stambaugh as "Overcoming Metaphysics," in *The End of Philosophy* (Chicago: University of Chicago Press, 2003).

52. Alain Badiou, *Manifeste pour la philosophie* (Paris: Seuil, 1989), 33–40; trans. Norman Madarash as *Manifesto for Philosophy* (Albany: State University of New York Press, 1999), 55.

53. Bernard Stiegler, *States of Shock: Stupidity and Knowledge in the 21st Century*, trans. Dan Ross (Cambridge: Polity Press, 2015), 228.

54. See Bernhard Waldenfels, "Technische Eingriffe in die Erfahrung," in *Bruchlinien der Erfahrung: Phänomenologie, Psychoanalyse, Phänomenotechnik* (Frankfurt am Main: Suhrkamp, 2002), 364–74.

55. Canguilhem, *Knowledge of Life*, 96.

56. The starting points of the research politics of the NBIC are well resumed by Gilbert Hottois in *Le transhumanisme est-il un humanisme* (Brussels: Académie Royale de Belgique, 2014).

57. This is the claim of Erich Hörl in *Sacred Channels: The Archaic Illusion of Communication*, trans. Nils Schott (Amsterdam: Amsterdam University Press, 2018), 33.

58. Waldenfels, *Bruchlinien der Erfahrung*, 370–71.

59. The most authoritative scientific sources are summarized by the IPCC (https://www.ipcc.ch/) and IPBES (https://ipbes.net/).

60. Science has understood itself as mediated by technics ever since Galileo's telescope, but today's more refined technologies have accentuated this phenomenon. See Ian Hacking, "Experimentation and Scientific Realism," *Philosophical Topics* 13 (1982): 154–72; Isabelle Stengers, *L'invention des sciences modernes* (Paris: La découverte, 1993); Bernadette Bensaude-Vincent, *Les vertiges de la technoscience: Façonner le monde atome par atome* (Paris: La découverte, 2009).

61. Susanna Lindberg, *Techniques en philosophie* (Paris: Hermann, 2020), chap. 6 "Les Techniques de la nature."

62. Gilles Deleuze and Félix Guattari, *Capitalisme et schizophrénie I: Anti-Œdipe* (Paris: Minuit, 1972); trans. Robert Hurley, Mark Seem, and Helen Lande as *Anti-Oedipus: Capitalism and Schizophrenia* (Minneapolis: University of Minnesota Press, 1983). The opposition between machine and structure was first sketched out by Félix Guattari in "Machine and Structure," trans. Rosemary Sheed, in *Molecular Revolution: Psychiatry and Politics* (New York: Penguin, 1984), 111–19; cf. Edward Thornton, "The Rise of the Machines: Deleuze's Flight from Structuralism," *Southern Journal of Philosophy* 55, no. 4 (2017).

63. A machine works according to the previous intercommunications of its structure and the positioning of its parts, but does not set itself into place any more than it forms or reproduces itself. This is even the point around which the

usual polemic between vitalism and mechanism revolves: the machine's ability to account for the workings of the organism, but its fundamental inability to account for its formations. From machines, mechanism abstracts a structural unity in terms of which it explains the functioning of the organism. Vitalism invokes an individual and specific unity of the living, which every machine presupposes insofar as it is subordinate to organic continuance, and insofar as it extends the latter's autonomous formations on the outside. But it should be noted that, in one way or another, the machine and desire thus remain in an extrinsic relationship, either because desire appears as an effect determined by a system of mechanical causes, or because the machine is itself a system of means in terms of the aims of desire. The link between the two remains secondary and indirect, both in the new means appropriated by desire and in the derived desires produced by the machines (Deleuze and Guattari, *Anti-Œdipe*, 337; Hurley et al., *Anti-Oedipus*, 283–84).

64. Deleuze and Guattari, 340; Hurley et al., 286.
65. Deleuze and Guattari, 341–42.
66. Deleuze and Guattari, 8–9.
67. N. Katherine Hayles, *How We Became Posthuman: Virtual Bodies in Cybernetics, Literature, and Informatics* (Chicago: University of Chicago Press, 1999).
68. Erich Hörl, *Sacred Channels*. See the original edition: *Die Heiligen Kanäle: Die archaische Illusion der Kommunikation* (Zürich: Diaphanes, 2005).
69. Hörl, *Sacred Channels*, 47.
70. Hörl, 47–54.
71. Hörl, 60.
72. Hörl, 55–56.
73. Hörl, 120–23, 251–97.
74. Hörl, 259.
75. Hörl, 280–92.
76. To quote the idea N. Katherine Hayles introduces in *Unthought: The Power of the Cognitive Nonconscious* (Chicago: University of Chicago Press, 2017), 9–40.
77. Arturo Rosenblueth, Norbert Wiener, and Julian Bigelow, "Behavior, Purpose and Teleology," *Philosophy of Science* 10 (1943): 18–24; Alan Turing, "Computing Machinery and Intelligence," *Mind* 236 (1950), 433–60.
78. Hayles, *How We Became Posthuman*, 8.
79. Humberto Maturana and Francisco Varela, *Autopoiesis and Cognition: The Realization of the Living* (London: Reidel, 1980), xii–xxiv.
80. Hayles, *How We Became Posthuman*, 11.
81. Hayles, 155.
82. See also Henri Atlan, "L'émergence du nouveau et du sens," in *Auto-Organisation: De la physique au politique*, eds. Paul Dumouchel and Jean-Pierre Dupuy (Paris: Seuil, 1983).
83. Yuk Hui, *Recursivity and Contingency* (London: Rowman and Littlefield, 2019), 4, 124–29.

84. Anne Alombert explains the critical position of allagmatics toward cybernetics very well in her unpublished doctoral thesis "Simondon et Derrida face aux questions de l'homme et de la technique. Ontogenèse et grammatologie dans le moment philosophique des années 1960" (unpublished PhD diss., Université de Nanterre, 2020), 194–209, see 328 for Simondon's view of the machine and organism. Alombert's work inspired some of the following reflections on Simondon's thought.

85. Gilbert Simondon, *Individuation à la lumière des notions de forme et d'information* (Grenoble: Jérôme Millon, 2013), 26–27; cf. Anne Sauvagnargues, "Crystals and Membranes: Individuation and Temporality," in *Gilbert Simondon: Being and Technology*, eds. Arne de Boever et al. (Edinburgh: Edinburgh University Press, 2012), 58–59.

86. Simondon, *Individuation à la lumière des notions de forme et d'information*, 32–34.

87. Simondon, 315.

88. Simondon, 307–25.

89. Gilbert Simondon, *Du mode d'existence des objets techniques* (Paris: Aubier, 2012), 11.

90. Simondon, 529–36.

91. David Gé Bartoli and Sophie Gosselin, *Le toucher du monde: Techniques du naturer* (Paris: Éditions Dehors, 2019).

92. Erich Hörl, "Introduction to General Ecology: The Ecologisation of Thinking," in *General Ecology: The New Ecological Paradigm*, eds. Erich Hörl and James Burton (London: Bloomsbury, 2017).

93. Stefan Lorenz Sorgner, "Pedigrees," in *Post- and Transhumanism: An Introduction*, Robert Ranisch and Stefan Lorenz Sorgner (Bern: Peter Lang, 2014), 29–47.

94. See Dierk Spreen, "Not Terminated: Cyborgized Men Still Remain Human Beings," in *Plessner's Philosophical Anthropology*, ed. Jos de Mul (Amsterdam: Amsterdam University Press, 2014), 425–42.

95. Arnold Gehlen, *Der Mensch: Seine Natur und seine Stellung in der Welt* (Wiebelsheim: AULA Verlag, 2009), 10.

Chapter 3

1. Bernard Stiegler, *Technics and Time 1: The Fault of Epimetheus*, trans. Richard Beardsworth and Richard Collins (Stanford: Stanford University Press, 1998), 183–203.

2. It is impossible to separate questions of anthropology and technology from their political interpretations and applications. This is why it is helpful to know that Scheler was a critic of both Nazism and Marxism, Heidegger and

Gehlen were cultural conservatives who were at one time involved with Nazism, and Plessner was a Jew who had to flee Germany. Cultural conservatism can to some extent explain Heidegger's dislike of modern technology (that is the main argument in his later criticism of Nazism) and Gehlen's defense of technology as a means of securing institutions, and an awareness of Plessner's democratic views helps to understand his view of technology as the means of constructing eccentric, utopic places. On Gehlen, see Olivier Agard, "Kulturkritik und Anthropologie: Adorno und Gehlen" in *Philosophische Anthropologie nach 1945: Rezeption und Fortwicklung*, eds. Guillaume Plas and Gerard Raulet (Nordhansen: Verlag Traugott Bautz Gmbh, 2014); on Plessner, see Alexis Dirakis, "Une anthropologie politique de la frontière, Réflexions à partir de l'anthropologie de Helmuth Plessner," *Le Débat* 1 (2016): 132–44. However, in the end, political preferences cannot account for the value of philosophical ideas. The same goes for the deplorable personal relations of these men. Although it is indeed helpful to know of these, they do not allow an evaluation of their works. Scheler accused Plessner of plagiarizing his ideas and Heidegger profited from spreading these rumors. Plessner, forced into exile, despised Gehlen, who made a fine career for himself under Nazism. Because of these academic rivalries and dramas, Philosophical anthropology never really became a philosophical school in its own right but rather just refers to a theoretical core shared by several authors. For a good explanation of this context, see Joachim Fischer, "Le noyau théorique propre à l'Anthropologie philosophique (Scheler, Plessner, Gehlen)," trans. Matthieu Amat and Alexis Dirakis, *Trivium* 25 (2017), 1–21. Fischer also introduces the use of a capital letter for Philosophical anthropology in order to distinguish this particular "school" from philosophical anthropology as a general subdiscipline of philosophy. I am adding these historical notes here because they are instructive, but I also want to emphasize that I find the crude reduction of twentieth-century German philosophy to political history short sighted, an approach that has recently become a common excuse for not reading the texts concerned. This is unfortunate because the period is very rich. While the historical contextualization of philosophy is always necessary, this is to the best of my knowledge the only period of philosophy where it has led to censorship, whereas I think that it should have led to particularly careful readings. Today, one has to repeat what should go without saying: philosophy cannot be judged ad hominem; good philosophers are not always nice persons and virtuous persons do not always make good philosophers.

3. Max Scheler, *Die Stellung des Menschen im Kosmos* (Darmstadt: Otto Reichl Verlag, 1982), 14; trans. Manfred S. Frings as *The Human Place in the Cosmos* (Evanston, IL: Northwestern University Press, 2009), 5. In a later essay, "*Unmenschlichkeit*," Helmuth Plessner comments on technical enhancements of the human being similar to those discussed by present-day transhumanists and includes them in "humanity"; inhumanity is a *human* possibility. Plessner, *Diesseits der Utopie* (Düsseldorf: Eugen Diederichs Verlag, 1966), 222, 227.

4. Plessner comments on Heidegger's rejection of philosophical anthropology and notes that it goes hand in hand with Heidegger's lack of understanding of bodily existence. Plessner, "Immer noch Philosophische Anthropologie?," in *Diesseits der Utopie* (Düsseldorf: Eugen Diederichs Verlag, 1966) 230–40. For an excellent comparison between Heidegger's and Plessner's points of view on the living being, see Christian Sommer, "Approches du vivant entre anthropologie et phénoménologie," in *Das Leben im Menschen oder der Mensch im Lebe: La vie dans l'homme, l'homme dans la vie*, eds. Thomas Ebke and Caterina Zanfi (Potsdam: Universitätsverlag Potsdam, 2017), 59–76.

5. On Kant's importance to Uexküll, see the excellent discussion in Brett Buchanan, *The Animal Environments of Uexküll, Heidegger, Merleau-Ponty, and Deleuze* (Albany: State University of New York Press, 2008), 7–38.

6. Jacob von Uexküll, *A Stroll through the Worlds of Animals and Men: A Picture Book of Invisible Worlds*, trans. Claire H. Schiller, Semiotica 89, 4 (1992): 319–91, 321.

7. "Time, which sways all happening, seems to us to be the only objectively stable thing in contrast to the colorful change of its contents, and now we see that the subject sways the time of his own world. Instead of saying, as heretofore, that without time, there can be no living subject, we shall now have to say that without a living subject, there can be no time. In the next chapter we shall see that the same is true of space: without a living subject, there can be neither space nor time. With this, biology has ultimately established its connection with the doctrine of Kant, which it intends to exploit in its Umwelt theory by stressing the decisive role of the subject" (Uexküll, *A Stroll through the Worlds of Animals and Men*, 326).

8. For example in *Man in the Age of Technology*, Arnold Gehlen compares the human being's circle of action with Norbert Wiener's cybernetical idea of feedback. Gehlen, *Man in the Age of Technology*, trans. Patricia Lipscomb (New York: Columbia University Press, 1980), 17.

9. Cf. Jussi Parikka, *Insect Media: An Archaeology of Animals and Technology* (Minneapolis: University of Minnesota Press, 2010), 68.

10. Scheler knew, e.g., Uexküll's *Umwelt und Innenwelt der Tiere* (1921) and *Bausteine zu einer biologischen Weltanschauung* (1913), as has been shown by Ralf Becker in "Creative Life and the Resentiment of Homo Faber: How Max Scheler Integrates Uexküll's Theory of Environment," in *Jacob Uexküll and Philosophy*, eds. Francesca Michelini and Kristian Köchy (New York: Routledge, 2020).

11. Scheler, *Die Stellung des Menschen im Kosmos*, 52; Frings, *The Human Place in the Cosmos*, 30.

12. Scheler, 21; Frings, 9.

13. Scheler, 53. The translator prefers the term *consciousness of self*; Frings, 30. See also Martin Nitsche, "Subjectivity and Eccentricity," in *Investigating Subjectivity: Classical and New Perspectives*, eds. Gert-Jan van der Heiden et al. (Leiden:

Brill, 2012), 268–70; and Heinz Witteriede, *Eine Einführung in die Philosophische Anthropologie: Max Scheler, Helmuth Plessner, Arnold Gehlen* (Frankfurt am Main: Peter Lang, 2009).

14. Scheler, *Die Stellung des Menschen im Kosmos*, 46–47, 63, 106; Frings, *The Human Place in the Cosmos*, 27, 37, 63.

15. Aristotle, *De anima*, II, 1, 412a–b.

16. Hans Jonas, the author of another bio-metaphysics—*Das Prinzip Leben: Ansätze zu einer philosophischen Biologie* (Frankfurt am Main: Suhrkamp, 1994)— says that his first experience of living philosophizing was in Heidegger's seminar on Aristotle's *De anima* in 1921. Hans Jonas, *Wissenschaft als persönliches Erlebnis* (Göttingen: Vandenhoeck and Ruprecht, 1987), 14.

17. We will see in a later chapter how Jacques Derrida and Giorgio Agamben in particular have criticized the metaphysical gap thus postulated between the animal and the human, especially because it leads to the defining of animality as a defective humanity. This is an important observation in all efforts for rethinking animality, but for the time being we start from humanity, whose continuity with the rest of nature is here at stake.

18. Arnold Gehlen, *Der Mensch: Seine Natur und seine Stellung in der Welt* (Wiebelsheim: AULA Verlag, 2009), 11.

19. "The living organic body, then, is distinguished from the inorganic body by its positional character or positionality." Helmuth Plessner, *Levels of Organic Life and the Human: An Introduction to Philosophical Anthropology*, trans. Millay Hyatt (New York: Fordham University Press, 2019), 121.

20. Plessner, 100 sqq. See also Pierre Osmo's introduction to the French translation of *Les degrés de l'organique et l'Homme* (Paris: Gallimard, 2017), 18–23.

21. Plessner, *Levels of Organic Life*, 116.

22. "Excentricity is the form of frontal positioning against the surrounding field that is characteristic of the human. As the I that makes possible the full return of the living system to itself, the human is no longer in the here/now but "behind" it, behind himself, without place, in nothingness, absorbed in nothingness, in a space- and time-like nowhere-never. Timeless and placeless, the human can experience himself and at the same time his timelessness and placelessness as a standing outside of himself, because the human is a living thing that no longer stands only in itself but whose "standing in itself" is the foundation of its standing. He is placed within his boundaries and therefore outside of these boundaries that confine him as living thing. He not only lives and experiences, but also experiences himself experiencing. The fact that he experiences himself as something that cannot be experienced, that does not inhabit the object position, as pure I (in distinction to the psychophysical individual I, which is identical to the "me" as an object of lived experience) is based solely in the particular way that the thing called "human" is set within its boundaries" (Plessner, 271).

23. Plessner, 288.

24. Scheler, *Die Stellung des Menschen im Kosmos*, 60; Frings, *The Human Place in the Cosmos*, 35.
25. Witteriede, *Eine Einführung in die Philosophische Anthropologie*, 79.
26. Plessner, *Les degrés de l'organique et l'Homme*, 486.
27. Plessner, *Levels of Organic Life*, 315.
28. Plessner, 314.
29. Plessner, 317.
30. The Soviet system aimed at the production of the new socialist man mainly by means of education/indoctrination and forceful administration of populations; the Nazi regime aimed at the physical elimination of unwanted population characteristics (Jews, homosexuals, the handicapped, etc.). However liberal eugenics too, for instance in the United States and Nordic countries, carried out forced sterilizations of the handicapped, among others. These policies do not amount to the same thing, and besides, it is dangerous to compare errors and horrors without careful historical contextualization. Nonetheless, all these approaches are based on similar assumptions concerning the metaphysical sense of human biology.
31. Plessner, *Levels of Organic Life*, 316. Thinking of human existence in terms of exile, Plessner comes close to Jan Patocka's *Heretical Essays in the Philosophy of History*, trans. Erazim Kohák (Chicago: Open Court, 1996).
32. Plessner, *Levels of Organic Life*, 318–19.
33. Plessner, 320.
34. Gehlen, *Man in the Age of Technology*, 4–5.
35. Gehlen, 13.
36. See Fischer, "Le noyau théorique propre à l'Anthropologie philosophique," 11–15.
37. Gehlen, *Man in the Age of Technology*, 19.
38. Gehlen, 117. Günther Anders makes the most elaborate interpretation of the nuclear bomb as the marker of a new era. Today, the digital appears to many people as more groundbreaking than nuclear technology, but the rupture brought about by the latter should not be forgotten.
39. Gehlen, *Man in the Age of Technology*, 10.
40. Martin Heidegger, *Sein und Zeit* (Tübingen: Niemeyer, 1984), 42; trans. John Macquarrie and Edward Robinson as *Being and Time* (Oxford: Blackwell, 1962). The original page numbers are indicated in the translation.
41. Heidegger, *Sein und Zeit*, 47–48.
42. Martin Heidegger, *Die Grundbegriffe der Metaphysik: Welt, Endlichkeit, Einsamkeit*. GA Bd 1929–1930 (Frankfurt am Main: Vittorio Klostermann, 1992), 283. Translated by William McNeill and Nicholas Walker as *The Fundamental Concepts of Metaphysics* (Bloomington: Indiana University Press, 2001). The original page numbers are indicated in the translation.
43. Heidegger, *Sein und Zeit*, 50.

44. Plessner, *Levels of Organic Life*, xxv.
45. Plessner, vii–xxxv; Plessner, "Immer noch Philosophische Anthropologie?," in *Diesseits der Utopie*, 230–40. See also Christian Sommer, "Approches du vivant entre anthropologie et phénoménologie."
46. Heidegger, *Sein und Zeit*, 54.
47. Heidegger, 58, 66.
48. Heidegger, 66.
49. Heidegger, 68.
50. Heidegger, 69.
51. Heidegger, *Die Grundbegriffe der Metaphysik*, 263.
52. Heidegger, 382–84.
53. Heidegger, 251–56.
54. Heidegger, 284.
55. Heidegger, 262–63.
56. Heidegger, *Aristoteles, Metaphysik theta 1–3, Vom Wesen und Wirklichkeit der Kraft*. GA Bd 33 (Frankfurt am Main: Vittorio Klostermann, 1990), 123 sqq.
57. Heidegger, *Die Grundbegriffe der Metaphysik*, 313–15.
58. Heidegger, 414; cf. 509.
59. Heidegger, 412.
60. Heidegger, 413.
61. Heidegger, 414.
62. Heidegger, 527.
63. Heidegger, 511.
64. Heidegger, 456–57.
65. Heidegger, 462, 518.
66. Heidegger, *Aristoteles*, 137–48.
67. Heidegger, *Die Grundbegriffe der Metaphysik*, 510.
68. Heidegger, 516–17.
69. Heidegger, 511–12.
70. For the criticism of Heidegger from the perspective of the phenomenology of life and a possible Heideggerian answer, see Sommer, "Approches du vivant entre anthropologie et phénoménologie."
71. Heidegger, "Die Frage nach der Technik," in *Vorträge und Aufsätze*, 9. The page number of the Neske edition is at the margin of the GA 7 edition. Translated by William Lovitt as "The Question Concerning Technology," in *The Question Concerning Technology and Other Essays*, 4.
72. Heidegger, "Vom Wesen und Begriff der *Physis*," in *Wegmarken* (Frankfurt am Main: Vittorio Klostermann, 1978), 248 sqq.
73. Heidegger, "Die Frage nach der Technik," 11, 15. My translation.
74. Heidegger, "Die Frage nach der Technik," 16; "The Question Concerning Technology," 12. In the following references to this article the first page number

refers to the German edition, the second one to the translation, in which the German page number is not visible.

75. Heidegger, 20/16.
76. Heidegger, 20/17.
77. Heidegger, 22/19.
78. Heidegger, 27/23–24.
79. That is, transhumanism is a dream characteristic of the epoch of technology.
80. Heidegger, "Die Frage nach der Technik," 30/27.
81. Heidegger, "Überwindung der Metaphysik" (1936–1946), in *Vorträge und Aufsätze*, 68; see translation "Overcoming Metaphysics," in *The End of Philosophy*, trans. Joan Stambaugh (Chicago: University of Chicago Press, 2003), 86.
82. Heidegger, "Überwindung der Metaphysik," 76, 95/93, 110.
83. Heidegger, "Die Frage nach der Technik," 37–40/33–36.
84. In "Letter on Humanism," Heidegger famously says that the most difficult thing for us to think of is our closest kin the living being, which is nonetheless separated from our existent being by an abyss. Heidegger, *Wegmarken*, 157, 323.
85. Plessner, *Levels of Organic Life*, 289.
86. Plessner, 293.
87. Plessner, 294.
88. Plessner, 298–99.
89. Heidegger, "Was ist metaphysik?," in *Wegmarken*, 111–113/101–103.

Chapter 4

1. As Jean-François Lyotard puts it in the introduction to his book *The Inhuman*: "In short, our contemporaries find it adequate to remind us that what is proper to humankind is its absence of defining property, its nothingness, or its transcendence, to display the sign of 'no vacancy.' I do not like this haste. What it hurries, and crushes, is what after the fact I have always tried, under diverse headings—work, figural, heterogeneity, dissensus, event, thing—to reserve: the unharmonizable. (And I am not the only one, which is why I write 'us.')" Jean-François Lyotard, *The Inhuman: Reflections on Time*, trans. Geoffrey Bennington and Rachel Bowlby (Cambridge: Polity Press, 1991), 4.

2. Philippe Lacoue-Labarthe, "The Echo of the Subject," in *Typography: Mimesis, Philosophy, Politics*, trans. Christopher Fynsk (Cambridge, MA: Harvard University Press, 1989).

3. Arianna Sforzini goes so far as saying that for Foucault, a phenomenology like that of Maurice Merleau-Ponty in *Phenomenology of Perception* is hardly anything other than positivism. Arianna Sforzini, *Michel Foucault, une pensée du corps* (Paris: PUF, 2014), 24. I would not go so far, and I rather agree

with Gilles Deleuze, whose explication of Foucault's relation to phenomenology can be generalized to apply to the poststructuralists' relation to phenomenology in general. According to Deleuze, Foucault is against "vulgar" phenomenology that is merely centered on intentionality, but he is inspired by Heidegger and Merleau-Ponty's late works in which intentionality tends toward Being, the fold of Being, ontology. Gilles Deleuze, *Foucault*, trans. Seán Hand (London: Athlone Press, 1988), 108–12.

4. Arthur Bradley, *Originary Technicity: The Theory of Technology from Marx to Derrida* (New York: Palgrave Macmillan, 2011), 2–3. This way of reading Derrida and other poststructuralists is not self-evident. For example, in 1999 Simon Critchley could still ask, hesitantly, if Derrida could be read as a thinker of originary technicity when he was generally not read as such. Simon Critchley, *Ethics, Politics, Subjectivity: Essays on Derrida, Levinas, and Contemporary French Thought* (London: Verso, 1999), 174–76.

5. David Wills, *Dorsality: Thinking Back through Technology and Politics* (Minneapolis: University of Minnesota Press, 2008), 3–4.

6. Michel Foucault, *The Order of Things: An Archaeology of the Human Sciences* (London: Routledge, 2005), 422; Foucault, *Les mots et les choses* (Paris: Gallimard, 1966), 398.

7. Alan Milchman and Alan Rosenberg, "The Aesthetic and Ascetic Dimensions of an Ethics of Self-Fashioning: Nietzsche and Foucault," *Parrhesia* 2 (2007): 45.

8. Jacques Derrida, "The Ends of Man," in *Margins of Philosophy*, trans. Alan Bass (Chicago: University of Chicago Press, 1982), 111; Derrida, "Les fins de l'homme," in *Marges de la Philosophie* (Paris: Minuit, 1972), 131.

9. Derrida, "The Ends of Man," 119, 141.

10. Derrida, 123, 147.

11. Derrida, 134, 161. See Ph. Lacoue-Labarthe and J.-L. Nancy, "Ouverture" to *Les fins de l'homme: À partir du travail de Jacques Derrida* (Paris: Galilée, 1981), 12–13.

12. Derrida, "The Ends of Man," 134–35.

13. The view that the human subject is a construction has sometimes led to strange misunderstandings, like when Judith Butler's Derrida- and Foucault-inspired descriptions of gender as construction led ignorant or malevolent defenders of conservative heteronormativity to claim that Butler defends the possibility of *voluntaristically* choosing, then producing, one's sexuality. This is of course not the case. The gender is the effect of extremely complex impersonal processes that characterize an entire historical situation and not a conscious individual will.

14. For example, Michel Foucault, *Surveiller et punir: Naissance de la prison* (Paris: Gallimard tel, 1975), 40. Trans. Alan Sheridan as *Discipline and Punish: The Birth of the Prison* (New York: Vintage Books, 1995), 31. The history of the present is thematized more precisely in the first chapter of Foucault, *The Government*

of Self and Others: Lectures at the Collège de France 1982–1983, trans. Graham Burchell (Basingstoke: Palgrave Macmillan, 2010), 11–21 (*Le gouvernement de soi et des autres: Cours au collège de France, 1982–1983* [Paris: Gallimard Seuil, 2008], 13–22); and Foucault, "Qu'est-ce que les lumières?," in *Dits et écrits II*, 1381–96; trans. Robert Hurley and others as "What Is Enlightenment?," in Michel Foucault, *Ethics, Subjectivity, and Truth: The Essential Works of Foucault 1954–1984*, 2nd ed., ed. Paul Rabinow (New York: New Press, 1997), 303–20.

15. For example, Yves Charles Zarka, "Foucault et l'idée d'une histoire de la subjectivité: Le moment moderne," *Archives de philosophie* 65 (2002): 255–67; Frédéric Gros, "Sujet moral et éthique chez Foucault," *Archives de philosophie* 65 (2002): 229–37. In his condescending account of Foucault, Allan Megill reads the *Order of Things* in particular as a structuralist book in which the epistemic function is thoroughly "subjectless, objectless" and marked by "all-embracing language." Allan Megill, *Prophets of Extremity: Nietzsche, Heidegger, Foucault, Derrida* (Berkeley: University of California Press, 1985), 204. If this were so, the late Foucault's interest in subjectivity would indeed be a break with the structuralist Foucault.

16. "I don't think there is a great difference between these books [*History of Sexuality*] and the earlier ones [*History of Madness, The Order of Things, Discipline and Punish*]. . . . You may have changed your point of view, you've gone round and round the problem, which is still the same, namely, the relations between the subject, truth, and the constitution of experience. I have tried to analyze how areas such as madness, sexuality, and deliquency may enter into a certain play of truth, and also how, through this insertion of human practice, of behavior, in the play of truth, the subject himself is affected" (Foucault, "Esthétique de l'existence," in *Dits et écrits II, 1976–1988* [Paris: Gallimard, 2017], 1550; trans. Alan Sheridan as "An Aesthetics of Existence," in *Politics, Philosophy, Culture: Interviews and Other Writings 1977–1984* [New York: Routledge, 1988], 48).

See also Sebastian Harrer, "The Theme of Subjectivity in Foucault's Lecture Series *L'Herméneutique du sujet*," *Foucault Studies* 2 (2005), 75–96; Milchman and Rosenberg, "The Aesthetic and Ascetic Dimensions," 54; Lynne Huffer, "Self," in *Cambridge Foucault Lexicon*, eds. Leonard Lawlor and John Nale (New York: Cambridge University Press, 2014), 443; Leonard Lawlor, *From Violence to Speaking Out: Apocalypse and Expression in Foucault, Derrida and Deleuze* (Edinburgh: Edinburgh University Press, 2016), 144.

17. See for instance Eve Tavor Bannet, *Structuralism and the Logic of Dissent: Barthes, Derrida, Foucault, Lacan* (Urbana: University of Illinois Press, 1989), 95, 162–63.

18. Michel Foucault, "Le sujet et le pouvoir," in *Dits et écrits II*, 1041–42; translated by Robert Harvey as "The Subject and Power" in Dreyfus and Rabinow's book *Michel Foucault: Beyond Structuralism and hermeneutics* (Chicago: University of Chicago Press, 1982), 208.

19. Foucault, "Esthétique de l'existence," in *Dits et écrits II*, 1552; trans. Sheridan, "An Aesthetics of Existence," 50–51.

20. Milchman and Rosenberg, "The Aesthetic and Ascetic Dimensions," 55.

21. "There are two meanings of the word 'subject': subject to someone else by control and dependence, and tied to his own identity by a conscience or self-knowledge." Foucault, "Le sujet et le pouvoir," *Dits et écrits II*, 1046; trans., "The Subject and Power," 212.

22. Judith Butler, "Subjection, Resistance, Resignification: Between Freud and Foucault," in *The Psychic Life of Power: Theories in Subjection* (Stanford: Stanford University Press, 1997), 90.

23. Judith Butler, "Subjection, Resistance, Resignification," 94, 100.

24. Judith Revel, "The Materiality of the Immaterial: Foucault, Against the Return of Idealisms and New Vitalisms," *Radical Philosophy* 149 (May–June 2008): 33–38.

25. In "The Subject and Power" Foucault explains that "something called Power, which is assumed to exist universally or in a concentrated or diffused form, does not exist. . . . In effect, what defines a relationship of power is that it is a mode of action which does not act directly and immediately on others. Instead it acts upon their actions: and action upon an action, on existing actions or on those which may arise in the present or the future." Michel Foucault, "Sujet et pouvoir," in *Dits et écrits II*, 1055; trans. "The Subject and Power," 219–20.

26. Bannett, *Structuralism and the Logic of Dissent*, 129.

27. Michel Foucault, *Histoire de la folie à l'âce classique* (Paris: Gallimard, 1972), 94; trans. Jonathan Murphy and Jean Khalfa as *History of Madness* (London: Routledge, 2006), 79–80.

28. Panopticon is a prison built as a circle around a central tower where the guards stay. The cells are all visible to the central tower but not to other cells, while the central tower's guards are not visible to the prisoners. In this way, the prisoners know that they can be constantly seen without seeing who sees them or whether there is somebody watching at all. According to the theory, this simple architectural structure makes the prisoners behave as if they were observed constantly: they watch themselves being watched. Michel Foucault, *Surveiller et punir*, 233 sqq; trans. Sheridan, *Discipline and Punish*, 119 sqq.

29. Foucault, 31, 34–38; trans. Sheridan, 24.

30. Foucault, 38; trans. Sheridan, 29.

31. Foucault, "La technologie politique des individus," in *Dits et écrits II*, 1632–35; see translation. Transcription by Rux Martin, "The Political Technology of Individuals," in *Technologies of the Self: A Seminar with Michel Foucault*, eds. Luther H. Hutton, Huck Gutman, and Patrick H. Hutton (Amherst: University of Massachusetts Press, 1988), 142–62.

32. Foucault, *Surveiller et punir*, 243; trans. Sheridan, *Discipline and Punish*, 208.

33. Foucault, 241–42; trans. Sheridan, 207.

34. Gilles Deleuze, "Post-scriptum sur les sociétés de contrôle," in *Pourparlers 1972-1991* (Paris: Minuit, 1990); see translation by Martin Joughin: Deleuze, "Postscript on the Societies of Control," *October* 59 (Winter 1992): 3–7.

35. Foucault, *Surveiller et punir*, 257; trans. Sheridan, *Discipline and Punish*, 221. Arianna Sforzini notes that the discovery of disciplinary techniques that steer and combine bodies reflected the idea of the body as itself a mechanism and machine, epitomized in La Mettrie's "Homme-machine." Sforzini, *Michel Foucault, une pensée du corps*, 49–56.

36. Foucault, *Surveiller et punir*, 251; trans. Sheridan, *Discipline and Punish*, 215. Disciplines are characterized as machines and technologies throughout the chapter "Panoptism."

37. In *The Will to Knowledge, Vol. I of the History of Sexuality*, Foucault also describes power contrary to its interpretation as an established organization or as a mode of subjugation as follows:

"It seems to me that power must be understood in the first instance as the multiplicity of force relations immanent in the sphere in which they operate and which constitute their own organization; as the process which, through ceaseless struggles and confrontations, transforms, strengthens, or reverses them; as the support which these force relations find in one another, thus forming a chain or a system, or on the contrary, the disjunctions and contradictions which isolate them from one another; and lastly, as the strategies in which they take effect, whose general design or institutional crystallization is embodied in the state apparatus, in the formulation of the law, in the various social hegemonies" (Michel Foucault, *The Will to Knowledge*, trans. Robert Hurley [New York: Pantheon Books, 1978], 92–93).

38. Foucault, *Surveiller et punir*, 34; trans. Sheridan, *Discipline and Punish*, 26, 28.

39. Foucault, 36; trans. Sheridan, 27.

40. Gilles Deleuze, *Foucault*, trans. Séan Hand (London: Athlone Press 1988), 23–44.

41. Deleuze, 27.

42. Foucault, *Surveiller et punir*, 239; trans. Sheridan, *Discipline and Punish*, 171, 205.

43. Deleuze, *Foucault*, 34.

44. Deleuze, 35–36.

45. Deleuze, 39.

46. Bannet, *Structuralism and the Logic of Dissent*, 144.

47. Foucault, "Le jeu de Michel Foucault," in *Dits et écrits II*, 300–1. English translation "The Confession of the Flesh," in *Power/Knowledge: Selected Interviews and Other Writings*, ed. Colin Gordon (New York: Pantheon Books, 1980), 194–228.

48. Foucault, "Le jeu de Michel Foucault," in *Dits et écrits II*, 299; Foucault, "The Confession of the Flesh," 194.

49. Some commentators have affirmed that Foucault negates Heidegger altogether in his work (Megill, *Prophets of Extremity*, 185), but actually, as Deleuze says, the late Heidegger especially was very important to Foucault (Deleuze, *Foucault*, 108 sqq). As McCumber notes, Foucault's acknowledgment of his debt to Heidegger comes late but intensely. John McCumber, *Philosophy and Freedom* (Bloomington: Indiana University Press, 2000), 110). In *Dits et écrits*, Foucault says very directly: "Heidegger has always been for me the essential philosopher. . . . My whole philosophical development was determined by my reading of him" ("Le retour de la morale," in *Dits et écrits II*, 1522, trans. cit. in McCumber, 110). In "Truth, Power and Self," Foucault adds that he finds the later work of Heidegger less enigmatic than *Being and Time* ("Vérité, pouvoir et soi" in *Dits et écrits II*, 1599). In "The Political Technology of Individuals" Foucault also inscribes himself in the same tradition of historical reflection on actuality as Heidegger and some others ("La technologie politique des individus," *Dits et écrits II*, 1633). "Truth, Power and Self" and "The Political Technology of Individuals" have been published in English as *Technologies of the Self: A Seminar with Michel Foucault*, 9-15; 142-62.

50. Foucault, "Qu'est-ce que les lumières?," in *Dits et écrits II*, 1381-96, 1394; trans. Hurley, "What Is Enlightenment?," 316. See also Foucault, *The Government of Self and Others*, 11-21 (*Le gouvernement*, 3-40). As Matthew Sharpe says in "Critique as Technology of the Self," *Foucault Studies* 2 (2005): 97-116, Foucault's reading of Kant's *What Is Enlightenment?* presents Kant's text as in line with Foucault's readings of antiquity: the aim each time is to reflect upon and work on one's own subjectivity such that one is capable of carrying one's role in the political community. This is also the task that Foucault gives to us today according to "What Is Enlightenment?": to develop a philosophical ethos that leads us to interrogate our own historical present and to work on our own freedom.

51. For a general explication of the idea of an aesthetics of existence, see for example Fabien Nègre, "L'esthétique de l'existence dans le dernier Foucault," *Raison Présente* 118 (1996): 47-71.

52. Foucault, *The Use of Pleasure, Vol. 2 of The History of Sexuality*, trans. Robert Hurley (New York: Vintage Books, 1990), 10-11.

53. Of course, Foucault's ethics can be compared to Aristotle's ethics, as Claire Colebrook does in "Ethics, Positivity, and Gender: Foucault, Aristotle, and Care of the Self," *Philosophy Today* 42 (1): 40-52. Here we are thinking of the theory of the four causes presented in Aristotle's *Metaphysics*, which was referred to by Judith Butler in *The Psychic Life of Power*, 90. Paul Patton also interprets Foucault's disciplinary, governmental, and ethical technologies in terms of Aristotle's four causes in "Technology," in *Cambridge Foucault Lexicon*, 503-8. Similarly Ilpo Helen interprets ethical work in terms of substance, form of subjectivation, ethical

work, and telos of the moral subject. Ilpo Helen, "Elämä seksuaalisuudessa," in Foucault, *Seksuaalisuuden historia*, trans. Kaisa Sivenius (Helsinki: Gaudeamus, 1998), 504. John McCumber interprets Foucault's view of discourse as a reworking of Aristotle's idea of the four causes in *Philosophy and Freedom*, 123. Gilles Deleuze, without ever mentioning Aristotle, appears to reproduce such a displacement of Aristotle in his explication of the four "folds"—body (that corresponds to the material cause), power (form), truth (agent), and expectation (telos)—that constitute Foucauldian subjectivation (Deleuze, *Foucault*, 104).

54. The following comes from Patton, "Technology," in *Cambridge Foucault Lexicon*, 507.

55. According to *The Government of Self and Others*, sexuality is just one example of subjectivation—but it is incontestably the most important one for late Foucault. Foucault, *The Government of Self and Others*, 5 (*Le gouvernement*, 6).

56. Foucault, *The Use of Pleasure*, 53 sqq.

57. Foucault, 38.

58. Foucault, 91.

59. Foucault, 92.

60. Foucault, *The Government of Self and Others*, 43 (*Le gouvernement*, 43).

61. Foucault, 104–5 (*Le gouvernement*, 98).

62. Foucault, 156, 173–74 (*Le gouvernement*, 144, 157–58).

63. Foucault, 157 (*Le gouvernement*, 145).

64. Foucault, *The Will to Knowledge*, 70.

65. Foucault, 90.

66. Foucault, 76.

67. Foucault, 113–31.

68. Foucault, 139.

69. Foucault, 143.

70. Deleuze, *Foucault*, 94–123.

71. Foucault's uses of the term *outside* draws particularly on Maurice Blanchot; see Foucault, *Maurice Blanchot: The Thought from Outside* / Blanchot: *Michel Foucault as I Imagine Him*, trans. Jeffrey Mehlman and Brian Massumi (New York: Zone Books, 1990); Foucault, *La pensée du dehors* (Paris: Fata morgana, 1986).

72. Deleuze, *Foucault*, 112.

73. Deleuze, 113.

74. Deleuze, 96–97.

75. Deleuze, 123.

76. Deleuze, 105–6.

77. Bernard Stiegler, *Taking Care of Youth and the Generations*, trans. Stephen Barker (Stanford: Stanford University Press, 2010), 23, 115, 122 sqq. However, Foucault emphasizes the art of listening and the art of writing as elements of Stoic self-techniques in a lecture on March 3, 1982, in *The Hermeneutics of the Subject*.

78. Stiegler, *Taking Care*, 122.

79. Antoinette Rouvroy and Guido Berns, "Le nouveau pouvoir statistique, Ou quand le contrôle s'exerce sur un réel normé, docile et sans événement car constitué de corps 'numériques,'" *Multitudes* 40 (2010): 88–103. See also Luiz Costa, *Virtuality and Capabilities in a World of Ambient Intelligence: New Challenges to Privacy and Data Protection*, Springer Law, Governance, and Technology Series, 2016. Éric Sadin has also studied what he calls "algorithmic life" in Éric Sadin, *Surveillance globale* (Paris: Climats, 2009); Sadin, *L'Humanité augmentée: L'administration numérique du monde* (Paris: L'échappée, 2013); Sadin, *La vie algorithmique: Critique de la raison numérique* (Paris: L'échappée, 2015). Bernhard Rieder resumes the discussion of the "algorithmic management" or "algorithmic accountability" of life in *Engines of Order: A Mechanology of Algorithmic Techniques* (Amsterdam: Amsterdam University Press, 2020), 9–14.

80. Rouvroy and Berns, "Détecter et prévenir: De la digitalisation des corps et de la docilité des normes," 2009. https://works.bepress.com/antoinette_rouvroy/30/.

81. Philip Alston, *Extreme Poverty and Human Rights*, report to the United Nations General Assembly, distr. 11, October 2019 (A/74/493). https://undocs.org/A/74/493.

82. Shoshana Zuboff, *The Age of Surveillance Capitalism: The Fight for a Human Future at the New Frontier of Power* (London: Profile Books, 2019).

83. Shoshana Zuboff, "Big Other: Surveillance Capitalism and the Prospects of an Information Civilization," *Journal of Information Technology* 30 (2015): 75–89.

84. Dominique Quessada calls "sousveillance" the constant automatic tracking of behavior. Dominique Quessada, "De la sousveillance: La surveillance globale, un nouveau mode de gouvernementalité," *Multitudes* 40 (2010): 54–59.

85. See Roy Boyne, *Foucault and Derrida: The Other Side of Reason* (London: Unwin Hyman, 1990), 4. Boyne gives the most balanced and detailed description of this important debate.

86. Jacques Derrida, "Cogito and the History of Madness," conference at Collège Philosophique in 1963 and later published in *Writing and Difference*, trans. Alan Bass (London: Routledge, 2003), 39–40.

87. Boyne, *Foucault and Derrida*, 77.

88. Foucault, "Mon corps, ce papier, ce feu" and "Réponse à Derrida," orig. in *Paideia* 11 (1972); republished in *Dits et écrits I*, 1113–36, 1149–63; trans by Foucault as "My Body, This Paper, This Fire," *Oxford Literary Review* 4, no. 1 (1979).

89. Foucault, "Réponse à Derrida," 1151.

90. Derrida, "Cogito and the History of Madness," 53.

91. Foucault, "Réponse à Derrida," 1152.

92. The place of the totally other is well marked by Eve Tavor Bannet, although her interpretation of the totally other in a straightforward Jewish framework strikes me as odd (Bannet, *Structuralism and the Logic of Dissent*, 184).

93. Derrida, "Tympan," in *Margins of Philosophy*, xiv.

94. Foucault, "Réponse à Derrida," 1163, trans. Geoff Bennington, "My Body, This Paper, This Fire," *Oxford Literary Review* 4, no. 1 (1979): 9–28, 27.

95. Derrida, "Artifactualities," in *Echographies of Television: Filmed Interviews*, trans. Jennifer Bajorek (Cambridge: Polity Press, 2002), 10.

96. Derrida, "Cogito and the History of Madness," 73.

97. As Derrida says of an oeuvre that functions, like every other, as a trace: "[It has] a sort of archival independence or autonomy that is quasi-machinelike (not machinelike but *quasi*-machinelike), a power of repetition, a repeatability iterability, serial and prosthetic substitution of self for self." Derrida, "Typewriter Ribbon: Limited Ink (2)" in *Without Alibi*, trans. Peggy Kamuf (Stanford: Stanford University Press, 2002), 133.

98. "My timid contribution," says Derrida, would consist in a displacement that will perhaps be "limited to underscoring 'materiality,' in place, so to say, of 'matter,' the insisting on 'thought of materiality,' or even 'material thought of materiality,' in place, if I may put it in this way, of 'materialist' thought, even within quotation marks." Derrida, "Typewriter Ribbon," in *Without Alibi*, 80.

99. This is why Mark Hansen prefers Stiegler's analysis of concrete technological systems to Derrida's abstract work on the aporia of the origin. Mark B. Hansen, "Realtime Synthesis and the Différance of the Body: Technocultural Studies in the Wake of Deconstruction," *Culturemachine* 6 (2004). Tracy Colony prefers Derrida's abstract thinking of the différance. Colony, "Epimetheus Bound: Stiegler on Derrida, Life, and the Technological Conditions," *Research in Phenomenology* 41 (2011): 72–89; and "The Future of Technics," *Parrhesia* 27 (2017): 64–87.

100. Derrida, "Différance," in *Margins of Philosophy*, 15.

101. "Hegel knew that this proper and animated body of the signifier was also a *tomb*. The association *soma/sema* is also at work in this semiology, which is in no way surprising. The tomb is the life of the body as the sign of death, the body as the other of the soul, the other of the animate psyche, of the living breath. But the tomb also shelters, maintains in reserve, capitalizes on life by marking that life continues elsewhere. The family crypt: *oikesis*. It consecrates the disappearance of life by attesting to the perseverance of life. Thus, the tomb also shelters from death. It warns the soul of possible death, warns (of) death of the soul, turns away (from) death. This double warning function belongs to the funerary monument. The body of the sign thus becomes the monument in which the soul will be enclosed, preserved, maintained, kept in presence, present, signified. At the heart of this monument the soul keeps itself alive, but it needs the monument only to the extent that it is exposed—to death—in its living relation to its own body. . . . The sign—the monument-of-life-in-death, the monument-of-life-in-death . . . is the *pyramid*" (Derrida, "The Pit and the Pyramid," in *Margins of Philosophy*, 82–83).

102. 'To write is to produce a mark that will constitute a kind of machine that is in turn productive, that my future disappearance in principle will not prevent from functioning. . . . For the written to be written, it must continue to 'act' and to be legible even if what is called the author of the writing no longer answers for what he has written, for what he seems to have signed, whether he is provisionally absent, or if he is dead, or if in general he does not support, with his absolutely current and present intention or attention, the plenitude of meaning, of that very thing which seems to be written 'in his name'" (Derrida, "Signature Event Context," in *Margins of Philosophy*, 316).

103. I have enumerated many of the technical objects mentioned by Derrida in my "Derrida's Quasi-Technique," *Research in Phenomenology* 46 (2016): 369–89.

104. Gilbert Simondon, *On the Mode of Technical Objects*, trans. Cécile Malaspina (Minneapolis: University of Minnesota Press, 2017). On the other hand, unlike Derrida but like Stiegler, Simondon proceeds to make a phenomenological study of concrete technical objects and of their associated milieus, while Derrida only studies the abstract machinic principle.

105. The infrastructural chain of quasi-transcendental structures is explained by Rodolphe Gasché in *The Tain of the Mirror: Derrida and the Philosophy of Reflection* (Cambridge, MA: Harvard University Press, 1988), chap. 9, whose transcendental interpretation of Derrida's early works frames this study far more than Richard Rorty's pragmatist readings or Simon Critchley's ethical readings. See Richard Rorty, *Essays on Heidegger and Others* (Cambridge: Cambridge University Press, 2010), 119–28; and Critchley, *Ethics, Politics, Subjectivity*, 83–102.

106. Derrida, "Différance," in *Margins of Philosophy*, 12.

107. Stiegler's ideas are presented in a later chapter. The idea of cyborg comes to philosophy especially with Donna Haraway's famous "A Cyborg Manifesto: Science, Technology, and Socialist-Feminism in the Late Twentieth Century," in *Simians, Cyborgs, and Women: The Reinvention of Nature* (New York: Routledge, 1991), 149–81.

108. For an earlier version of these considerations, see my "On Prosthetic Existence: What Differentiates Deconstruction from Transhumanism and Posthumanism," in *Humanism and Its Discontents: The Rise of Transhumanism and Posthumanism*, ed. Paul Jorion (London: Palgrave Macmillan, 2022).

109. Jean-Luc Nancy, "The Intruder," in *Corpus*, trans. Richard Rand (New York: Fordham University Press, 2008), 163.

110. Derrida, "Plato's Pharmacy," 104–5.

111. Derrida, 92.

112. Derrida, 94. Socrates, too, is accused of toxic, demonic, narcotic speech (Derrida, 118).

113. Derrida, 91.

114. Derrida, 78.

115. "Writing (or, if you will, the *pharmakon*) is thus presented to the king . . . as a finished work presented to his appreciation. And this work is itself an art, a capacity to work, a power of operation. The artefactum is an art." Derrida, 75–76.

116. Derrida, 111.

117. Derrida, 98–99.

118. Derrida, 108. Phrase adapted to the sentence.

119. André Leroi-Gourhan, *Gesture and Speech*, trans. Anna Bostock Berger (Cambridge, MA: MIT Press, 1993), 91.

120. Leroi-Gourhan, 106; cited and commented in Wills, *Dorsality*, 8.

121. Derrida, *Of Grammatology*, trans. Gayatri Chakravorty Spivak (Baltimore: Johns Hopkins University Press, 1997), 84. If Leroi-Gourhan provides Derrida with a paleoanthropological explication of the originary technicity of the human, Freud provides him with an analogical psychological explication. In "Freud and the Scene of Writing," Derrida reads Freud's "Note on the Mystic Writing Pad" as a text in which the "*structure* of the psychical *apparatus* is *represented* by a writing machine." Derrida, "Freud and the Scene of Writing," in *Writing and Difference*, 250. "Here the question of *technology* (a new name must perhaps be found in order to remove it from its traditional problematic) may not be derived from an assumed opposition between the psychical and the nonpsychical, life and death. Writing, here, is *techne* as the relation between life and death, between present and representation, between the two apparatuses. It opens up the question of technics: of the apparatus in general and of the analogy between the psychical apparatus and the nonpsychical apparatus." Derrida, 287. Derrida returns to the magical writing block as Freud's technical model for hypomnesic memory in *Mal d'Archive* (Paris: Galilée, 1995), 29. Both Leroi-Gourhan and Freud think of the human in terms of a memory and memory in terms of writing.

122. Derrida, *Of Grammatology*, 244–45.

123. As Arthur Bradley says, "According to the famous logic that is first developed in the context of his reading of Rousseau in the *Grammatology*, what apparently serves to *supplement* the state of nature—culture, language, writing, and the entire field of technics—actually exposes to an originary lack within what should be the integrity or plenitude of the human being itself which calls for supplementation." Arthur Bradley, *Originary Technicity*, 97.

124. Derrida, *Of Grammatology*, 106.

125. Derrida, 8.

126. Derrida, 11.

127. "[In Saussure] Writing would thus have the exteriority that one attributes to utensils; to what is even an imperfect tool and a dangerous, almost maleficent, technique. . . . It is less a question of outlining than of protecting, and even of restoring the internal system of the language in the purity of its concept against the gravest, most perfidious, most permanent contamination . . . affecting the

language and befalling it *from without* . . . the contamination of writing. . . . This tone began to make itself heard when, at the moment of already tying the *epistémè* and the *logos* within the same possibility the *Phaedrus* denounced writing as the intrusion of an artful technique, a forced entry of a totally original sort, an archetypal violence: eruption of the *outside* within the *inside*, breaching into the interiority of the soul, the living presence of the soul within the true *logos*, the help that speech lends to itself.

Writing, sensible matter and artificial exteriority: a "clothing." . . . For Saussure it is even a garment of debauchery, a dress of corruption and disguise, a festival mask that must be exorcised, that is to say warded off, by the good word" (Derrida, *Of Grammatology*, 34).

128. Derrida, 82.
129. Derrida, 144.
130. Derrida, 145.
131. Although "it is difficult to avoid mechanist, technicist and teleological language at the very moment when it is precisely a question of retrieving the origin and the possibility of the movement, or the machine, of the *technè*, of orientation in general." Derrida, *Of Grammatology*, 84. Derrida's comment on Leroy-Gourhan's paleoanthropological description of the liberation of the hand for writing applies to Derrida's own project as well.
132. Derrida, 79.
133. Derrida, "Signature Event Context" in *Margins of Philosophy*, 315.
134. Derrida, "Tympan," in *Margins of Philosophy*, x.
135. Derrida, "The Pit and the Pyramid," in *Margins of Philosophy*, 107.
136. Derrida, *Of Grammatology*, 209.
137. Philippe Lacoue-Labarthe, *Typography: Mimesis, Philosophy, Politics* (Stanford: Stanford University Press), 1998.
138. Derrida, "Psyché: Invention of the Other," in *Psyche: Inventions of the Other Vol. 1*, trans. Catherine Porter (Stanford: Stanford University Press, 2007), 10.
139. Derrida, 24.
140. Derrida, 46.
141. The difference between invention and messianicity also explains the difference between Stiegler's and Derrida's conceptions of the future: Stiegler thinks the future in terms of invention, whereas Derrida thinks it in terms of messianicity. See Tracy Colony, "The Future of Technics," 64–87, 66.
142. Derrida, *Of Grammatology*, 44.
143. Derrida, 46. The trace is not only associated only with human language but also with traces left by animal and nonliving beings. Derrida, 47.
144. Derrida, 75.
145. This is why, according to the famous and famously misunderstood dictum, "there is nothing outside of the text (there is no outside-text), *il n'y a pas de hors-texte.*" Derrida, *Of Grammatology*, 158. Apart from the fact that this

remark is used to characterize Rousseau's autobiographical writings, this does not mean in general that there would not be anything "real" but that the real takes on "meaning from a trace and from an invocation of a supplement." Derrida, 159.

146. Derrida, 46–47.

147. Derrida, 61; cf. 110. "*The trace itself does not exist.*" Derrida, 166. Also the trace, or the différance, "has no meaning and is not." Derrida, "Différance," in *Margins of Philosophy*, 22.

148. Derrida, *Of Grammatology*, 65.

149. Derrida, 68; cf. "The synthesis of marks, traces of retentions and protentions . . . that I propose to call archi-writing, archi-trace or *différance*. Which (is) (simultaneously) spacing (and) temporization." Derrida, "Différance," in *Margins of Philosophy*, 13.

150. Plato, *Timaeus*, 48e–51b, quoted in Derrida, "Plato's Pharmacy," in *Dissemination*, trans. Barbara Johnson (Chicago: University of Chicago Press, 1981), 160. John Sallis develops Derrida's indications into a detailed reading of Plato's khora in *Chorology: On Beginning in Plato's Timaeus* (Bloomington: Indiana University Press, 1995). For a careful presentation of Derrida's later developments of the *khora* and its relation to messianicity without messianism (that we will examine later), see Michael Naas, *Miracle and Machine: Jacques Derrida and the Two Sources of Religion, Science, and Media* (New York: Fordham University Press, 2012), 152–96.

151. Derrida, "Plato's Pharmacy," 159–60.

152. Derrida, "Khora," in *On the Name*, trans. Ian McLeod (Stanford: Stanford University Press, 1993), 90.

153. Derrida, "Khora," 99.

154. Jacques Derrida, *La vie la mort: Séminaire (1975–1976)* (Paris: Seuil Bibliothèque Derrida, 2019), 122; Pascale-Anne Brault and Michael Naas, trans., *Life Death* (Chicago: University of Chicago Press, 2020). As the page numbers of the French edition are marked in the margins of the English translation, I quote those from the French edition.

155. Derrida, *La vie la mort*, 115.

156. Derrida, 21.

157. Derrida, 27–31.

158. Derrida, 106, 109.

159. Derrida, 100; see also 106.

160. Derrida, 44.

161. Derrida, 32–33.

162. Derrida, 33–41.

163. Derrida, 126–28.

164. Derrida, 168.

165. Derrida, 167.

166. Derrida, 32.
167. Derrida, 119.
168. Derrida, 123.
169. In today's machine learning, the principle of retroaction is applied to computer programs. However, this still does not amount to making the program in the first place but only to adapting an existing program to changing circumstances. On the challenges of artificial intelligence, see Yuk Hui, *Recursivity and Contingency* (London: Rowman and Littlefield, 2019).
170. This and the other quotes of this paragraph come from Jacques Derrida, *Faith and Knowledge*, in *Acts of Religion*, ed. Gil Anidjar (New York: Routledge, 2002), 80, note 27.
171. Derrida, *Faith and Knowledge*, 82.
172. "Indeed, without autoimmunity, without this breach in the immunitary and self-protective systems of the organism, there would be no possibility for the supplement that might destroy *or* save it, bring it to an end or allow it to live on. Without autoimmunity, the organism would have, in short, no future before it." Michael Naas, *Miracle and Machine: Jacques Derrida and the Two Sources of Religion, Science, and Media* (New York: Fordham University Press, 2012), 82.
173. Derrida, *Faith and Knowledge,*. 81.
174. Derrida, 82.
175. For a clear presentation of the theory of autopoiesis from a Derridian point of view, see Cary Wolfe, "In the Shadow of Wittgenstein's Lion," in *Zoontologies: The Question of the Animal*, ed. Cary Wolfe (Minneapolis: University of Minnesota Press, 2003), 35–48. For a rich explication of the discussion on the complementarity of Derrida's and Luhmann's thinking, see Cary Wolfe, *What Is Posthumanism?* (Minneapolis: University of Minnesota Press, 2010), 3–29. Luhmann's idea of system is well presented by Hanna Lukkari in chapter 3 "The Art of Not Being Paralyzed: Niklas Luhmann's Evolutionary Theory of Legal Paradox" of her PhD dissertation "Law, Politics and Paradox: Orientations in Legal Formalism" (PhD diss., University of Helsinki, 2020).
176. Niklas Luhmann, "Deconstruction as Second-Order Observing," *New Literary History* 24, no. 4 (1993): 771.
177. This is the main principle of cybernetics. One sees easily that the so-called artificial intelligence, that is, machine learning systems, is thought along the principles of autopoiesis. The importance to artificial intelligence of the principle of recursivity instead of simple machinic repetition is emphasized by Yuk Hui in *Recursivity and Contingency* (London: Rowman and Littlefield, 2019).
178. Luhmann, "Deconstruction as Second-Order Observing," 767.
179. Willis Santiago Guerra Filho, "Luhmann and Derrida: Immunology and Autopoiesis," in *Luhmann Observed*, eds. Anders La Cour and Andreas Philippopoulos-Mihailopoulos (London: Palgrave Macmillan, 2013), 233.

180. Wolfe, *What Is Posthumanism?*, 13, 24.

181. Gunther Teubner, "Economics of Gift—Positivity of Justice: The Mutual Paranoia of Jacques Derrida and Niklas Luhmann," *Theory, Culture & Society* 18, no. 1 (2001): 29–47.

182. Derrida and Stiegler, *Echographies of Television*, 38–39.

183. Derrida, *Faith and Knowledge*, 67. Derrida is aware of the bias that is introduced when (Catholic) Christianity is declared to be the most important of the global religions, but he nonetheless adapts to this "Roman" horizon. Now this article has led to a debate concerning Derrida's relation to religion where some claim that in his later texts Derrida has turned toward religion, whereas others answer that his thinking is fundamentally atheist and rather describes technology as the framework of the contemporary world (the main positions in the debate are summarized by Bradley in *Originary Technicity*, 115, who thinks that in the last instance it is not about religion but about the historical vs. the transcendental. Naas's *Miracle and Machine* is a useful minute reading of *Faith and Knowledge* also from the viewpoint of the question of religion). I believe that if this was Derrida's aim, the second option would be closer to truth, but that in reality this way of setting the debate misses Derrida's point, which is not that of faith *in* anything but faith as an act of trusting or believing as part of any act of thinking whatsoever, whether faith or knowledge.

184. Jacques Derrida, *Specters of Marx: The State of the Debt, the Work of Mourning and the New International*, trans. Peggy Kamuf (New York: Routledge, 1994), 210.

185. Derrida, *Faith and Knowledge*, 82. Furthermore:
"The same movement that renders indissociable religion and tele-techno-scientific reason in its most critical aspect reacts inevitably *to itself.* It secretes its own antidote but also its own power of auto-immunity. We are here in a space where all self-protection of the unscathed, of the safe and sound, of the sacred (*heilig*, holy) must protect itself against its own protection, its own police, its own power of rejection, in short against its own, which is to say, against its own immunity. It is this terrifying but fatal logic of the *auto-immunity of the unscathed* [*l'auto-immunité de l'indemne*] that will always associate Science and Religion" (Derrida, *Faith and Knowledge*, 79–80).

186. Derrida, 43.

187. Derrida and Stiegler, *Echographies of Television*, 33. "It makes no sense to be against TV, journalists and media in general." Derrida, *Papier Machine* (Paris: Galilée, 2001), 236. My translation.

188. In the second essay of *The Genealogy of Morals*, Nietzsche presents a remarkable analysis of promise.

"Forgetfulness is no mere *vis inertiæ*, as the superficial believe, rather is it a power of obstruction, active and, in the strictest sense of the word, positive. . . . But this very animal who finds it necessary to be forgetful, in whom, in fact, forgetfulness represents a force and a form of *robust* health, has reared for himself an opposition-power, a memory, with whose help forgetfulness is, in certain instances, kept in check—in the cases, namely, where promises have to be made;—so that it is by no means a mere passive inability to get rid of a once indented impression, not merely the indigestion occasioned by a once pledged word, which one cannot dispose of, but an *active* refusal to get rid of it, a continuing and a wish to continue what has once been willed, an actual *memory of the will*. . . . How thoroughly, in order to be able to regulate the future in this way, must man have first learnt to distinguish between necessitated and accidental phenomena, to think causally, to see the distant as present and to anticipate it, to fix with certainty what is the end, and what is the means to that end; above all, to reckon, to have power to calculate—how thoroughly must man have first become *calculable, disciplined, necessitated* even for himself and his own conception of himself, that, like a man entering into a promise, he could guarantee himself *as a future*. . . . This is simply the long history of the origin of *responsibility*. That task of breeding an animal which can make promises, includes, as we have already grasped, as its condition and preliminary, the more immediate task of first *making* man to a certain extent, necessitated, uniform, like among his like, regular, and consequently calculable" (Friedrich Nietzsche, *The Genealogy of Morals*, trans. Horace B. Samuel [Edinburgh: Foulis, 1913], 61–63).

189. For example, the entire system of money is based on the promise of conservation of money's value in the future—and on people's faith that money will not suddenly lose its value nor turn out to be counterfeit, although this is of course in principle possible. This idea is notably developed in Derrida, *Given Time 1: Counterfeit Money*, trans. Peggy Kamuf (Chicago: University of Chicago Press, 1992); and commented by Christian Arnsperger, Egide Berns, Simon Critchley, Jacques Derrida, Marcel Drach, and Jean-Jacques Goux, "L'esprit de l'argent. Autour des ecrits de Jacques Derrida sur l'argent," in *L'argent: Croyance, mesure, spéculation*, ed. Marcel Drach (Paris: La découverte, 2004), 199–232.

190. Derrida, *Faith and Knowledge*, 83.

191. Derrida, 83.

192. The German word *Geschlecht* is derived from *schlagen*, strike, stamp; it means blood relations, species, race, and sex, among others, and it is used to translate all senses of the Latin *genus*. In a series of articles that are all subtitled *Geschlecht*, Derrida shows how the politically charged problematics of *Geschlecht* accompanies Heidegger's thinking of historicity, *Geschichtlichkeit*, and spirit, *Geist*. Derrida's studies of this problematics include Derrida, *Of Spirit, Heidegger and the Question*, trans. Rachel Bowlby (Chicago: University of Chicago Press, 2017); "Sexual Difference, Ontological Difference, *Geschlecht 1*," trans. Ruben Berezdiwin and Elizabeth Rottenberg and "The Hand of Heidegger (*Geschlecht 2*)," trans. John P. Leavey and Elizabeth Rottenberg; both in *Psyche 2: Inventions of the Other*, ed. Peggy Kamuf (Stanford: Stanford University Press, 2008), 7–26; *Geschlecht 3: Sex,*

Race, Nation, Humanity, eds. Geoffrey Bennington et al., trans. Katie Chenoweth and Rodrigo Therezo (Chicago: University of Chicago Press, 2020); "Heidegger's Ear: Philopolemology (*Geschlecht* 4)," trans. John P. Leavey Jr., in *Reading Heidegger: Commemorations*, ed. John Sallis (Bloomington: Indiana University Press, 1993), 163–218. For a meticulous presentation of Derrida's series of articles, see David Farrell Krell, *Phantoms of the Other: Four Generations of Derrida's Geschlecht* (Albany: State University of New York Press, 2015).

193. Heidegger, *Being and Time*, § 9, 71.

194. Bernard Stiegler, "Derrida and Technology: Fidelity at the Limits of Deconstruction and the Prosthesis of Faith," in *Jacques Derrida and the Humanities: A Critical Reader*, ed. Tom Cohen (Cambridge: Cambridge University Press, 2001).

195. In *Given Time*, Derrida studies the *gift* as "the impossible" because it is what gives time and being without having any presence: "The 'present' of the gift is no longer thinkable as a now, that is, as a present bound up in a temporal synthesis." Derrida, *Given Time 1*, 7, 9. The gift can be gift only if it does not appear presently as such.

196. "In all these cases, the gift can certainly keep its phenomenality, or if one prefers, its appearance as a gift. But its very appearance, the simple phenomenon of the gift annuls it as gift, transforming the apparition into a phantom and the operation into a simulacrum." Derrida, *Given Time 1*, 14; Derrida, *Dissemination*, 165; Derrida, *Of Hospitality*, trans. Rachel Bowlby (Stanford: Stanford University Press, 2000), 37.

197. Derrida, *Specters of Marx*, 10.

198. Derrida, 166.

199. Derrida, xvi.

200. Derrida, 125. In *Ecographies of Television*, when Stiegler presses Derrida to explicate the inheritance of Heidegger's notion of Dasein in Derrida's spectrology, Derrida admits this inheritance in the very concept of inheritance but points out that Heidegger never goes as far as evoking specters. Derrida and Stiegler, *Echographies of Television*, 130–31. One could say that this is because the specter refers finally to the *other Dasein*, which Heidegger always had difficulties including in his thinking.

201. Derrida, *Specters of Marx*, xviii.

202. Derrida, *Specters of Marx*, 16.

203. Derrida finds *The Communist Manifesto* particularly lucid concerning the irreducibility of the technical and the media. Derrida, *Specters of Marx*, 14.

204. Derrida, xvii.

205. Derrida, 67.

206. Derrida, 114. "Inheriting does not consist in receiving goods or capital that would be in one place, already and once for all, localized in a bank, a data bank, an image bank, or whatever. Inheritance implies decision, responsibility,

response and, consequently, critical selection, choice. There is always choice, whether one likes it or not, whether it is or isn't conscious." Derrida and Stiegler, *Echographies of Television*, 68–69.

207. Derrida and Stiegler, 86–87.
208. Derrida, *Specters of Marx*, 125.
209. Derrida, 6–7.
210. Derrida, 18.
211. Derrida, 25.
212. Derrida, 26. As Michael Naas says, "Messianicity without messianism is thus the name of an opening to what is radically other." Naas, *Miracle and Machine*, 161. "Derrida links messianicity not only to absolute surprise—to the possibility of the best as well as the worst—but to a 'phenomenal form' of peace and justice" (163). Naas shows very well the link between messianicity, *khora*, and the motive of faith to which we will come shortly. "Derrida is trying to think a link that must be understood only in relationship to a kind of interruption, to a certain *without* (in the case of messianicity without messianism) or a certain *withdrawal* (in the case of *khora*) that opens up the possibility of religion without being reducible to any determinate form of religion." According to Naas, this will allow Derrida to "criticize—to deconstruct if you will—all determinate religious traditions" (154). While Naas interprets *Faith and Knowledge* in function of the question of religion, I read it in function of the question of tele-technology.
213. Derrida, *Specters of Marx*, 33; cf. 93–94.
214. Derrida, 46.
215. Derrida, 74.
216. Derrida, 212.
217. Derrida, 211.
218. Derrida, *Of Hospitality*, 19.
219. This frustrating double bind is explained in Derrida, *Force of Law: The Metaphysical Foundation of Authority*, in *Deconstruction and the Possibility of Justice*, eds. Drucilla Cornell et al. (Abingdon: Routledge, 1992).
220. Derrida, *Specters of Marx*, 121.
221. Derrida, 63.
222. Derrida, *On Touching: Jean-Luc Nancy*, trans. Christine Irizarry (Stanford: Stanford University Press), 2005.
223. Derrida and Stiegler, *Echographies of Television*, 5.
224. Stiegler, "Fidelity at the Limits of Deconstruction," 251.
225. Derrida, "Typewriter Ribbon: Limited Ink (2)," 71–160.
226. Derrida, *Faith and Knowledge*, 56.
227. Derrida, 58.
228. I analyze these in the context of contemporary digital remote meeting technologies in my "Four Transcendental Illusions of the Digital World: A Derridian Approach," *Research in Phenomenology* 3 (2021), 394–413.

229. Contemporary information ordering is not only programmed by others to us, but it also follows logics proper to machines but that, unlike human deduction, human programmers cannot necessarily follow. The technological process is well explained by Rieder in *Engines of Order*. "One could argue that data mining techniques embody forms of cognition or enunciation that are, on the one hand, nonanthropomorphic in the sense that they consist of procedures that can only be enacted by fast computing machinery and, on the other side, thoroughly entangled in operational arrangements." Rieder, *Engines of Order*, 255.

230. Jussi Parikka, "Deep Times and Media Mines: A Descent into Ecological Materiality of Technology" in *General Ecology: The New Ecological Paradigm*, Erich Hörl and James Burton (London: Bloomsbury, 2017).

231. N. Katherine Hayles, *How We Became Posthuman*, 7.

232. N. Katherine Hayles, *Unthought: The Power of the Cognitive Nonconscious*, 9–40.

Chapter 5

1. Jean-François Lyotard, *The Inhuman: Reflections on Time*, trans. Geoffrey Bennington and Rachel Bowlby (Cambridge: Polity Press, 1991), 2, 7.

2. Bernard Stiegler, *Taking Care of Youth and the Generations*, trans. Stephen Barker (Stanford: Stanford University Press, 2010), chap. 54.

3. Gilles Deleuze and Félix Guattari, *A Thousand Plateaus: Capitalism and Schizophrenia*, trans. Brian Massumi (Minneapolis: University of Minnesota Press, 1987), 399.

4. As Derrida says, the development of tele-technologies is "producing a practical deconstruction of the traditional and dominant concepts of the state and the citizen." Jacques Derrida and Bernard Stiegler, *Echographies of Television: Filmed Interviews*, trans. Jennifer Bajorek (Cambridge: Polity Press, 2002), 36. He even says that the technical transformation brought about by the telephone, television, email, and the Internet may have done more to democratization than all the discourses on behalf of human rights. Derrida and Stiegler, 71.

5. According to Derrida, contemporary tele-technology excels in *displacing places* such that "the link between the political and the local, the *topolitical*, is as it were *dislocated.*" Derrida and Stiegler, 57. I develop the idea of a displace further in my "Technics of Space, Place and Displace," in *Azimuth* 10, Special Issue "Intersections: At the Technophysics of Space" (2018): 27–44.

6. This is also because of the economic and political conditions of telecommunication globalization, as discussed, for example, by Derrida and Stiegler, *Echographies of Television*, 4–5.

7. Derrida and Stiegler, 81, 17.

8. Bernard Stiegler, *Technics and Time 1: The Fault of Epimetheus*, trans. Richard Beardsworth and Richard Collins (Stanford: Stanford University Press, 1998), 134.

9. This and the following: Stiegler, 183. "The prosthesis is not a mere extension of the human body; it is the constitution of the body qua 'human.'" Stiegler, 152–53.

10. "Derrida bases his own thought of différance as a general history of life, that is, as a general history of the *gramme*, on the concept of program insofar as it can be found on both sides of such divides. Since the *gramme* is older than the specifically human written forms, and because the letter is nothing without it, the conceptual unity that différance is contests the opposition animal/human and, in the same move, the opposition nature/culture. 'Intentional consciousness' finds the origin of its possibility before the human; it is nothing else but 'the emergence that has the *gramme* appearing *as such*.' We are left with the question *of determining what the conditions of such an emergence of the "gramme as such" are, and the consequences as to the general history of life and/or of the gramme. This will be our question*" (Stiegler, 137).

11. The following in Stiegler, 150–53, 183–87.

12. Stiegler, 177.

13. Derrida and Stiegler, *Echographies of Television*, 36–39.

14. Bernard Stiegler, "Derrida and Technology: Fidelity at the Limits of Deconstruction and the Prosthesis of Faith," in *Jacques Derrida and the Humanities: A Critical Reader*, ed. Tom Cohen (Cambridge: Cambridge University Press, 2001). Mark B. Hansen and Ben Roberts follow Stiegler's preference for the analysis of concrete technological systems, while Tracy Colony finds Derrida's transcendental approach more fruitful. Mark B. Hansen, "'Realtime Synthesis' and the Différance of the Body: Technocultural Studies in the Wake of Deconstruction," *Culturemachine* 6 (2004), https://culturemachine.net/deconstruction-is-in-cultural-studies/realtime-synthesis-and-the-differance-of-the-body/; Ben Roberts, "Stiegler Reading Derrida: The Prosthesis of Deconstruction in Technics," in *Postmodern Culture* 16, no. 1 (2005), https://muse.jhu.edu/article/192267. Tracy Colony, "Epimetheus Bound: Stiegler on Derrida, Life, and the Technological Conditions," *Research in Phenomenology* 41 (2011): 72–89; and Colony, "The Future of Technics," *Parrhesia* 27 (2017): 64–87.

15. Stiegler, *Technics and Time 1*, 205.

16. Stiegler, 204.

17. Stiegler, 234; cf. 17–18.

18. Stiegler, 172.

19. Heidegger's examples of technical objects are so few that they cannot be said to constitute anything like a history. The difference between the old bridge in Heidelberg and the power plant over the Rhine can at most indicate

the difference between two epochs. Contrary to this, Gilbert Simondon calls for a detailed history of technology and even sketches it out in his lecture courses *Psychosociologie de la technicité* and especially *Naissance de la technologie*, published in Gilbert Simondon, *Sur la technique (1953–1983)* (Paris: PUF, 2014). In the first part of *Technics and Time 1* where Stiegler exposes the fundamental principles of history of technology, he also draws upon Bernard Gille's history of technology.

20. Stiegler studies these different cases in all of his books; a summary can be found for example in Bernard Stiegler, *Automatic Society Vol. 1: The Future of Work*, trans. Dan Ross (Cambridge: Polity Press, 2016), 19.

21. Calculative machine memory differs from human memory because the machine does not forget but the forgetful human can invent new solutions where old ones were forgotten or because the machine calculates much more quickly but the human can figure out new patterns of reasoning. Gilbert Simondon explains very clearly the difference between machine and human memory in "Critique of the Relation between Man and the Technical Object as it Is Presented by the Notion of Progress Arising from Thermodynamics and Energetics: Recourse to Information Theory," in *On the Mode of Existence of Technical Objects*, trans. Cécile Malaspina and John Rogove (Minneapolis: Univocal Publishing, 2017), 135.

22. That is, both living and technical organisms are defined by the recursivity of their own operation and by the contingency of the external impulsion, as Yuk Hui describes in his important *Recursivity and Contingency* (London: Rowman and Littlefield, 2019). Another important account of the use of systems theory in the explication of both living and machinic beings is Erich Hörl's *Sacred Channels: The Archaic Illusion of Communication*, trans. Nils Schott (Amsterdam: Amsterdam University Press, 2018).

23. N. Katherine Hayles, *Unthought: The Power of the Cognitive Nonconscious* (Chicago: University of Chicago Press, 2017), chap. 1. The idea of nonconscious cognition goes beyond the so-called extended mind, which emphasizes that cognition is a situated activity that includes both the human cognizers and diverse material objects around them, because in addition to this, Hayles's idea of nonconsciousness also shows that these "external" things are not only extensions of human cognition but have their own, distinctive ways of cognizing. On "extended mind," see Michael L. Anderson, "Embodied Cognition: A Field Guide," in *Artificial Intelligence* 149 (Amsterdam: Elsevier Science, 2003): 91–130.

24. Stiegler, *Technics and Time 1*, 2–4.
25. Stiegler, *Automatic Society 1*, 116.
26. Stiegler, 38–39.
27. Simondon, *On the Mode of Existence of Technical Objects*, 64–68.
28. Stiegler, *Automatic Society 1*, 56.
29. Stiegler, 80.
30. Stiegler, 10–11.
31. Stiegler, 11.

32. The evolution of the term *anthropocene*, which stands at the center of an enormous discussion, is presented by Bruno Latour, *Facing Gaia: Eight Lectures on the New Climatic Regime*, trans. C. Porter (Cambridge: Polity Press, 2017). The debate is much more general, however; for a good general presentation of the debate, see Agostino Cero, "The Anthropocene as an Epistemic Hyperobject: The Stratigraphic Fallacy," *Acta Philosophica Fennica* 97 (2021): 129–52.

33. Stiegler, *Automatic Society 1*, 15, 121.

34. Félix Guattari, *The Three Ecologies*, trans. Ian Pindar and Paul Sutton (London: Athlone Press, 2000).

35. Stiegler, *Automatic Society 1*, 32–33.

36. Stiegler, *Technics and Time 1*, 148. There are no logical reasons for limiting the "who?" to just *Homo sapiens*, but Stiegler does not in fact analyze versions of the "who?" other than the human. Although he accepts a very open definition of the human starting from the early ancestors of the *Homo sapiens*, he, like the entire philosophical tradition, takes it for granted that the use of technics defines its user as a human. Both Derrida and Agamben question the priority given to *Homo sapiens*—especially given that many animal species use technics as epiphylogenetic memory (anthills, beehives, etc.).

37. Stiegler, *Technics and Time 1*, 183.

38. Stiegler, 232, 258.

39. Stiegler, 193.

40. Stiegler, 196–97.

41. Stiegler, 240. Both quotations are from Heidegger, *Being and Time*, 12.

42. Stiegler, 234.

43. Stiegler, 267.

44. Stiegler, 65.

45. Stiegler, 46–49.

46. Stiegler, *Technics and Time 3: Cinematic Time and the Question of Malaise*, trans. Stephen Barker (Stanford: Stanford University Press, 2011), 90.

47. Stiegler, 90.

48. Stiegler, 88.

49. Stiegler, 94.

50. Gilbert Simondon, *Individuation à la lumière des notions de forme et d'information* (Grenoble: Jérôme Millon, 2013), 24–26.

51. Stiegler, *Technics and Time 3*, 95. Jason Read explains clearly that Stiegler's principal critique of Simondon is his assertion that the pre-individual, the language, habits, and customs that form the basis of individuation, primarily exist in the form of texts, artefacts, and other technological inscriptions of memory. The pre-individual is not nature, existing as some inchoate set of drives and potentials, nor is it simply mediated by culture, by its historical organisation in transindividual relations, but it exists in the form of writing, tools, and record-

ings that form the basis of our experience and history (Jason Read, *The Politics of Transindividuality* [Leiden: Brill, 2016], 119).

52. Stiegler, *Technics and Time 3*, 95.

53. On why Simondon's transindividuality is not intersubjectivity, see Read, *The Politics of Transindividuality*, 113; and Stiegler, *La technique et le temps 1* (Paris: Galilée, 1994), 27: "*(nous abandonons ici Habermas et le concept d'intersubjectivité, extrêmement fragile, sur lequel reposent ses analyses)*" (this parenthesis was dropped in the English translation).

54. Stiegler, *Technics and Time 2*, trans. Stephen Barker (Stanford: Stanford University Press, 2009), 84.

55. Stiegler, *Technics and Time 2*, 3.

56. Derrida and Stiegler, *Echographies of Television*, 3–9, 41–46.

57. Stiegler, *Automatic Society*, 19, 98, 106, chap. "Overtaken."

58. By "displace," I mean a space produced by technological dispositifs, which consists of informational of physical transits. See my "Technics of Space, Place and Displace," in *Azimuth* 10, Special Issue "Intersections: At the Technophysics of Space" (2018): 27–44.

59. On Heidegger and Leninism, see Timothy C. Campbell, *Improper Life: Technology and Biopolitics from Heidegger to Agamben* (Minneapolis: University of Minnesota Press, 2011), 6–9. However, especially in *Black Notebooks*, Heidegger emphasizes that as expressions of technological "machination" (*Machenschaft*) Bolshevism and national socialism amount to the same. In the *Notebooks* from 1931–1941, it is clear that Nazism no longer represents for him the possibility of a new beginning but is rather the most brutal realization of the epoch of technics that formats people to catastrophic technological modernity. See Martin Heidegger, *Schwarze Hefte 1939–1941, GA 96* (Frankfurt am Main: Klostermann, 2014), 109, 127, 190, 195, 256–57.

60. Simondon, *On the Mode of Existence of Technical* Objects, 55–66.

61. Michal Foucault, *Society Must Be Defended: Lectures at the Collège de France, 1975–1976*, eds. Mauro Bertani and Alessandro Fontana, trans. David Macey (New York: Picador, 2003), 34.

62. Foucault, *Society Must Be Defended*, 35–36.

63. Foucault, 241–43.

64. Foucault, 241.

65. Foucault, 245.

66. Giorgio Agamben, "The Invention of an Epidemic," *European Journal of Psychoanalysis* (February 26, 2020). Republished in *Where are We Now? The Epidemic as Politics*, trans. Valeria Dani (New York: Rowman and Littlefield, 2021), 20–22.

67. Foucault, *Society Must Be Defended*, 43.

68. Foucault, 45.

69. Foucault, 46.

70. Giorgio Agamben, "What Is an Apparatus?" in *What Is an Apparatus? and Other Essays*, trans. David Kishik and Stefan Pedatella (Stanford: Stanford University Press, 2009), 8, 12.
71. Foucault, *Society Must Be Defended*, 13–15.
72. Foucault, 248.
73. Foucault, 61, 81–82, 254–63.
74. This motif has been further developed by Achille Mbembe, *Necropolitics*, trans. Steve Corcoran (Durham: Duke University Press, 2019), chap. 3 "Necropolitics."
75. Giorgio Agamben, *Homo Sacer: Sovereign Power and Bare Life*, in *The Omnibus Homo Sacer*, trans. Daniel Heller-Roazen (Stanford: Stanford University Press, 2017), 99–102, 117, 137–47.
76. Giorgio Agamben, *Language and Death: The Place of Negativity*, trans. Karen Pinkus and Michel Hardt (Minneapolis: University of Minnesota Press, 1991). Agamben's *Auseinandersetzung* with Derrida is difficult; Leland de la Durantaye lists its twists in *Giorgio Agamben: A Critical Introduction* (Stanford: Stanford University Press, 2009), 184–91. In *Language and Death*, Agamben reads the same metaphysical sources as Derrida but downplays the value of Derrida's deconstructive approach:

> Although we must certainly honor Derrida as the thinker who has identified with the greatest rigor . . . the original status of the *gramma* and of meaning in our culture, it is also true that he believed he had opened a way to surpassing metaphysics, while in truth he merely brought the fundamental problem of metaphysics to light. For metaphysics is not simply the primacy of voice over *gramma*. . . . Metaphysics is always already grammatology and this is fundamentology in the sense that *gramma* (or the Voice) functions as the negative ontological foundation." (Agamben, *Language and Death*, 39)

In *Stanzas*, he continues: "The metaphysics of the writing and the signifier is but the reverse face of the metaphysics of the signified and the voice, and not, surely, its transcendence." Agamben, *Stanzas: Word and Phantasm in Western Culture*, trans. Ronald L. Martinez (Minneapolis: University of Minnesota Press, 1993), 156. It is surely wrong to situate Derrida within metaphysics, for he has always affirmed that he is neither inside nor outside of metaphysics but works at its margins. However, the reason for Agamben's assessment of Derrida's early thought as within metaphysics is less truthful than strategic, for it allows Agamben to underline the real originality of his own thinking of the Voice as taking-place-of language. In truth, however, it appears clear that Agamben's philosophical position, especially when it comes to a Heideggerean reading of metaphysics, could only take place within an implicit dialogue with Derrida, as well as with Jean-Luc Nancy and Philippe Lacoue-Labarthe, and it is a pity for us readers that he has never generously spelled out what he has learned from these thinkers. See also Naomi Waltham-Smith, "The Use of Ears: Agamben Overhearing Derrida Overhearing Heidegger," in *Parrhesia* 33 (2020): 113–49, 117; and Virgil W. Brower, "Jacques Derrida," in *Agamben's Philosophical Lineage*, eds. Adam Kotsko and Carlo Salzani (Edinburgh: Edinburgh University Press, 1917). This being said, it is true that Derrida has not been any more tender with Agamben. See, for example, how he mocks Agamben's desire to always designate the first moments in history and how he contests the distinction between *bios* and *zoe* that is so important to Agamben in *Homo Sacer*. Jacques Derrida, *The Beast and the Sovereign I*, trans. Geoffrey Bennington (Chicago: University of Chicago Press, 2009), 93–95, 316 sqq. Whether deliberately or not, Derrida ignores the fact that in the *Homo Sacer* series Agamben will situate the key concepts of use and habitude precisely at the threshold where *bios* and *zoe* are indistinguishable. An impartial comparison of the *Homo Sacer* project and Derrida's *The Best and the Sovereign* project remains to be done, their subjects are obviously related.

77. Agamben, *Homo Sacer*, 9. "The production of bare life is the originary activity of sovereignty," Agamben, 21.

78. For example, "We have seen, then, in Nazi society something that is really quite extraordinary: this is a society which has generalized biopower in an absolute sense, but which has also generalized the sovereign right to kill." Foucault, *Society Must Be Defended*, 260.

79. Agamben, *Homo Sacer*, 17.

80. Agamben, 19. For an excellent comparative analysis of Agamben's conception of the ground of law, see Hanna Lukkari, *Law, Politics and Paradox: Orientations in Legal Formalism* (PhD diss., University of Helsinki, 2020).

81. Agamben, *Homo Sacer*, 10.

82. Agamben, 27.

83. Agamben, *The Open: Man and Animal*, trans. Kevin Attell (Stanford: Stanford University Press, 2004), 37, 38. In this book, Agamben shows through drawing on the history of philosophy how "the originary political conflict is . . . that between the humanity and the animality of man. The animal is the Undisclosable which man keeps and brings to light as such" (73). The difference between natural life and bare life is also explained by Carlo Salzani in "From Benjamin's *bloßes Leben* to Agamben's *nuda vita*: A Genealogy," in *Towards the Critique of Violence: Walter Banjamin and Giorgio Agamben*, eds. Brendan Moran and Carlo Salzani (London: Bloomsbury, 2015), 109–118.

84. Agamben, *Homo Sacer*, 5; cf. Agamben, *Means without End: Notes on Politics*, trans. Vincenzo Binetti and Cesare Cesarino (Minneapolis: University of Minnesota Press, 2000), 1.

85. As Arne de Boever also puts it in *The Agamben Dictionary*, ed. Alex Murray and Jessica Whyte (Edinburgh: Edinburgh University Press, 2011), 30.
86. Agamben, *Homo Sacer*, 70.
87. Agamben, 88–92.
88. Agamben, 151.
89. Agamben, 116–18.
90. Agamben, 132–36.
91. Agamben, *State of Exception*, in *Omnibus Homo Sacer*, 169.
92. For example, Agamben, *Homo Sacer*, part 1 of the collected volume *Omnibus Homo Sacer* (Stanford: Stanford University Press, 2017), 148; Agamben, *Means without End*, 37.
93. Mika Ojakangas, "Impossible Dialogue on Bio-power: Agamben and Foucault," *Foucault Studies* 2 (May 2005): 11–15, 25.
94. Ojakangas, 27.
95. Matthew Abbott, *The Figure of This World: Agamben and the Question of Political Ontology* (Edinburgh: Edinburgh University Press 2014), 3, 17.
96. Abbott, 22–23. The sense of Agamben's paradigms reflects his genealogical method. A paradigm is not a particular case incarnating a universal law but a singular case detached from its context and used to make a group of phenomena intelligible. It does not explain, but it helps in interpretation. See also Vanessa Lemm, "Michel Foucault" in *Agamben's Philosophical Lineage*, eds. Adam Kotsko and Carlo Salzani (Edinburgh: Edinburgh University Press 1917), chap. 4, 54–55. Jeffrey Bussolini shows how Agamben's idea of paradigm draws from Foucauldian method, which also regularly departs from paradigms (great confinement, Panopticon, etc.). Moving from singularity to singularity, paradigms avoid inductive and deductive modes of explication and open immanent ways of explication. Bussolini, "Critical Encounter Between Giorgio Agamben and Michel Foucault: Review of Recent Works of Agamben: *Il Regno e la Gloria: per una genealogia dell'economia e del governo*; *Il sacramento del linguaggio: Archeologia del giuramento*, and *Signatura rerum: Sul Metodo*," *Foucault Studies* 10 (November 2010): 108–43.
97. Agamben, *Means without End*, 4.
98. Agamben, 3–4.
99. Agamben, *The Use of Bodies*, in the collected volume *Omnibus Homo Sacer*, 1214–15.
100. "Glenn Gould, to whom we attribute the habit of playing the piano, does nothing but make use-of-himself insofar as he plays and knows habitually how to play the piano. He is not the title holder and master of the potential to play, which he can put to work or not, but constitutes-himself as having use of the piano, independently of his playing it or not playing it in actuality. *Use, as habit, is a form-of-life and not the knowledge or faculty of a subject.*" Agamben, *The Use*

of Bodies, in *Omnibus Homo Sacer*, 1084; cf. Agamben, *The Coming Community*, trans. Michael Hardt (Minneapolis: University of Minnesota Press, 2007), 6, 36. See also Agamben, *Homo Sacer*, *Omnibus Homo Sacer*, 40–42; and for Agamben's intensive analysis of the Aristotelian notion of potentiality, see Agamben, "On Potentiality," in *Potentialities*, trans. Daniel Heller-Roazen (Stanford: Stanford University Press, 1999), 177–84. As David Bleeden emphasizes, "Potentiality is not simply a potentiality to, a 'can,' but also a potentiality not-to, a 'cannot.'" Bleeden, "One Paradigm, Two Potentialities: Freedom, Sovereignty and Foucault in Agamben's Reading of Aristotle's *dynamis*," *Foucault Studies* 10 (2010): 68–84, 71.

101. Agamben, *The Use of Bodies*, in *Omnibus Homo Sacer*, 1217.

102. Agamben, *The Highest Poverty*, in *Omnibus Homo Sacer*, 963, 967; Agamben, *Use of Bodies*, 1223–24. For a careful presentation of the monastic form-of-life, see Valeria Bonacci, "Form-of-Life and Use in Homo Sacer," in *The Politics, Ethics and Aesthetics of Inoperativity*, eds. Giovanni Marmont and German E. Primera, *Journal of Italian Philosophy* 3 (2020): 217–45.

103. Giorgio Agamben, *The Coming Community*. As Anke Snoek says: "Agamben searches for the nexus in which the two powers [political techniques and the technologies of the self] intertwine, and he finds them in the production of 'bare life': life that is subjected to power through the exclusion of its essential element. But this point is also the place in which Agamben develops his notion of resistance and ethics. Bare life is closely related to the 'art of life,' or what Agamben calls a form-of-life." Anke Snok, "Agamben's Foucault: An Overview," *Foucault Studies* 10 (2010): 44–67, 54.

104. Timothy Campbell, *Improper Life: Technology and Biopolitics from Heidegger to Agamben* (Minneapolis: University of Minnesota Press, 2011), 32, 35, 36.

105. Agamben, "What Is an Apparatus?," 14.

106. Agamben, 16.

107. Agamben, 19–20.

108. Agamben, 16, 23.

109. Agamben, 21.

110. Agamben, 15–16, 20.

111. On the way in which contemporary art diverts technics, see for example chapter 4 of my book *Techniques en philosophie* (Paris: Hermann, 2020). In his article "The Allegory of the Cave: Foucault, Agamben, and the Enlightenment," *Foucault Studies* 10 (2010): 44–67, Arne de Boever sides with Stiegler against Agamben, claiming that although Agamben continues Foucault's work on biopower, Foucault's shift to the technics of the self remains unthought in Agamben's work (15) and his call for a profanation of apparatuses tends to mysticism (9). This article was written in 2010, before the publication of *The Use of Bodies* in 2014, which is not only a vast work on technics of the self but also shows why forms-of-life naturally lead to profanation.

112. Agamben, *The Use of Bodies*, in *Omnibus Homo Sacer*, 1117–18.
113. Agamben, 1123.
114. Agamben, 1125.
115. Agamben, 1277–78.
116. Agamben, 1242.
117. Of course, Agamben is not against "freedom, equality, and fraternity" as such, but he points at the terror latent in the total rule of freedom, equality, and fraternity, such as it was described by Hegel in his reading of the period of terror during the French Revolution. Modern political subjectivity must be universal, but for this precise reason it is also exclusionary.
118. Agamben, *The Use of Bodies*, in *Omnibus Homo Sacer*, 1078.
119. Agamben, 1065.
120. Agamben, 1056–57.
121. Artistic practice is one of Agamben's essential examples of an ethical relation to self: "And the painter, the poet, the thinker—and in general, anyone who practices a *poiesis* and an activity—are not the sovereign subjects of a creative operation and of a work. Rather, they are anonymous living beings who, by always rendering inoperative the works of language, of vision, of bodies, seek to have an experience of themselves and to constitute their life as form-of-life." Agamben, 1251.
122. Agamben, 1053.
123. On the "cyborg body" of the musician, see Peter Szendy, *Phantom Limbs: On Musical Bodies* (New York: Fordham University Press, 2015).
124. Naomi Waltham-Smith, "The Use of Ears: Agamben Overhearing Derrida Overhearing Heidegger," 138.
125. Agamben, *The Use of Bodies*, in *Omnibus Homo Sacer*, 1053.
126. Agamben, 1030.
127. Agamben, 1063.
128. Agamben, 1039–40.
129. For a useful summary of Agamben's reading of Aristotle's theory of slavery, see Arthur Bradley, "In the Sovereign Machine: Sovereignty, Governmentality, Automaticity," *Journal for Cultural Research* 22, no. 3 (2018): 209–23, 212–15, 221.
130. Gert-Jan van der Heiden, "Exile, Use, and Form-of-Life: On the Conclusion of Agamben's *Homo Sacer* Series," *Theory, Culture & Society* 37, no. 2 (2020): 61–78, 64.
131. Agamben, *The Use of Bodies*, in *Omnibus Homo Sacer*, 1039–40.
132. Agamben, 1045.
133. Agamben, 1099.
134. Agamben, 1055.
135. Agamben, 1101.
136. Agamben, *The Coming Community*, 28–29.

137. Martin Heidegger, *Elucidations of Hölderlin's Poetry*, trans. Keith Hoeller (New York: Humanity Books, 2000), 112.

138. Heidegger explains: "What is proper to the Greeks is the fire of heaven. . . . In order to appropriate this proper character, they must pass through what is foreign to them. This is *clarity of presentation*. . . . What is *natural* to the Germans, on the contrary, is *clarity of presentation*. . . . What the Germans must encounter as foreign to them, and what they must become experienced with in the foreign land, is the *fire of heaven*." Heidegger, *Elucidations of Hölderlin's Poetry*, 112–13.

139. Heidegger, 173.

140. However, Agamben's newspaper article against the COVID-19 measures was shortsighted and irresponsible, as the ensuing debate showed. Original debate in *Inscriptions* 3 (July 2020): art 72. The interventions have now been published as *Coronavirus, Psychoanalysis, and Philosophy: Conversations on Pandemics, Politics, and Society*, eds. Fernando Castillón and Thomas Marchevsky (London: Routledge, 2021).

Chapter 6

1. Such an ontology can be found in Friedrich Kittler, as shown beautifully by Frédérique Vargoz in her unpublished PhD dissertation "Média et systèmes d'inscription dans la pensée de Friedrich Kittler: une lecture allemande de Jacques Lacan, Michel Foucault et Jacques Derrida" (PhD diss., Université de Grenoble Alpes, 2021).

Bibliography

Abbott, Matthew. *The Figure of This World: Agamben and the Question of Political Ontology*. Edinburgh: Edinburgh University Press, 2014.

Agamben, Giorgio. *The Coming Community*. Translated by Michael Hardt. Minneapolis: University of Minnesota Press, 2007.

———. *Homo Sacer: Sovereign Power and Bare Life*. Translated by Daniel Heller-Roazen. Stanford: Stanford University Press, 1998.

———. *Infancy and History: On the Destruction of Experience*. Translated by Liz Heron. London: Verso, 2007. First published 1978.

———. "The Invention of an Epidemic," *European Journal of Psychoanalysis*, February 26, 2020. Republished in *Where Are We Now? The Epidemic as Politics*, translated by Valeria Dani, 20–22. New York: Rowman and Littlefield, 2021.

———. *Language and Death: The Place of Negativity*. Translated by Karen Pinkus and Michel Hardt. Minneapolis: University of Minnesota Press, 1991.

———. *Means without End: Notes on Politics*. Translated by Vincenzo Binetti and Cesare Casarino. Minneapolis: University of Minnesota Press, 2000.

———. *The Omnibus Homo Sacer*. Stanford: Stanford University Press, 2017.

———. "On Potentiality." In *Potentialities*, translated by Daniel Heller-Roazen, 177–84. Stanford: Stanford University Press, 1999.

———. *The Open: Man and Animal*. Translated by Kevin Attell. Stanford: Stanford University Press, 2004. First published 2002.

———. *Stanzas: Word and Phantasm in Western Culture*. Translated by Ronald L. Martinez. Minneapolis: University of Minnesota Press, 1993.

———. *The Use of Bodies*. Translated by Adam Kotsko. Stanford: Stanford University Press, 2016.

———. *What Is Apparatus? and Other Essays*. Translated by David Kishik and Stefan Pedatella. Stanford: Stanford University Press, 2009.

Alombert, Anne. *Simondon et Derrida face aux questions de l'homme et de la technique: Ontogenèse et grammatologie dans le moment philosophique des années 1960*. Paris: Université de Nanterre, 2020.

Alston, Philip. *Extreme Poverty and Human Rights.* Report to the United Nations General Assembly, distr. 11. October 2019, A/74/493. https://undocs.org/A/74/493.

Anderson, Michael L. "Embodied Cognition: A Field Guide." *Artificial Intelligence* 149 (2003): 91–130.

Angier, Tom, and Anne Balansard. *Techné dans les dialogues de Platon: L'empreinte de la sophistique.* Sankt Augustin: Academia Verlag, 2001.

Aristotle. *The Complete Works of Aristotle.* Edited by Jonathan Barnes. 2 vols. Princeton, NJ: Princeton University Press, 1984.

Bacon, Francis. *The New Organon.* Edited by Lisa Jardine and Michael Silverstone. Cambridge Texts in the History of Philosophy. Cambridge: Cambridge University Press, 2000.

Badiou, Alain. *Manifeste pour la philosophie.* Paris: Seuil, 1989. Translated by Norman Madarash as *Manifesto for Philosophy* (Albany: State University of New York Press, 1999).

Bannet, Eve Tavor. *Structuralism and the Logic of Dissent: Barthes, Derrida, Foucault, Lacan.* Urbana: University of Illinois Press, 1989.

Bartoli, David Gé, and Sophie Gosselin. *Le toucher du monde: Techniques du naturer.* Paris: Dehors, 2019.

Bennett, Jane. *Vibrant Matter: A Political Ecology of Things.* Durham, NC: Duke University Press, 2010.

Bensaude-Vincent, Bernadette. *Les vertiges de la technoscience: Façonner le monde atome par atome.* Paris: La découverte, 2009.

Besnier, Jean-Michel. *Demain les posthumains: Le futur a-t-il encore besoin de nous?* Paris: Pluriel, 2012.

Bleeden, David. "One Paradigm, Two Potentialities: Freedom, Sovereignty and Foucault in Agamben's Reading of Aristotle's *dynamis*." *Foucault Studies* 10 (2010): 68–84.

de Boever, Arne. "The Allegory of the Cave: Foucault, Agamben, and the Enlightenment." *Foucault Studies* 10 (2010): 44–67.

de Boever, Arne, Alex Murray, John Roffe, and Ashley Woodward, eds. *Gilbert Simondon: Being and Technology.* Edinburgh: Edinburgh University Press, 2012.

Bonacci, Valeria. "Form-of-Life and Use in Homo Sacer." In *The Politics, Ethics, and Aesthetics of Inoperativity*, edited by Giovanni Marmont and German E. Primera. *Journal of Italian Philosophy* 3 (2020): 217–24.

Boyne, Roy. *Foucault and Derrida: The Other Side of Reason.* London: Unwin Hyman, 1990.

Bradley, Arthur. "In the Sovereign Machine: Sovereignty, Governmentality, Automaticity." *Journal for Cultural Research* 22, no. 3 (2018): 209–23.

———. *Originary Technicity: The Theory of Technology from Marx to Derrida.* New York: Palgrave Macmillan, 2011.

Braidotti, Rosi. *The Posthuman*. Cambridge: Polity Press, 2013.
Buchanan, Allen, Dan W. Brock, Norman Daniels, and Daniel Wikler. *From Chance to Choice: Genetics and Justice*. Cambridge: Cambridge University Press, 2000.
Buchanan, Brett. *The Animal Environments of Uexküll, Heidegger, Merleau-Ponty, and Deleuze*. Albany: State University of New York Press, 2008.
Bussolini, Jeffrey. "Critical Encounter Between Giorgio Agamben and Michel Foucault: Review of Recent Works of Agamben: *Il Regno e la Gloria: per una genealogia dell'economia e del governo*; *Il sacramento del linguaggio: Archeologia del giuramento*, and *Signatura rerum: Sul Metodo*." *Foucault Studies* 10 (November 2010): 108–43.
Butler, Judith. "Subjection, Resistance, Resignification: Between Freud and Foucault." In *The Psychic Life of Power: Theories in Subjection*. Stanford: Stanford University Press, 1997.
Campbell, Timothy. *Improper Life: Technology and Biopolitics from Heidegger to Agamben*. Minneapolis: University of Minnesota Press, 2011.
Canguilhem, Georges. *Knowledge of Life*. Translated by Stefanos Geroulanos and Daniela Ginsburg. New York: Fordham University Press, 2008. First published 1965.
Castillón, Fernando, and Thomas Marchevsky, eds. *Coronavirus, Psychoanalysis, and Philosophy: Conversations on Pandemics, Politics, and Society*. London: Routledge, 2021.
Cerullo, Michael A. "Uploading and Branching Identity." *Minds and Machines* 25 (2015): 17–36.
Choong, Kartina. "Death Before Organ Donation: Exploring the Chasm between Lay and Professional Knowledge." Unpublished conference presentation at the International Conference: Phenomenology of Medicine and Bioethics. Sweden: Södertörn University, June 13–15, 2018.
Clark, Andy, and David Chalmers. "The Extended Mind." *Analysis* 58 (1998): 1.
Colebrook Claire. "Ethics, Positivity, and Gender: Foucault, Aristotle, and Care of the Self." *Philosophy Today* 42, no. 1 (Spring 1998): 40–52.
Colony, Tracy. "Epimetheus Bound: Stiegler on Derrida, Life, and the Technological Conditions." *Research in Phenomenology* 41 (2011): 72–89.
———. "The Future of Technics." *Parrhesia* 27 (2017): 64–87.
Costa, Luiz. *Virtuality and Capabilities in a World of Ambient Intelligence: New Challenges to Privacy and Data Protection*. Cham: Springer Law, Governance and Technology Series, 2016.
Critchley, Simon. *Ethics, Politics, Subjectivity: Essays on Derrida, Levinas, and Contemporary French Thought*. London: Verso, 1999.
Deleuze, Gilles. *Foucault*. Translated by Seán Hand. London: Athlone Press, 1988.
———. "Post-scriptum sur les sociétés de contrôle." *L'autre journal* 1 (May 1990).
Deleuze, Gilles, and Félix Guattari. *Capitalisme et schizophrénie I: Anti-Œdipe*. Paris: Minuit, 1972. Translated by Robert Hurley, Mark Seem, and Helen Lande

as *Anti-Oedipus: Capitalism and Schizophrenia* (Minneapolis: University of Minnesota Press, 1983).

———. *A Thousand Plateaus: Capitalism and Schizophrenia*. Translated by Brian Massumi. Minneapolis: Minneapolis University Press, 1987.

van der Heiden, Gert-Jan. "Exile, Use, and Form-of-Life: On the Conclusion of Agamben's *Homo Sacer* Series." *Theory, Culture & Society* 372 (2020): 61–78.

van der Heiden, Gert-Jan, Karel Novotny, Inga Römer, and Lazlo Tengelyi, eds. *Investigating Subjectivity: Classical and New Perspectives*. Leiden: Brill, 2012.

Derrida, Jacques. *The Animal that Therefore I Am*. Translated by David Wills. New York: Fordham University Press, 2008. First published 2006.

———. *Dissemination*. Translated by Barbara Johnson. Chicago: University of Chicago Press, 1981.

———. *Faith and Knowledge*. In *Acts of Religion*, edited by Gil Anidjar. New York: Routledge, 2002.

———. *Force of Law: The Metaphysical Foundation of Authority*. In *Deconstruction and the Possibility of Justice*, edited by Drucilla Cornell, Michel Rosenfeld, and David Carlson. London: Routledge, 1992.

———. *Geschlecht 3: Sex, Race, Nation, Humanity*. Edited by Geoffrey Bennington, Katie Chenoweth, and Rodrigo Therezo. Translated by Katie Chenoweth and Rodrigo Therezo. Chicago: University of Chicago Press, 2020.

———. *Given Time 1: Counterfeit Money*. Translated by Peggy Kamuf. Chicago: University of Chicago Press, 2004.

———. "Heidegger's Ear: Philopolemology *Geschlecht 4*." Translated by John P. Leavey Jr. In *Reading Heidegger: Commemorations*, edited by John Sallis, 163–218. Bloomington: Indiana University Press, 1993.

———. *La vie la mort: Séminaire 1975–1976*. Paris: Seuil Bibliothèque, 2019. Translated by Pascale-Anne Brault and Michael Naas as *Life Death*. Chicago: University of Chicago Press, 2020.

———. *Mal d'Archive*. Paris: Galilée, 1995.

———. *Marges de la Philosophie*. Paris: Minuit, 1972. Translated by Alan Bass as *Margins of Philosophy*. Chicago: University of Chicago Press, 1982.

———. *Of Grammatology*. Translated by G. C. Spivak. Baltimore: Johns Hopkins University Press, 1976. First published 1967.

———. *Of Hospitality*. Translated by Rachel Bowlby. Stanford: Stanford University Press, 2000.

———. *Of Spirit, Heidegger, and the Question*. Translated by Rachel Bowlby. Chicago: University of Chicago Press, 2017.

———. *On the Name*. Translated by Ian McLeod. Stanford: Stanford University Press, 1993.

———. *On Touching: Jean-Luc Nancy*. Translated by Christine Irizarry. Stanford: Stanford University Press, 2005.

———. *Papier Machine*. Paris: Galilée, 2001.

———. *Psyche: Inventions of the Other, Vol. 1*. Translated by Catherine Porter. Stanford: Stanford University Press, 2007.
———. *Psyche 2: Inventions of the Other*. Edited by Peggy Kamuf. Stanford: Stanford University Press, 2008.
———. *Specters of Marx: The State of the Debt, the Work of Mourning, and the New International*. Translated by Peggy Kamuf. New York: Routledge, 1994.
———. *Without Alibi*. Translated by Peggy Kamuf. Stanford: Stanford University Press, 2002.
———. *Writing and Difference*. Translated by Alan Bass. London: Routledge, 2003. First published 1967.
Derrida, Jacques, and Bernard Stiegler. *Echographies of Television: Filmed Interviews*. Translated by Jennifer Bajorek. Cambridge: Polity Press, 2002.
Descartes, René. *Discourse on the Method and Meditations on First Philosophy*. Translated by Elizabeth F. Haldane. Edited by David Weissman. New Haven, CT: Yale University Press, 1996.
———. *Meditations on First Philosophy*. In *The Philosophical Works of Descartes*, translated by Elizabeth S. Haldane. Cambridge: Cambridge University Press, 1911.
———. *Œuvres et lettres*. Paris: Bibliothèque de la Pléiade, 1953.
———. *The World and Other Writings*. Translated and edited by Stephen Gaukroger. Cambridge: Cambridge University Press, 1998.
Descola, Philippe. *Beyond Nature and Culture*. Translated by Janet Lloyd. Chicago: University of Chicago Press, 2013.
Diderot, Denis. *D'Alembert's Dream*. In *Rameau's Nephew and D'Alembert's Dream*, edited and translated by Leonard Tancock. London: Penguin Classics, 1976.
Dirakis, Alexis. "Une anthropologie politique de la frontière, Réflexions à partir de l'anthropologie de Helmuth Plessner." *Le Débat* 1 (2016): 132–44.
Drach, Marcel, ed. *L'argent: Croyance, mesure, spéculation*. Paris: La découverte, |2004.
Dreyfus, Hubert. *What Computers Still Can't Do: A Critique of Artificial Reason*. Rev. ed. of the original 1972 edition. Cambridge, MA: MIT Press, 1992.
Dumouchel, Paul, and Jean-Pierre Dupuy. *Auto-organisation: De la physique eu politique*. Paris: Seuil, 1983.
Ellul, Jacques. *The Technological System*. Translated by Joachim Neugroschel. New York: Continuum, 1980. First published 1977.
———. *The Technological Society*. Translated by John Wilkinson. New York: Knopf, 1964.
"EU Guidelines on Ethics in Artificial Intelligence: Context and Implementation." European Parliamentary Research Service, PE 640.163, September 2019.
Fischer, Joachim. "Le noyau théorique propre à l'Anthropologie philosophique Scheler, Plessner, Gehlen." Translated by Matthieu Amat and Alexis Dirakis. *Trivium* 25 (2017).

de Fontenay, Elisabeth. *Le silence des bête: La philosophie à l'épreuve de l'animalité.* Paris, Fayard, 1998.

Foucault, Michel. *The Care of the Self, Vol. 3 of the History of Sexuality.* Translated by Robert Hurley. New York: Pantheon Books, 1986.

———. *Dits et écrits I 1954–1975.* Paris: Gallimard, 2001.

———. *Dits et écrits II 1976–1988.* Paris: Gallimard, 2017.

———. *Ethics: Subjectivity and Truth. The Essential Foucault 1.* Edited by Paul Rabinow. Translated by Robert Hurley and others. New York: New Press, 1997.

———. *Histoire de la folie à l'âce classique.* Paris: Gallimard, 1972.

———. *Introduction to Kant's Anthropology.* Edited and translated by Roberto Nigro and Kate Briggs. Semiotexte Foreign Agents Series. Cambridge, MA: MIT Press, 2008.

———. *La pensée du dehors.* Paris: Fata morgana, 1986. Translated by Jeffrey Mehlman and Brian Massumi as *Maurice Blanchot: The Thought from Outside*; Blanchot: *Michel Foucault as I Imagine Him.* New York: Zone Books, 1990.

———. *Le gouvernement de soi et des autres: Cours au collège de France, 1982–1983.* Paris: Gallimard Seuil, 2008. Translated by Graham Burchell as *The Government of Self and Others: Lectures at the College of France 1982–1983.* Basingstoke: Palgrave Macmillan, 2010.

———. *Les mots et les choses.* Paris: Gallimard, 1966. Translated by Alan Sheridan as *The Order of Things: An Archaeology of the Human Sciences.* London: Routledge, 2005.

———. *L'herméneutique du sujet: Cours au collège de France 1981–1982.* Paris: Gallimard, 2001. Translated by Graham Burchell as *The Hermeneutics of the Subject: Lectures at the College of France 1918–1982.* New York: Palgrave Macmillan, 2005.

———. *Michel Foucault: Beyond Structuralism and Hermeneutics.* Edited by Hubert L. Dreyfus and Paul Rabinow. Chicago: University of Chicago Press, 1982.

———. *Michel Foucault, Politics, Philosophy, Culture: Interviews and Other Writings 1977–1984.* Edited by Lawrence D. Kritzman. New York: Routledge, 1990.

———. "Society Must Be Defended." *Lectures at the Collège de France 1975–1976.* London: Penguin Classics, 2020.

———. *Surveiller et punir: Naissance de la prison.* Paris: Gallimard tel, 1975.

———. *Technologies of the Self: A Seminar with Michel Foucault.* Edited by Luther H. Hutton, Huck Gutman, and Patrick H. Hutton. Amherst: University of Massachusetts Press, 1988.

———. *The Use of Pleasure, Vol. 2 of the History of Sexuality.* Translated by Robert Hurley. New York: Vintage Books, 1990.

———. *The Will to Knowledge, Vol. I of the History of Sexuality.* Translated by Robert Hurley. New York: Pantheon Books, 1978.

Freud, Sigmund. *Totem and Taboo.* London: Ark Paperback, 1983. First published 1913.

Friedman, R. Z. "Hypocrisy and the Highest Good: Hegel on Kant's Transition from Morality to Religion." *Journal of the History of Philosophy* 24, 4 (1986): 503–33.
Garrido Wainer, Juan Manuel. *La formation des formes*. Paris: Galilée, 2008.
Gasché, Rodolphe. *The Tain of the Mirror: Derrida and the Philosophy of Reflection*. Cambridge, MA: Harvard University Press, 1988.
Gehlen, Arnold. *Der Mensch: Seine Natur und seine Stellung in der Welt*. Wiebelsheim: AULA Verlag, 2009. First published 1950.
———. *Man in the Age of Technology*. Translated by Patricia Lipscomb. New York: Columbia University Press, 1980.
Gros, Frédéric. "Sujet moral et éthique chez Foucault." *Archives de philosophie* 2 (2002): tome 65, 229–37.
Gros, Frédéric, and Carlos Lévy, eds. *Foucault et la philosophie antique*. Paris: Kimé, 2003.
Guattari, Félix. "Machine and Structure." In *Molecular Revolution: Psychiatry and Politics*, translated by Rosemary Sheed, 111–19. New York: Penguin, 1984.
———. *The Three Ecologies*. Translated by Ian Pindar and Paul Sutton. London: Athlone Press, 2000. First published 1989.
Guerra Filho, Willis Santiago. "Luhmann and Derrida: Immunology and Autopoiesis." In *Luhmann Observed*, edited by Anders La Cour and Andreas Philippopoulos-Mihailopoulos. London: Palgrave Macmillan, 2013.
Habermas, Jürgen. *Die Zukunft der menschlichen Natur: Auf dem Weg zu einer liberalen Eugenik?* Frankfurt am Main: Suhrkamp, 2001.
Hacking, Ian. "Experimentation and Scientific Realism." *Philosophical Topics* 13 (1982): 154–72.
Hagner, Michael, and Erich Hörl. *Die Transformation des Humanen: Beiträge zur Kulturgeschichte der Kybernetik*. Frankfurt am Main: Suhrkamp, 2008.
Hansen, Mark B. "'Realtime Synthesis' and the Différance of the Body: Technocultural Studies in the Wake of Deconstruction." *Culturemachine* 6 (2004).
Haraway, Donna. *The Companion Species Manifesto: Dogs, People, and Significant Otherness*. Chicago: Prickly Paradigm Press, 2003.
———. *Simians, Cyborgs, and Women: The Reinvention of Nature*. New York: Routledge, 1991.
Harrer, Sebastian. "The Theme of Subjectivity in Foucault's Lecture Series *L'Herméneutique du sujet*." *Foucault Studies* 2 (2005): 75–96.
Hayles, N. Katherine. *How We Became Posthuman: Virtual Bodies in Cybernetics, Literature, and Informatics*. Chicago: University of Chicago Press, 1999.
———. *Unthought: The Power of the Cognitive Nonconscious*. Chicago: University of Chicago Press, 2017.
Hegel, G. W. F. *The Phenomenology of Spirit*. Translated by Terry Pinkard. Cambridge: Cambridge University Press, 2018.
Heidegger, Martin. *Aristoteles, Metaphysik theta 1–3, Vom Wesen und Wirklichkeit der Kraft*. GA Bd 33. Frankfurt am Main: Vittorio Klostermann, 1990.

———. *Die Grundbegriffe der Metaphysik: Welt, Endlichkeit, Einsamkeit*. GA Bd 1929–1930. Frankfurt am Main: Vittorio Klostermann, 1992. First published 1913. Translated by William McNeill and Nicholas Walker as *The Fundamental Concepts of Metaphysics*. Bloomington: Indiana University Press, 2001.

———. "Die Zeit des Weltbildes." In *Holzwege*. Frankfurt am Main: Vittorio Klostermann, 1980. Translated by William Levitt as "The Age of the World Picture." In *The Question Concerning Technology and Other Essays*. New York: Garland Publishing, 1977.

———. *Elucidations of Hölderlin's Poetry*. Translated by Keith Hoeller. New York: Humanity Books, 2000.

———. *Introduction to Metaphysics*. Translated by Gregory Fried and Richard Polt. New Haven, CT: Yale University Press, 2000.

———. *Pathmarks*. Translated by Thomas Sheehan and William McNeill. Cambridge: Cambridge University Press, 1998.

———. *Sein und Zeit*. Tübingen: Max Niemeyer, 1984. Translated by John Macquarrie and Edward Robinson as *Being and Time*. Oxford: Blackwell, 1985.

———. "Überwindung der Metaphysik." In *Vorträge und Aufsätze*. Stuttgart: Neske, 1994. Translated by Joan Stambaugh as "Overcoming Metaphysics." In *The End of Philosophy*. Chicago: University of Chicago Press, 2003.

Helen, Ilpo. "Elämä seksuaalisuudessa." In Foucault, *Seksuaalisuuden historia*, translated by Kaisa Sivenius. Helsinki: Gaudeamus, 1998.

Herbrechter, Stefan. *Posthumanism: A Critical Analysis*. London: Bloomsbury, 2013.

Herbrechter, Stefan, and Ivan Callus, eds. *Discipline and Practice: The (Ir)resistibility of Theory*. Lewisburg, PA: Bucknell University Press, 2004.

Horkheimer, Max, and Theodor Adorno. *Dialectic of Enlightenment: Philosophical Fragments*, translated by Edmund Jephcott. Stanford: Stanford University Press, 2001. First published 1947.

Hörl, Erich. *Sacred Channels: The Archaic Illusion of Communication*. Translated by Nils Schott. Amsterdam: Amsterdam University Press, 2018. The German original edition: *Die Heiligen Kanäle: Die archaische Illusion der Kommunikation*. Zurich: Diaphanes, 2005.

———. "The Technological Condition." *Parrhesia* 22 (2015): 1–15.

Hörl, Erich, and James Burton. *General Ecology: The New Ecological Paradigm*. London: Bloomsbury, 2017.

Hottois, Gilbert. *Le transhumanisme est-il un humanism*. Brussels: Académie Royale de Belgique, 2014.

———. *Species technica*, suivi d'un *Dialogue vingt ans plus tard*. Paris: Vrin, 2002.

Hui, Yuk. *Recursivity and Contingency*. London: Rowman and Littlefield, 2019.

Husserl, Edmund. *Ideas Pertaining to a Pure Phenomenology and to Phenomenological Philosophy. First Book: General Introduction to a Pure Phenomenology*. Translated by F. Kersten. The Hague: Nijhoff, 1982. First published 1913.

Jonas, Hans. *Das Prinzip Leben: Ansätze zu einer philosophischen Biologie*. Frankfurt am Main: Suhrkamp, 1994.

———. *Technik, Medizin und Ethik*. Frankfurt am Main: Suhrkamp, 1985.
———. *Wissenschaft als persönliches Erlebnis*. Göttingen: Vandenhoeck and Ruprecht, 1987.
Kant, Immanuel. *Idée d'une histoire universelle au point de vue cosmopolitique*. Paris: Bordas, 1988.
———. *Kritik der reinen Vernunft 1–2*. Frankfurt am Main: Suhrkamp, 1995.
———. *Political Writings*. Translated by H. B. Nisbet. Cambridge: Cambridge University Press, 1991.
Kapp, Ernst. *Elements of a Philosophy of Technology 1877*. Minneapolis: University of Minnesota Press, 2018.
Kotsko, Adam, and Carlo Salzani, eds. *Agamben's Philosophical Lineage*. Edinburgh: Edinburgh University Press, 2017.
Koyré, Alexandre. *From the Closed World to the Infinite Universe*. Baltimore: Johns Hopkins University Press, 1968.
Krell, David Farrell. *Phantoms of the Other: Four Generations of Derrida's Geschlecht*. Albany: State University of New York Press, 2015.
Krijnen, Christian. "Kants Subjektstheorie und die Grundlegung einer Philosophischen Anthropologie." *Zeitschrift für philosophische Forschung*, Bd 62, H 2 (April–June 2008): 254–73.
Kurzweil, Ray. *The Singularity Is Near: When Humans Transcend Biology*. Penguin Books, 2005.
de la Durantaye, Leland. *Giorgio Agamben: A Critical Introduction*. Stanford: Stanford University Press, 2009.
de la Mettrie, Julien Offray. In *Machine Man and Other Writings*, edited by Ann Thomson. Cambridge: Cambridge University Press, 2012.
Lacoue-Labarthe, Philippe. *Typography: Mimesis, Philosophy, Politics*. Translated by Christopher Fynsk. Cambridge, MA: Harvard University Press, 1989.
Lacoue-Labarthe, Philippe, and Jean-Luc Nancy, eds. *Les fins de l'homme: À partir du travail de Jacques Derrida*. Paris: Galilée, 1981.
Latour, Bruno. *Facing Gaia: Eight Lectures on the New Climatic Regime*. Translated by C. Porter. Cambridge: Polity Press, 2017.
Lawlor, Leonard. *From Violence to Speaking Out: Apocalypse and Expression in Foucault, Derrida, and Deleuze*. Edinburgh: Edinburgh University Press, 2016.
———. *This Is Not Sufficient: An Essay on Animality and Human Nature in Derrida*. New York: Columbia University Press, 2007.
Lawlor, Leonard, and John Nale, eds. *The Cambridge Foucault Lexicon*. New York: Cambridge University Press, 2014.
Leroi-Gourhan, André. *L'homme et la matière*. Paris: Albin Michel, 1943.
———. *Le geste et la parole I: Technique et langage*. Paris: Albin Michel, 1964. Translated by Anna Bostock Berger as *Gesture and Speech*. Cambridge, MA: MIT Press, 1993.
———. *Le geste et la parole II: La mémoire et les rythmes*. Paris: Albin Michel, 1965.
———. *Milieu et technique*. Paris: Albin Michel, 1945.

Lévi-Strauss, Claude. *The Elementary Structures of Kinship*. London: Eyre and Spottiswoode, 1970. First published 1949.
Lilley, Stephen. *Transhumanism and Society: The Social Debate over Human Enhancement*. SpringerBriefs in Philosophy. Dordrecht: Springer, 2013. https://doi.org/10.1007/978-94-007-4981-8.
Lindberg, Susanna. "Derrida's Quasi-Technique." *Research in Phenomenology* 46 (2016): 369–89.
———. "On Prosthetic Existence: What Differentiates Deconstruction from Transhumanism and Posthumanism." In *Humanism and Its Discontents: The Rise of Transhumanism and Posthumanism*, edited by Paul Jorion. New York: Palgrave Macmillan, 2022.
———. "Technics of Space, Place, and Displace." In *Azimuth* 10, Special Issue "Intersections: At the Technophysics of Space" (2018): 27–44.
———. *Techniques en Philosophie*. Paris: Hermann, 2020.
Lock, Margaret. *Twice Dead: Organ Transplantation and the Reinvention of Death*. Berkeley: University of California Press, 2002.
Louden, Robert B. *Kant's Human Being: Essays on His Theory of Human Nature*. Oxford: Oxford University Press, 2011.
Luhmann, Niklas. "Deconstruction as Second-Order Observing." *New Literary History* 24, no. 4 (1993).
Lukkari, Hanna. "Law, Politics, and Paradox: Orientations in Legal Formalism." PhD diss., University of Helsinki, 2020.
Lyotard, Jean-François. *The Inhuman: Reflections on Time*. Translated by Geoffrey Bennington and Rachel Bowlby. Cambridge: Polity Press, 1991. First published 1988.
Manchev, Boyan. *La métamorphose et l'instant: La désorganisation de la vie*. Paris: La Phocide, 2009.
Maturana, Humberto, and Francisco Varela. *Autopoiesis and Cognition: The Realization of the Living*. London: Reidel, 1980.
Mbembe, Achille. *Necropolitics*. Translated by Steve Corcoran. Durham, NC: Duke University Press, 2019.
McCumber, John. *Philosophy and Freedom*. Bloomington: Indiana University Press, 2000.
Megill, Allan. *Prophets of Extremity: Nietzsche, Heidegger, Foucault, Derrida*. Berkeley: University of California Press, 1985.
Michelini, Francesca, and Kristian Köchy, eds. *Jacob Uexküll and Philosophy*. New York: Routledge, 2020.
Milchman, Alan, and Alan Rosenberg. "The Aesthetic and Ascetic Dimensions of an Ethics of Self-Fashioning: Nietzsche and Foucault." *Parrhesia* 2 (2007).
Mills, Nicolás García. "Realizing the Good: Hegel's Critique of Kantian Morality." *European Journal of Philosophy* 26, 1 (2017): 195–212.
More, Max, and Natascha Vita-More, eds. *The Transhumanist Reader*. Oxford: Wiley-Blackwell, 2013.

de Mul, Jos, ed. *Plessner's Philosophical Anthropology*. Amsterdam: Amsterdam University Press, 2014.
Mumford, Lewis. *Technics and Civilization*. New York: Harbinger Books, 1964.
Murray, Alex, and Jessica Whyte, eds. *The Agamben Dictionary*. Edinburgh: Edinburgh University Press, 2011.
Naas, Michael. *Miracle and Machine: Jacques Derrida and the Two Sources of Religion, Science, and Media*. New York: Fordham University Press, 2012.
Nancy, Jean-Luc. *L'intrus*. Paris: Galilée, 2000. Translated by Richard Rand as *Corpus*. New York: Fordham University Press, 2008. Translated by Susan Hanson: http://www.maxvanmanen.com/files/2014/10/Nancy-LIntrus.pdf.
Nègre, Fabien. "L'esthètique de l'existence dans le dernier Foucault." *Raison Présente* 118 (1996): 47–71.
Neyrat, Frédéric. *Homo labyrinthus: Humanisme, antihumanisme, posthumanisme*. Paris: Éditions Dehors, 2015.
Nietzsche, Friedrich. *The Genealogy of Morals*. Translated by Horace B. Samuel. Edinburgh: Foulis, 1913.
Ojakangas, Mika. "Impossible Dialogue on Bio-power: Agamben and Foucault." *Foucault Studies* 2 (May 2005).
Osmo, Pierre. Introduction to the French Translation of *Les degrés de l'organique et l'Homme*. Paris: Gallimard, 2017.
Parikka, Jussi. *Insect Media: An Archaeology of Animals and Technology*. Minneapolis: University of Minnesota Press, 2010.
Patocka, Jan. *Heretical Essays in the Philosophy of History*. Translated by Erazim Kohák. Chicago: Open Court, 1996.
Pichot, André. *L'eugénisme, ou les généticiens saisis par la philantropie*. Paris: Hatier, 1995.
———. *La société pure: De Darwin a Hitler*. Paris: Flammarion, 2000.
Pilsch, Andrew. *Transhumanism: Evolutionary Futurism and the Human Technologies of Utopia*. Minneapolis: University of Minnesota Press, 2017.
Plas, Guillaume, and Gerard Raulet, eds. *Philosophische Anthropologie nach 1945: Rezeption und Fortwicklung*. Nordhansen: Verlag Traugott Bautz Gmbh, 2014.
Plato. *Complete Works*. Edited by John M. Cooper. Indianapolis: Hackett, 1997.
Plessner, Helmuth. *Diesseits der Utopie*. Düsseldorf: Eugen Diederichs Verlag, 1966.
———. *Levels of Organic Life and the Human: An Introduction to Philosophical Anthropology*. Translated by Millay Hyatt. New York: Fordham University Press, 2019.
Quessada, Dominique. "De la sousveillance: La surveillance globale, un nouveau mode de gouvernementalité." *Multitudes* 40 (2010): 54–59.
Ranisch, Robert, and Stefan Lorenz Sorgner. *Post- and Transhumanism: An Introduction*. Bern: Peter Lang, 2014.
Read, Jason. *The Politics of Transindividuality*. Leiden: Brill, 2016.

Revel, Judith. "The Materiality of the Immaterial: Foucault, Against the Return of Idealisms and New Vitalisms." *Radical Philosophy* 149 (May-June 2008): 33-38.

———. "Michel Foucault: repenser la technique." In *Tracés: Revue des sciences humaines* 16 (2009): 139-49.

Rieder, Bernhard. *Engins of Order: A Mechanology of Algorithmic Techniques*. Amsterdam: Amsterdam University Press, 2020.

Roberts, Ben. "Stiegler Reading Derrida: The Prosthesis of Deconstruction in Technics." *Postmodern Culture* 16, no. 1 (2005).

Rogozinski, Jacob. *Kanten: Esquisses kantiennes* Paris: Kimé, 1996.

Rorty, Richard. *Essays on Heidegger and Others*. Cambridge: Cambridge University Press, 2010.

Rosenblueth, Arturo, Norbert Wiener, and Julian Bigelow. "Behavior, Purpose and Teleology." *Philosophy of Science* 10 (1943): 18-24.

Rossi, Paolo. *The Birth of Modern Science*. Oxford: Blackwell, 2001.

Rouvroy, Antoinette, and Thomas Berns. "Détecter et prévenir: De la digitalisation des corps et de la docilité des normes." 2009. http://www.crid.be/pdf/public/6243.pdf.

———. "Le nouveau pouvoir statistique, Ou quand le contrôle s'exerce sur un réel normé, docile et sans événement car constitué de corps 'numériques.'" *Multitudes* 40 (2010): 88-103.

Sadin, Éric. *L'Humanité augmentée: L'administration numérique du monde*. Paris: L'échappée, 2013.

———. *La vie algorithmique: Critique de la raison numérique*. Paris: L'échappée, 2015.

———. *Surveillance globale*. Paris: Climats, 2009.

Sallis, John. *Chorology: On Beginning in Plato's Timaeus*. Bloomington: Indiana University Press, 1995.

Scheler, Max. *Die Stellung des Menschen im Kosmos*. Darmstadt: Otto Reichl Verlag, 1930. Translated by Manfred S. Frings as *The Human Place in the Cosmos* (Evanston, IL: Northwestern University Press, 2009).

Searle, John. "Minds, Brains, and Programs." *Behavioral and Brain Sciences* 3 (1980): 417-57.

Sforzini, Arianna. *Michel Foucault, une pensée du corps*. Paris: PUF, 2014.

Sharp, Lesley A. *Strange Harvest: Organ Transplants, Denatured Bodies, and the Transformed Self*. Berkeley: University of California Press, 2006.

Sharpe, Matthew. "Critique as Technology of the Self." *Foucault Studies* 2 (2005): 97-116.

Shildrick, Margrit. "Microchimerism, Immunity, and Temporality: Rethinking the Ecology of Life and Death." *Australian Feminist Studies* 34, no. 99 (2019): 10-24.

Simondon, Gilbert. *Du mode d'existence des objets techniques*. Paris: Aubier, 2012. Translated by Cécile Malaspina and John Rogove as *On the Mode of Existence of Technical Objects* (Minneapolis: Univocal Publishing, 2017).

———. *Individuation à la lumière des notions de forme et d'information*. Grenoble: Jérôme Millon, 2013.

———. *Sur la technique 1953–1983*. Paris: PUF, 2014.

Sloterdijk, Peter. *Regeln für den Menschenpark*. Frankfurt am Main: Suhrkamp Verlag, 1999.

———. *You Must Change Your Life*. Malden, MA: Polity Press, 2013.

Snok, Anke. "Agamben's Foucault: An Overview." *Foucault Studies* 10 (2010): 44–67.

Sommer, Christian. "Approches du vivant entre anthropologie et phénoménologie." https://publishup.uni-potsdam.de/opus4-ubp/frontdoor/deliver/index/docId/39590/file/ebke_S59-76.pdf.

Stengers, Isabelle. *L'invention des sciences modernes*. Paris: La découverte, 1993.

Stiegler, Bernard. *Automatic Society Vol. 1: The Future of Work*. Translated by Dan Ross. Cambridge: Polity Press, 2016.

———. "Derrida and Technology: Fidelity at the Limits of Deconstruction and the Prosthesis of Faith." In *Jacques Derrida and the Humanities: A Critical Reader*, edited by Tom Cohen. Cambridge: Cambridge University Press, 2001.

———. *States of Shock: Stupidity and Knowledge in the 21st Century*. Translated by Dan Ross. Cambridge: Polity Press, 2015.

———. *Taking Care of Youth and the Generations*. Translated by Stephen Barker. Stanford: Stanford University Press, 2010.

———. *Technics and Time 1: The Fault of Epimetheus*. Translated by Richard Beardsworth and Richard Collins. Stanford: Stanford University Press, 1998.

———. *Technics and Time 2*. Translated by Stephen Barker. Stanford: Stanford University Press, 2009.

———. *Technics and Time 3: Cinematic Time and the Question of Malaise*. Translated by Stephen Barker. Stanford: Stanford University Press, 2011.

Svenaeus, Fredrik. "To Die Well: The Phenomenology of Suffering and End of Life Ethics." *Medicine, Health Care, and Philosophy* 23 (2020): 335–42.

Szendy, Peter. *Kant in the Land of Extraterrestrials: Cosmopolitical Philosofictions*. New York: Fordham University Press, 2013.

———. *Phantom Limbs: On Musical Bodies*. New York: Fordham University Press, 2015.

Teubner, Gunther. "Economics of Gift: Positivity of Justice; The Mutual Paranoia of Jacques Derrida and Niklas Luhmann." *Theory, Culture & Society* 181 (2001): 29–47.

Thornton, Edward. "The Rise of the Machines: Deleuze's Flight from Structuralism." *Southern Journal of Philosophy* 55, no. 4 (2017).

Turing, Alan. "Computing Machinery and Intelligence." *Mind* 59 (1950): 433–60.

von Uexküll, Jacob. *A Stroll through the Worlds of Animals and Men: A Picture Book of Invisible Worlds*. Translated by Claire H. Schiller. *Semiotica* 89, no. 4 (1992): 319–91.

Varela, Francisco. "Intimate Distances: Fragments for a Phenomenology of Organ Transplantation." *Journal of Consciousness Studies* 8, no. 5–7 (2001): 259–71.

Vargoz, Frédérique. "Média et systèmes d'inscription dans la pensée de Friedrich Kittler: Une lecture allemande de Jacques Lacan, Michel Foucault et Jacques Derrida." PhD diss., Université de Grenoble Alpes, 2021.

Villani, Cédric. "For a Meaningful Artificial Intelligence: Towards a French and European Strategy." Parliamentary Mission assigned by the French Government, 2018.

Viveiros de Castro, Eduardo. *Cannibal Metaphysics*. Translated by Peter Skafish. Minneapolis: Univocal Publishing, 2014. First published 2011.

———. "Perspectivisme et multinaturalisme en Amérique indigène." *Journal des anthropologues* (2014): 161–81.

Waldenfels, Bernhard. *Bruchlinien der Erfahrung*. Frankfurt am Main: Suhrkamp, 2002.

Waltham-Smith, Naomi. "The Use of Ears: Agamben Overhearing Derrida Overhearing Heidegger." *Parrhesia* 33 (2020): 113–49.

Warminski, Andrzej. "Spectre Shapes: The Body of Descartes?" *Qui Parle* 6, no. 1 (1992): 93–112.

Wills, David. *Dorsality: Thinking Back through Technology and Politics*. Minneapolis: University of Minnesota Press, 2008.

Witteriede, Heinz. *Eine Einführung in die Philosophische Anthropologie: Max Scheler, Helmuth Plessner, Arnold Gehlen*. Frankfurt am Main: Peter Lang, 2009.

Wolfe, Cary. *What Is Posthumanism?* Minneapolis: University of Minnesota Press, 2010.

———, ed. *Zoontologies: The Question of the Animal*. Minneapolis: University of Minnesota Press, 2003.

Zarka, Yves Charles. "Foucault et l'idée d'une histoire de la subjectivité: Le moment moderne." *Archives de philosophie* 2 (2002): tome 65, 255–67.

Zittel, Claus, ed. *Philosophes of Technology. Francis Bacon and His Contemporaries*. Leiden: Brill, 2008.

Zuboff, Shoshana. *The Age of Surveillance Capitalism: The Fight for a Human Future at the New Frontier of Power*. London: Profile Books, 2019.

———. "Big Other: Surveillance Capitalism and the Prospects of an Information Civilization." *Journal of Information Technology* 30 (2015): 75–89.

Index

affectivity, 23-24
Agamben, Giorgio, 11-12, 48, 202, 228-257, 268
algorithms, 21, 139-140, 210-212, 225-226, 283n17, 309n79
animal, animality, 25-31, 64, 82-88, 93-97, 134, 145-147, 151-154, 235, 247, 257, 268, 285n37, 285n39
anthropology, 24-30, 36, 66; as philosophical question, 32-34, 151-152, 155-156, 203-207; see also Philosophical Anthropology
anthropotechnics, 2-3, 15-24, 2-3, 36, 88, 263-264, 266, 269, 281-282n2
antihumanism, 12, 267-268
apparatus, 122, 126, 228-229, 232-233, 239-243, 249-250, 312n121
Aristotle, 41-42, 44-46, 85, 95-98, 129, 245-246, 307-308n53
artificial intelligence, 22, 29, 69-70, 139-140, 195-196, 211
autoimmunization, 169-173, 178-179
autopoiesis 68-69, 104, 170-172, 194

Bacon, Francis, 52-54
bare life, 234-238, 141-142 147, 250
bioethics, 16-20
biology, 68-69, 80-85, 88, 90, 164-174

bio-technics, 2-3, 12, 40, 62, 73-76, 113, 192, 228, 259261-263, 269-277
biotechnology, 4, 259, 264
birth, 18-20

calculation, 42, 66, 106, 208, 210-212
Canguilhem, Georges, 56, 61, 164-165
code, 64-65, 67-68, 73-74, 159, 165, 191, 211, 261, 267
cognition, 21-23, 67, 195; embodied, 22; situated, 22
community, 88, 104-105, 131-132, 200-202, 204, 219, 220, 222, 230, 235, 238, 242, 245, 249, 253, 256
computation, 225
cybernetics, 66, 68-70, 72, 83, 99, 111, 152, 167, 170, 194-195

Darwin, Charles, 82-83, 95, 164
death, 16-18, 148-149, 166-168, 170, 179, 182, 217, 219, 231-232
death of man, *see* end of man
Derrida, Jacques, 11-12, 22-23, 27-28, 43, 110-114, 140-195, 207, 234, 240, 257, 267-268
Descartes, René, 30-31, 51-52, 54-58, 89, 140-142
Descola, Philippe, 25-27
digitality, 21, 23-24, 139, 192-194, 201-202, 214-216, 224-227

345

discipline, 49, 59, 116, 118–122, 128, 137
dispositif, 9, 115, 124, 126–128, 133–135, 147, 154, 193, 232–233, 243

education, 19, 41, 46–49, 61, 75, 88, 116, 130, 134, 139, 182, 193, 199, 223, 227, 263; *Bildung*, 96, 100
end of man, 8, 112–113, 125
environmentality, 1, 65, 73–75
extended mind, 22, 322n23

form-of-life, 236–238, 240–242
Freud, Sigmund, 8, 25
Foucault, Michel, 8–9, 12, 32–33, 35, 41, 46–49, 110–144, 147, 155, 190–196, 226, 228–239, 241, 243, 145–147, 267

Gehlen, Arnold, 11, 77, 80, 82, 88–89, 104–106, 296–297n2

Hayles, N. Katherine, 10, 23, 64, 68–69, 72, 194–196, 211
Hegel, Georg Wilhelm Friedrich, 32–34, 85, 112, 159–160
Heidegger, Martin, 3, 11–12, 25, 40–41, 60, 80–82, 89–110, 266–267; *Being and Time*, 81, 89–93, 102, 105; *Dasein*, 17, 24, 34–35, 90–97, 105–107, 147, 175, 180–182, 207–209, 216–219, 237, 266, 318n200; *Ge-Stell*, 9, 60, 81, 98–101, 107, 126–127, 147, 230, 232, 257, 272–273; *The Fundamental Concepts of Metaphysics*, 93–99; "The Question Concerning Technology," 98–101, 105; *Zeug*, 92, 95, 97, 147, 207, 216, 243
homelessness, unhomeliness, 81, 86, 106–107

human being, the human, 1–12, 15–35, 75–77, 79–81, 197–200, 260–268; according to Agamben, 229–231, 235, 238, 251, 257; according to Derrida, 143–144, 152–156, 164; according to Foucault, 115, 138; according to Heidegger, 93–103; according to Stiegler, 202–207, 217, 228, 251, 257; and bio-technics, 79; and machine, 59; in Philosophical Anthropology, 85–91, 97, 103–109; in poststructuralism, 109–114, 190–192
humanism, 7–12, 29, 109, 143, 197–200
humanity, humankind, 1–12, 15–37, 75–77, 143–144, 197–198, 259–260
Husserl, Edmund, 34
hyperindustrial society, 61, 203, 224, 256
hypermodernity 61–62

information, 67–71, 111, 165–166, 171–172, 202, 211, 261
information technologies, 2, 61–70, 201, 263–264
inhumanity, the inhuman, 12, 29, 76, 197–198, 235, 251–252, 263
instrument, 1–2, 49–55, 57, 93, 99, 122, 200, 206, 229, 243–250, 268–269, 274–275
instrumentality, 1, 10, 12, 22, 39, 48, 106, 144, 147, 157–158, 208, 229, 243
it, instead of he, she or they, 4–5

Kant, Emmanuel, 30–34
khora, 12, 145–146, 162–163, 184–185, 189, 191, 193, 212, 268

Index   347

Leroi-Gourhan, André, 15, 22, 205–206
Lévi-Strauss, Claude, 25, 66
life, 2, 11–13, 12–13, 19–21, 46–49,
    61–63, 68–69, 72, 80–81, 111,
    173–174, 269–276
Lyotard, Jean-François, 29, 76,
    197–198, 251, 263

machine, 21, 29, 39–61, 63–70, 75,
    126–127, 142–145, 154, 170–171,
    194–196, 201, 230, 260–261,
    276; Agamben's concept of,
    235–238, 249–250; autopoietic,
    171; computational, 23, 66–70,
    210–212, 225; Deleuze's idea of,
    63–65, 123–124; Derrida's metaphor
    of, 143–145, 151, 155, 157–161,
    176, 178; in comparison with
    organism, 52, 55–59, 70, 72, 83,
    111; in comparison with society,
    59–61, 124, 127, 137–138, 236,
    238, 240; Stiegler's use of, 210–215;
    thermodynamic 214–215;
Mumford, Lewis, 50–52, 60

Nancy, Jean-Luc, 148, 262–263, 272
negativity, 11–12, 79, 81, 105–106,
    110, 136–137, 269
nothingness, 81, 86, 104–107, 109–
    111, 137, 154–155, 190, 266–267,
    269, 299n22; *nichtigkeit*, 86–87

ontology, 12, 27, 58, 64–65, 70,
    90–91, 128, 145, 162, 181, 184, 191,
    234–239, 261–262, 268
organ, 16–18, 20, 50–52, 57, 83,
    147–148, 246
organism, 50–52, 62–63, 68–72,
    82–89, 152, 167–170, 178–179,
    260–261
organon, organum, 50–52

originary technicity, 3, 11, 37, 77, 79,
    81, 88, 106, 109, 111, 143, 145–146,
    155, 158, 163–164, 168, 174,
    188–192, 252–253, 262, 265–266,
    269–272

personal identity: affected by digital
    technologies, 20–21
pharmakon, 138, 145, 149–151, 170,
    176, 210, 214, 231, 144, 253, 268,
    273
philosophical anthropology 25, 82–89,
    97–98, 101–107, 109–110, 190, 204,
    266, 196–197n2
Plato, 22–23, 41–49, 79, 114, 149–150,
    162–164, 195
Plessner, Arnold, 80, 82–83, 85–88,
    90–91, 93–94, 98, 101–109, 155
positionality, 85–87, 91, 94
posthumanism, 5–11, 35–36
postindustrial society, 61
poststructuralism, 6, 110–113
prosthesis, 23, 146–155, 172, 175,
    178–179, 205, 208–209, 217–223,
    268

robot, 29, 59, 72

Scheler, Max, 11, 80, 82–86, 90, 93,
    296–297n2
self-techniques, 127–132, 139
Simondon, Gilbert, 70–74, 144, 153,
    209, 212–216, 221–222, 264, 275
skill, 2, 12, 40–49, 59, 75, 130,
    228–130, 260–263, 270, 272, 276,
    288n1
Stiegler, Bernard, 11–12, 23, 41, 61,
    79, 138–140, 174, 188, 198–199,
    202–228, 250–257, 268
subject, subjectivity, 3–4, 30–34, 37,
    60, 63, 72, 75–76, 89–90, 109–113,

subject, subjectivity *(continued)*
 115–116, 122–123, 125, 142–144,
 164, 174, 181, 203, 216, 228,
 231–232, 259–265
subjection, 115–127
subjectivation, 114–119, 124, 127–139,
 147, 155, 229, 231–233, 240–242,
 249–250
supplementarity, 147, 152–156, 164,
 167–168, 179

technics: concept of, 39–40
techne, 39–49, 58, 75, 97–101, 107,
 114, 229, 253, 262, 288–289n1;
 techne tou biou, 41, 46, 48–49, 114,
 117, 127–132
technology: term, 39–40, 288–289n1
techno-nature, 62, 74–75, 272
technical or technological humanity,
 1–3, 40, 264–266
tele-technology, 113, 146, 169,
 174–179, 186–187, 189, 193, 207

television, 187, 224–225
transhumanism, 5–11, 35–36
Turing, Alan, 21, 67, 194

Uexküll, Jacob von, 82–84, 91, 93–95,
 164, 271
unthought, 23, 68, 135–136, 194–196,
 211–212
use, 11–13, 30, 99–100, 130–132, 134,
 147–148, 151–153, 193, 200, 210,
 213–215, 227, 229–233, 239–250,
 254–257, 260, 268, 270, 273

Viveiros de Castro, Eduardo, 26

Wiener, Norbert, 67–70
writing, 22–23, 43, 49, 67, 113–114,
 138, 140–146, 149–169, 173–175,
 179–180, 182, 187–188, 192,
 194–195, 205, 207, 209, 210,
 224, 234

www.ingramcontent.com/pod-product-compliance
Ingram Content Group UK Ltd.
Pitfield, Milton Keynes, MK11 3LW, UK
UKHW041922140426
5217IPUK00014B/272